T0128482

Springer Undergraduate Mathematics Series

The Springer Undergraduate Mathematics Series (SUMS) is a series designed for undergraduates in mathematics and the sciences worldwide. From core foundational material to final year topics, SUMS books take a fresh and modern approach. Textual explanations are supported by a wealth of examples, problems and fully-worked solutions, with particular attention paid to universal areas of difficulty. These practical and concise texts are designed for a one- or two-semester course but the self-study approach makes them ideal for independent use.

More information about this series at http://www.springer.com/series/3423

Steven T. Dougherty

Combinatorics and Finite Geometry

 Springer

Steven T. Dougherty
Department of Mathematics
University of Scranton
Scranton, PA, USA

ISSN 1615-2085   ISSN 2197-4144   (electronic)
Springer Undergraduate Mathematics Series
ISBN 978-3-030-56394-3   ISBN 978-3-030-56395-0   (eBook)
https://doi.org/10.1007/978-3-030-56395-0

Mathematics Subject Classification: 05, 06, 51, 52, 94

This Springer imprint is published by the registered company Springer Nature Switzerland AG
The registered company address is: Gewerbestrasse 11, 6330 Cham, Switzerland

This book is dedicated to my family.

Preface

The mathematics that will be discussed in this text falls into the branch of mathematics known as combinatorics. Combinatorics is a very old and very large branch of mathematics encompassing numerous topics. One nice feature about combinatorics is that you need to know very little mathematics to understand the open questions. Unlike many branches of mathematics, there is very little technical language and very few definitions standing in the way of understanding interesting unsolved problems in the area. At present, there are numerous problems which no one has answered yet, but that can be explained to almost anyone. In fact, amateurs have, on occasion, made interesting discoveries in combinatorics. We shall talk about one such example in the third chapter.

On the other hand, these open problems are often very, very, difficult to solve. In a strange twist, it is often true that the easier it is to state a problem, the harder it is to solve it. Many problems in mathematics or science seem hopelessly difficult because it requires a good deal of study just to understand the language of the problem. However, it is often true that the standard techniques of the field can be applied to solve the problem. So even though the problems may seem very difficult, they are not. When problems are easy to state and have been around for a while, then you can be fairly certain that the standard techniques do not apply. These problems are like infections for which known antibiotics simply do not work. To solve them you need something new and original. In this same vein, they are often very infectious, once you begin to think about them you crave to know their solution. In a very real sense, the best in mathematics is like unrequited love, the more the discipline refuses to reveal its secrets the more you desire them.

Combinatorics is also interesting because it has a wide overlap with other branches of mathematics including abstract algebra, number theory, coding theory, and topology. These and other branches of mathematics often have techniques that are useful in solving the problems that arise in combinatorics and often problems in these other areas have combinatorial solutions. Combinatorics also has a very wide variety of applications in science. In fact, many parts of combinatorics, which were purely abstract at their birth, were later vital to scientific applications. In fact, the dawn of the electronic age has brought forth a great interest in combinatorial matters because of their connection to computers and the communication of information.

While combinatorics is a very rich and diverse subject, we shall focus our attention to some specific areas, for example, finite geometry and design theory. Finite geometry, as the name suggests, is closely related to infinite geometry. They share many of the same ideas and definitions. Geometry, along with number theory, is one of the oldest branches of mathematics. In every culture that developed mathematics, geometry developed first. The reason for this is clear. Namely, geometry is probably the most intuitive branch of mathematics. The ideas behind geometry are very natural to consider and have applications from the most elementary aspects of life to the most advanced science. In finite geometry, we study these geometric ideas in a finite setting and relate these intuitive notions to combinatorial problems.

Geometry has often formed the basis for the development of the rest of mathematics. Modern mathematics was built on the base of ancient Greek mathematics, which phrased every idea (even algebraic ones) in terms of geometry. Greek mathematics greatly influenced the focus, notation, and tone of all modern mathematics. It was on this intuitive, geometric foundation that mathematics was built.

In the nineteenth century, mathematicians began to look at all mathematics including geometry in a very abstract way and created models for different geometries using algebraic methods. It was noticed that these same techniques could be used to create geometries, which were finite in that they contained only finitely many points and lines. These are geometries where many of the fundamental properties of standard geometry were true, but there were only a finite number of points and lines. It was soon realized that finite geometry had numerous connections with interesting questions from combinatorics. Design theory arose from this connection. A design can be thought of as a geometric understanding of a combinatorial problem.

Later, in the second half of the twentieth century, it was realized that these very abstract ideas in finite geometry could be used in the very concrete problem of electronic communication. Many of the ideas in this text have applications both in the mathematics of coding theory (making sure that a message is received correctly) and in cryptography (making sure that secret messages are not read by undesired recipients). One small example of how finite geometry can be used in coding theory will be shown in the text. Numerous other applications of combinatorics exist in mathematics, statistics, and science. There has always been a healthy exchange of ideas from those who study combinatorics as pure mathematics and others who apply it. Namely, those who apply it give the researcher interesting problems to think about, and the researcher gives solutions to the combinatorial questions raised by others.

Many combinatorial problems, on the other hand, have their origins in recreational mathematics and from questions arising in other branches of mathematics. Often, some interesting questions in mathematics have arisen in very odd circumstances. For example, the Kirkman schoolgirl problem was raised as a recreational problem as was the question of the existence of the mutually orthogonal Latin squares. We shall give some examples, including these, in the text of how some interesting recreational mathematics gave rise to some of the most well-known and important questions in combinatorics.

Notations

We shall establish some notations that will be used throughout the text. We begin with the natural numbers. We shall denote the natural numbers by $\mathbb{N}$ and we assume that 0 is a natural number. Therefore, $\mathbb{N} = \{0, 1, 2, 3, \ldots\}$. We denote the integers by $\mathbb{Z}$ and $\mathbb{Z} = \{\ldots, -3, -2, -1, 0, 1, 2, 3, \ldots\}$. We denote the rationals by $\mathbb{Q}$ where $\mathbb{Q} = \{\frac{a}{b} \mid a, b \in \mathbb{Z}, b \neq 0\}$. We denote the greatest integer function of x by $\lfloor x \rfloor$, namely, $\lfloor x \rfloor = \max\{n \mid n \leq x, n \in \mathbb{Z}\}$.

Guide for Lecturers

An effort has been made to make each chapter independent and self-contained to allow it to be read without having to refer back constantly to earlier chapters. The topics in Sects. 1.1–1.4, 2.1, and 2.2 are really the only ones that must be understood to read the remaining chapters. (It should be pointed out that while the major results of Sect. 2.2, for example, the existence of finite fields of all prime power orders, must be understood, it is not necessary to have a complete understanding of the algebraic techniques used to obtain these results.) Given a knowledge of these basic chapters, it should be possible for a student to learn the material independently or to make an interesting course by choosing topics from the remaining sections.

It is possible to make several different courses from this text. For students who have not had a course in discrete mathematics, nor a bridge course, the author suggests that Chaps. 1–6 make a nice introductory course on the subject. In this scenario, it is probably best to do the first three sections of Chap. 6. It is also possible to skip the topics from Sect. 2.3 as they are not used in the remainder of the text, but it is probably best to choose one of these types of numbers to include. The topics from Chap. 13 can be introduced throughout to allow students to get some concrete experience with the combinatorial objects. In this case, one can make two courses, where the second course consists of the remaining sections of Chap. 6 along with Chaps. 7–13. This configuration also works well if the first course is a standard course and the second one is an independent study course.

For students who have had a course in discrete mathematics or a bridge course, it is possible to skip Sects. 1.1–1.3, and 2.1. Given that these students are more likely to be mathematically mature, one can then choose from topics in Chaps. 7 and 8 or for students whose interest is in computer science one might wish to choose Chap. 11 or Chap. 12 (or both).

For more advanced students, one can start with Chap. 3 and make a course from Chaps. 3–10, choosing topics from the remaining 3 chapters based on student interest.

Scranton, USA   Steven T. Dougherty

Acknowledgements

The author is grateful to Dr. Cristina Fernández-Córdoba, Dr. Charles Redmond, and Dr. Mercè Villanueva for their comments on earlier drafts of this text. The author is particularly grateful to Dr. Esengül Saltürk who carefully read the entire text and offered numerous suggestions.

Contents

Foundational Combinatorial Structures

1.1 Basic Counting

Counting is at the core of combinatorics. In general, what we mean by this seemingly simple term is that we wish to be able to count the number of objects when they are described in an abstract way. For example, one might want to count the possible outcomes in a lottery game in order to determine the probability of winning or one might want to count the possible number of arrangements of symbols to make a secret message.

To aid us in counting, we use the foundational objects and relations of mathematics, namely, sets and functions. A set is simply a collection of objects. In this text, we are largely concerned with finite sets which simplify matters a great deal. A set can be described by its elements, for example, $\{a_1, a_2, \ldots, a_n\}$ or it can be described as $\{x \mid p(x)\}$, where $p(x)$ is a statement about x. As examples, we can have the set $\{1, 2, 3, 4\}$ or we can have the set $\{x \mid x \text{ is an integer and } 1 \leq x \leq 4\}$. These sets are equal as they contain precisely the same elements.

We say that $a \in A$, if the element a is in A. In the previous example, we have $1 \in \{1, 2, 3, 4\}$, but $5 \notin A$ since 5 is not an element of A. The set B is a subset of A, denoted $B \subseteq A$, if for every $b \in B$ we have $b \in A$. It follows immediately that $A = B$ if and only if $A \subseteq B$ and $B \subseteq A$.

A function $f : A \rightarrow B$ is a subset of $A \times B = \{(a, b) \mid a \in A, b \in B\}$ where for each $a \in A$ there exists a unique $b \in B$ with (a, b) in the function; we write this as $f(a) = b$. For example, if A is the set of people in the United States and B is the set of social security numbers, then the function that maps each person to their social security number can be thought of as the set of ordered pairs (a, b) where a is a person and b is the social security number assigned to a. In the same way, we can simply write $f(a) = b$. The key property of a function, meaning the one that distinguishes it from a general relation, is that each element a has a unique b with $f(a) = b$. Therefore, if we let $A = B$ be the set of real numbers, and f be the relation

S. T. Dougherty, *Combinatorics and Finite Geometry*, Springer Undergraduate Mathematics Series,

that maps a to its square root, i.e., $f(a) = \sqrt{a}$, then this is not a function since $2^2 = 4$ and $(-2)^2 = 4$. It is possible to turn this into a function by changing it to $f(a) = |\sqrt{a}|$ which gives the positive square root for each number. When $f(x) = \sqrt{x}$ is used in mathematics, it is generally assumed that it is the function as described here.

We denote the cardinality of a set A by $|A|$. Specifically $|A| = n$ if and only if there is a bijection between A and the set $\{1, 2, 3, \ldots, n\}$. If $A = \emptyset$ then $|A| = 0$ and if no n exists with such a bijection then the set is infinite.

There are three types of functions that are often used in counting arguments, namely, injections, surjections, and bijections. A function is injective if $f(a) = f(a')$ implies $a = a'$. Intuitively this means that no element in B is mapped more than once. A function is surjective if for every $b \in B$ there exists an $a \in A$ with $f(a) = b$. In other words, every element of B is the image of some element in A. A function is bijective if it is both injective and surjective. A bijection makes a one-to-one correspondence between the elements of A and B.

To make an inverse function, it is necessary for a function to be injective. If the function were not injective then there would exist a_1, a_2 with $a_1 \neq a_2$ such that $f(a_1) = b$ and $f(a_2) = b$. Then an inverse function could not be defined, since the element b would have to be mapped to two distinct elements, violating the definition of a function.

For an injective function $f : A \to B$, define the inverse function $f^{-1} : B \to A$ by $f^{-1}(b) = a$ if and only if $f(a) = b$. If follows from the definition that if $f : A \to B$ is a bijection then $f^{-1} : B \to A$ is also a bijection.

Lemma 1.1 *Let $f : A \to B$ and $g : B \to C$ with f and g bijections. Then the composition $g \circ f : A \to C$ is a bijection.*

Proof Assume $(g \circ f)(a_1) = (g \circ f)(a_2)$ then $g(f(a_1)) = g(f(a_2))$ which implies that $f(a_1) = f(a_2)$ since g is an injection. Then $a_1 = a_2$, since f is an injection. Therefore, $g \circ f$ is an injection.

Let $c \in C$. Then, since g is a surjection, there exists a $b \in B$ with $g(b) = c$. Since f is a surjection, we have that there exists an $a \in A$ with $f(a) = b$. Then $(g \circ f)(a) = g(f(a)) = g(b) = c$ and $g \circ f$ is a surjection. Therefore $g \circ f$ is a bijection and we have our result. $\square$

A very important technique for comparing the cardinalities of sets is to construct maps from one set to another that are either injective, surjective, or both and use the results that follow.

Theorem 1.1 *If A and B are finite sets and there exists a bijection $f : A \to B$, then $|A| = |B|$.*

Proof Assume $|B| = n$, then there exists a bijection $g : B \to \{1, 2, \ldots, n\}$. Then $g \circ f$ is a bijection from A to $\{1, 2, \ldots, n\}$. Then, by definition, $|A| = n$ giving that $|A| = |B|$. $\square$

In the next three theorems, functions are used to compare the relative sizes of sets. The statements are true for infinite sets as well but we shall only prove them for finite sets.

Theorem 1.2 *If A and B are finite sets and there is an injection from A to B then* $|A| \leq |B|$.

Proof Assume $|A| > |B|$ and f is an injection from A to B with $|A| = n$, $|B| = m$. We have $n < m$. We have a bijection from A to $\{1, 2, \ldots, n\}$. Let a_i be the element of A that corresponds to i under this bijection. Then we can write $A = \{a_1, a_2, \ldots, a_n\}$. Let $b_i = f(a_i)$. Since f is an injection, $f(a_{m+1}) \neq f(a_i)$ for $1 \leq i \leq m$. But $B - \{b_1, b_2, \ldots, b_m\}$ is empty which gives a contradiction. □

Theorem 1.3 *If A and B are finite sets and there is a surjection from A to B then* $|A| \geq |B|$.

Proof Assume $|A| < |B|$ and f is a surjection from A to B with $|A| = n$, $|B| = m$. We have $n > m$. Just as in the previous theorem, we have a bijection from A to $\{1, 2, \ldots, n\}$ and we let a_i be the element of A that corresponds to i under this bijection, with $b_i = f(a_i)$. Then b_{m+1} is not the image of any element under the function f, which contradicts that f is a surjection. □

The following theorem is known as the Schröder-Bernstein Theorem. It is an extremely important theorem in set theory and fairly difficult to prove for infinite sets, where the proof requires some form of the Axiom of Choice. For finite sets the proof is easy.

Theorem 1.4 *If A and B are finite sets and there exists an injection* $f : A \to B$ *and an injection* $g : B \to A$ *then* $|A| = |B|$.

Proof The injection f gives $|A| \leq |B|$ by Theorem 1.2 and the injection f gives $|B| \leq |A|$ by the same theorem. It follows that $|A| = |B|$. □

The most basic counting principle is the multiplicative principle.

Multiplicative Counting Principle: If there are n ways of choosing an object from a set A and there are m ways of choosing an object from a set B then there are nm ways of choosing objects from $A \times B$.

This is the principle stated in its most general form. Of course, it extends by induction to the following.

Generalized Multiplicative Counting Principle: If there are n_i ways of choosing an object from the set A_i, then there are $\prod_{i=1}^{k} n_i = n_1 n_2 \ldots n_k$ ways of choosing elements from $A_1 \times A_2 \times \cdots \times A_k$.

Essentially, we are simply determining the cardinalityof

$$A_1 \times A_2 \times \cdots \times A_k = \{(a_1, a_2, \ldots, a_k) | a_i \in A_i\}.$$

Example 1.1 As a simple example, consider the sets $A = \{1, 2, 3, 4\}$ and $B = \{a, b, c\}$. There are $4 \cdot 3 = 12$ pairs, namely,

$$\{(1, a), (1, b), (1, c), (2, a), (2, b), (2, c), (3, a), (3, b), (3, c), (4, a), (4, b), (4, c)\}.$$

Example 1.2 Assume someone has 5 pairs of pants, 8 shirts, and 3 pairs of shoes. Then, this person has $5(8)(3) = 120$ possible outfits by the multiplicative principle.

We shall now examine some other counting techniques.

Recall the recursive definition of $n!$, that is $0! = 1$ and $n! = n(n-1)!$. It follows immediately from this definition that for $n \neq 0$ we have that $n! = \prod_{i=1}^{n} i$.

Theorem 1.5 *The number of ways of arranging n items is $n!$.*

Proof Consider the n spaces into which the n items will be arranged. In the first space there are n choices. In the second there are $n-1$ choices since you can choose any of the n objects except the one that was placed in the first space. At the k-th space there are $n - k + 1$ choices since you can choose any of the n objects except those used in the first $k - 1$ spaces. The generalized multiplicative counting principle gives that there are

$$n(n-1)(n-2)\cdots(2)1 = \prod_{i=1}^{n} i = n!$$

ways of arranging the n items.   □

Example 1.3 Consider the ways of arranging the elements a, b, and c. They are as follows:

$$a\ b\ c$$
$$a\ c\ b$$
$$b\ a\ c$$
$$b\ c\ a$$
$$c\ a\ b$$
$$c\ b\ a$$

Here $3! = 6$ and there are 6 ways of arranging the elements.

The value of $n!$ grows wildly in comparison to n. For example, when n is 10 we have $10! = 3, 628, 800$. As another example of how large this number is, consider a standard set of playing cards. There are 52 cards in such a deck. Hence there are 52!

ways of arranging this deck. The number 52! is greater than $8 \cdot 10^{67}$. Consider that there are $31,557,600$ seconds in a year. If we assume the universe has existed at most 10 billion years, then there have been less than 10^{18} seconds since the beginning of time. Assuming that at every second someone shuffled a million decks of cards, this would still give only 10^{24} different arrangements of a deck that have ever occurred, assuming of course that there were no repeated formations. Take a deck of cards and shuffle it. Given this scenario, the chance that there has ever been a deck of cards arranged in this manner is less than $\frac{1}{10^{43}}$. You have a much better chance of winning the lottery billions of times! The point here is that the number of ways of arranging elements grows extremely quickly. This makes it quite difficult to use computers to solve combinatorial problems, since the number of different arrangements to check is often far too great for a computer.

Of course, there are more complicated ways of picking elements of a set. For example, if you wanted to pick three people from a class to be president, secretary, and treasurer, then the order in which they are picked makes quite a difference. Whereas if you are simply picking three students to be a committee then the order does not matter.

Let $P(n, k)$ denote the number of ways of choosing k elements from n elements where the order of the choice matters. Let $C(n, k)$ denote the number of ways of choosing k elements from n elements where the order of the choice does not matter. We can determine both of these in the following theorem.

Theorem 1.6 *For* $0 \le k \le n$, $P(n, k) = \frac{n!}{(n-k)!}$ *and* $C(n, k) = \frac{P(n,k)}{k!} = \frac{n!}{k!(n-k)!}$.

Proof To choose k items from n items, consider the first space. There are n choices for that space. Then there are $n - 1$ for the second as before. For the kth space there are $n - k + 1$ choices and hence by the generalized multiplicative counting principle there are $n(n-1)\cdots(n-k+1) = \frac{n!}{(n-k)!}$ ways of choosing k items from n where the order of the choosing matters.

To determine the number of ways of choosing k items from n where order does not matter, we know by Theorem 1.5 that each choice of k items can be arranged in $k!$ ways. So each choice in counting $P(n, k)$ is in a set of size $k!$ consisting of the same k elements permuted. Therefore, $C(n, k) = \frac{P(n,k)}{k!} = \frac{n!}{k!(n-k)!}$. □

The number $P(n, k)$ is often referred to as the number of permutations. The number $C(n, k)$ is often called a combination and is sometimes denoted $\binom{n}{k}$ or ${}_nC_k$.

Example 1.4 A standard deck of cards consists of 52 cards with 4 suits each containing 13 cards. A standard poker hand consists of 5 cards. There are $C(52, 5) = \frac{52!}{5!47!} = 2,598,960$ possible poker hands. A flush is a set of 5 cards that are all in the same suit. The number of poker hands that are a flush can be found by choosing 5 cards from any suit and multiplying by the number of suits. Namely, there are $4C(13, 5) = 4(1287) = 5148$ hands that are a flush. A straight consists of 5 cards in a row from any suit. Any card can start a straight except for the face cards. Namely, there are 40 cards that can start a straight. Given any of those cards, there are $C(4, 1) = 4$ choices for the second, third, fourth, and fifth cards. Then the multiplicative principle gives that there are $40(4)(4)(4)(4) = 10,240$ poker hands that are straights. Since there are fewer flushes than straights a flush beats a straight in poker.

Example 1.5 A lottery game has 40 balls in a bin and 6 balls are pulled out. A winner is anyone who correctly picks all 6 numbers. There are $C(40, 6) = \frac{40!}{6!34!} = 3,838,380$ possible outcomes. Therefore, someone with a single ticket has a 1 out of $3,838,380$ chance of winning this game.

Theorem 1.7 *For* $0 \le k \le n$, $C(n, k) = C(n - 1, k) + C(n - 1, k - 1)$.

Proof First technique: Pick one element a out of a set A of cardinality n. Every subset of size k of A either contains a or does not contain a. The number of those that do not contain a is $C(n - 1, k)$ since you are choosing k objects out of $A - \{a\}$. The number of those that do contain a is $C(n - 1, k - 1)$ since you are choosing $k - 1$ objects out of $A - \{a\}$ and adding the element a. Therefore, $C(n, k) = C(n - 1, k) + C(n - 1, k - 1)$.

  Second technique: We simply add $C(n - 1, k) + C(n - 1, k - 1)$ using the formula given in Theorem 1.6:

$$
\begin{aligned}
C(n - 1, k) + C(n - 1, k - 1) &= \frac{(n - 1)!}{k!(n - 1 - k)!} + \frac{(n - 1)!}{(k - 1)!(n - 1 - (k - 1))!} \\
&= \frac{(n - 1)!(n - k)}{k!(n - 1 - k)!(n - k)} + \frac{(n - 1)!k}{k(k - 1)!(n - k)!} \\
&= \frac{(n - 1)!(n - k) + (n - 1)!k}{k!(n - k)!} \\
&= \frac{(n - 1)!n}{k!(n - k)!} \\
&= \frac{n!}{k!(n - k)!} = C(n, k). \qquad \square
\end{aligned}
$$

  This recursion, together with the obvious fact that $C(n, 0) = C(n, n) = 1$, allows us to compute any value of $C(n, k)$. The values of $C(n, k)$ are often written in the following table, which is known as Pascal's triangle. It was known prior to Pascal in ancient China and to the Italian mathematician Tartaglia, so it is sometimes referred to as Tartaglia's triangle. The table is written so that the recursive nature of the numbers is evident.

$$
\begin{array}{ccccccccccccccc}
 & & & & & & & 1 & & & & & & & \\
 & & & & & & 1 & & 1 & & & & & & \\
 & & & & & 1 & & 2 & & 1 & & & & & \\
 & & & & 1 & & 3 & & 3 & & 1 & & & & \\
 & & & 1 & & 4 & & 6 & & 4 & & 1 & & & \\
 & & 1 & & 5 & & 10 & & 10 & & 5 & & 1 & & \\
 & 1 & & 6 & & 15 & & 20 & & 15 & & 6 & & 1 & \\
1 & & 7 & & 21 & & 35 & & 35 & & 21 & & 7 & & 1 \\
\end{array}
$$

1 8 28 56 70 56 28 8 1

.
.
.

The elements in each place are determined by adding the two numbers immediately above them. This uses the result in Theorem 1.7. Notice that the rows and columns are both indexed starting at 0 not 1. So the element in row 6 and column 2 is 15. Of course

$$
C(6, 2) = \frac{6!}{4!}2! = \frac{6(5)}{2} = 15.
$$

The table is also symmetric; this follows from the next theorem.

Theorem 1.8 *For $0 \leq k \leq n$, $C(n, k) = C(n, n - k)$.*

Proof **First technique:** By choosing k objects out of n objects to be included in a set, you are also choosing $n - k$ out of n objects to be excluded from the set. Therefore, $C(n, k) = C(n, n - k)$.

Second technique: We can simply apply the formula given in Theorem 1.6 which gives $C(n, k) = \frac{n!}{k!(n-k)!}$ and $C(n, n - k) = \frac{n!}{(n-k)!(k)!}$ which are equal. $\square$

Theorem 1.9 (The Binomial Theorem) *Let $n \geq 0$, then*

$$
(x + y)^n = \sum_{k=0}^{n} C(n, k)x^{n-k}y^k.
$$

Proof We shall prove the theorem by induction. For $n = 0$, we have $(x + y)^0 = 1$ and $\sum_{k=0}^{0} C(n, k)x^{n-k}y^k = 1$.

Now assume $\sum_{k=0}^{n} C(n, k)x^{n-k}y^k = (x + y)^n$. Then we have the following (assuming that $C(n, k)$ is 0 when k is negative):

$$
\sum_{k=0}^{n+1} C(n + 1, k)x^{n+1-k}y^k = \sum_{k=0}^{n+1} (C(n, k) + C(n, k - 1))x^{n+1-k}y^k
$$

$$
= \sum_{k=0}^{n+1} C(n, k)x^{n+1-k}y^k + \sum_{k=0}^{n+1} C(n, k - 1)x^{n+1-k}y^k
$$

$$
= \sum_{k=0}^{n} C(n, k)x^{n+1-k}y^k + \sum_{k=1}^{n+1} C(n, k - 1)x^{n+1-k}y^k
$$

$$
= \sum_{k=0}^{n} C(n, k)x^{n+1-k}y^k + \sum_{k=0}^{n} C(n, k)x^{n-k}y^{k+1}
$$

$$
= x \sum_{k=0}^{n} C(n, k)x^{n-k}y^k + y \sum_{k=0}^{n} C(n, k)x^{n-k}y^k
$$

$$
= x(x + y)^n + y(x + y)^n = (x + y)^{n+1}.
$$

Therefore by mathematical induction we have the result.   □

This theorem allows us to use Pascal's triangle to determine $(x + y)^n$. For example, the coefficients of $(x + y)^5$ come from the numbers in the row corresponding to 5 in the table, namely

$$
(x + y)^5 = x^5 + 5x^4y + 10x^3y^2 + 10x^2y^3 + 5xy^4 + y^5.
$$

Let S be a set. We define the power set of S to be

$$
\mathcal{P}(S) = \{B \,|\, B \subseteq S\}.
$$

That is, the power set is a set containing all of the subsets of S as elements.

Theorem 1.10 *Let S be a set with n elements, then $|\mathcal{P}(S)| = 2^n$.*

Proof For each of the n elements there are 2 choices, namely, it is either in a subset or not in a subset. Hence there are 2^n possible subsets.   □

Example 1.6 Let $S = \{a, b, c\}$. Then

$$
\mathcal{P}(S) = \{\emptyset, \{a\}, \{b\}, \{c\}, \{a, b\}, \{a, c\}, \{b, c\}, \{a, b, c\}\}.
$$

So S has cardinality 3 and the power set has cardinality $2^3 = 8$.

Notice that the sum of the elements in the nth row of Pascal's triangle is 2^n. We can prove this by applying the Binomial Theorem. That is, if $\sum_{k=0}^{n} C(n, k)x^{n-k}y^k = (x + y)^n$ then let $x = y = 1$ and we have $\sum_{k=0}^{n} C(n, k) = (2)^n$. You can also prove this using Theorem 1.10. Namely, if there are 2^n subsets of a set of cardinality n, then count the number of subsets of each cardinality k with $0 \leq k \leq n$, namely, $C(n, k)$. This gives $\sum_{k=0}^{n} C(n, k) = 2^n$. Either proof gives the following corollary.

Corollary 1.1 *For all $n \geq 0$, we have that $\sum_{k=0}^{n} C(n, k) = 2^n$.*

As an example, if $n = 5$, then $1 + 5 + 10 + 10 + 5 + 1 = 2^5 = 32$.

The final counting scenario is if we are choosing k objects out of n where order does not matter but we are allowed to replace elements.

Theorem 1.11 *The number of ways of choosing k objects out of n, where order does not matter, but repetitions are allowed is $C(n + k - 1, k)$.*

Proof Assume we have n objects and place them in a line as $o_1, o_2, \ldots, o_n$. In front of the o_i you place a number of 1s indicating how many of object 1 you want. Then you end this sequence of 1s with a 0 to indicate to move on to the next one. Hence there are $n + k - 1$ places with a 1 or 0 in it, noting that there are precisely k occurrences of 1. Therefore, the number of ways of choosing k objects out of n where order does not matter but repetitions are allowed is $C(n + k - 1, k)$. □

Example 1.7 Assume there are 7 flavors of ice cream and you are allowed to take 3 scoops of any flavors that you want. How many possible combinations are there. Here, there are 7 objects and 3 to choose from. Then by Theorem 1.11 the possible number of combinations is $C(7 + 3 - 1, 3) = C(9, 3) = \frac{9!}{3!6!} = 84$.

We summarize these counting techniques as follows.

Objects	Chosen	Repetitions	Order	Number of choices
n	k	Allowed	Matters	n^k
n	k	Not Allowed	Matters	$P(n, k) = \frac{n!}{(n-k)!}$
n	k	Allowed	Does Not Matter	$C(n + k - 1, k) = \frac{(n+k-1)!}{(n-1)!k!}$
n	k	Not Allowed	Does Not Matter	$C(n, k) = \frac{n!}{(n-k)!k!}$

Exercises

1. Let $A_n = \{1, 2, \ldots, n\}$. Prove that if n is even, then the number of even numbers and odd numbers in A_n are equal and if n is odd, then the number of even numbers and odd numbers in A_n are not equal. Prove which one is greater in the second case.

2. Give examples of functions f from the set $\{1, 2, 3, \ldots, n\}$ to the set $\{1, 2, 3, \ldots, m\}$ such that

    a. f is injective but not surjective
    b. f is surjective but not injective
    c. f is bijective but not the identity
    d. f is neither injective nor surjective

    That is, find n, m, and f satisfying the conditions in each part.
3. Prove that if f is an injection from $\{1, 2, 3, \ldots, n\}$ to $\{1, 2, 3, \ldots, n\}$, then f must be a surjection and thus a bijection.
4. Prove that if $n > m$ then there is no injection from $\{1, 2, 3, \ldots, n\}$ to $\{1, 2, 3, \ldots, m\}$.
5. Prove that if there exists an injection from A to A that is not a surjection then A must be infinite.
6. Recall that the set of integers denoted by $\mathbb{Z}$ is the set

$$\{\ldots, -4, -3, -2, -1, 0, 1, 2, 3, 4, \ldots\}.$$

    Give an example of a function from $\mathbb{Z}$ to $\mathbb{Z}$ that is an injection but not a surjection. Give an example of a function from $\mathbb{Z}$ to $\mathbb{Z}$ that is a surjection but not an injection.
7. Prove that if $|A| = n$ and $|A| = m$ then $n = m$, that is, prove that the cardinality of a finite set is unique.
8. Let $S = \{a, b, c, d\}$. Find all 16 elements of $\mathcal{P}(S)$.
9. Prove that if p is a prime then $C(p, i)$ is divisible by p for $0 < i < p$.
10. Prove that $\sum_{i=1}^{n} iC(n, i) = 2^{n-1}n$.
11. Prove that $C(2n, n)$ is even for $n > 0$.
12. How many vectors of length n are there where the coordinates are chosen from a set of order k.
13. Let a be an element in a set A of size n. Determine the number of subsets of A of size k containing a.
14. A standard pair of dice consists of two cubes with the numbers 1 through 6 written on the faces. Determine the number of ways a player can roll the dice and obtain a total of 2, 3, $\ldots$, 12. Which is the most likely roll?
15. Assume a license plate has six places with a number between 1 and 9 or a letter of the English alphabet in it. Determine the number of possible license plates.
16. Count the number of possible wins a player can make in a tic-tac-toe game played on a 3×3 board, a 4×4 board, and a $3 \times 3 \times 3$ board.
17. How many ways can you choose 6 books from 12 choices if repeats are allowed and order does not matter?
18. On an 8 by 8 chessboard a rook attacks another rook if they share a row or column. Determine how many rooks can be placed with no rook attacking another rook on this board.
19. How many ways can you get 5 scoops of ice cream from 12 flavors?

20. On an 8 by 8 chessboard a bishop attacks another bishop if they share a diagonal. Determine how many bishops can be placed with no bishop attacking another on this board.

21. Determine how many tic-tac-toe games can be played on a 3 by 3 board. Determine how many final configurations are there on this board.

22. Let A be a set with cardinality n and let B be a set with cardinality m. Let f be the bijection between A and $\{1, 2, 3, \ldots, n\}$ and let g be the bijection between B and $\{1, 2, 3, \ldots, m\}$. Prove that the function $h : (A \times B) \to \{1, 2, 3, \ldots, nm - 1, nm\}$ defined by $h(a, b) = f(a) + (g(b) - 1)n$ is a bijection.

23. Assume $|A_i| = n_i$. Prove by induction that $|A_1 \times A_2 \times \cdots \times A_k| = \prod_{i=1}^{k} n_i$. Namely, prove that there exists a bijection from

$$A_1 \times A_2 \times \cdots \times A_k$$

to

$$\{1, 2, 3, \ldots, \prod_{i=1}^{k} n_i\}$$

using the fact that $A_1 \times A_2 \times \cdots \times A_k$ can be viewed as $(A_1 \times A_2 \times \cdots \times A_{k-1}) \times A_k$.

24. Determine the number of poker hands that are a straight flush (a straight all in the same suit), 4 of a kind, 3 of a kind, 2 of a kind, and a full house (3 of a kind with 2 of a different kind).

25. The German army Enigma machine had three rotors which permuted letters and a plugboard that interchanged six pairs of letters. Each rotor had 26 different initial positions and the three worked in tandem. Count the number of possible initial configurations for this machine.

26. A blackjack hand consists of two cards from a standard deck. Each card has the value of its number, face cards have a value of 10 and an ace has the value of either 1 or 11. Count the number of possible blackjack hands that sum to 12, 18, or 21.

27. In the classic Monty Hall problem there are three curtains. Behind one is a prize. A player picks one curtain and then the host opens a curtain that does not have a prize. The player is offered the choice of switching their choice of curtain. Count the number of ways that switching gets the prize and count the number of ways that switching does not win and deduce what strategy the player should adopt.

28. The game of Morra is an Italian game that goes back to Roman times. The game is played with two people who each put out a number of fingers on one hand from 0 to 5. Simultaneously, the player calls out a guess for the sum of the two numbers (saying Morra if the guess is 0). Count the number of possible games. Then solve the harder problem of counting the number of games where

each player plays rationally, that is, a player only guesses a sum that is greater than or equal to the number of fingers they put out. For example, a player puts out a 4 then they would only guess sums from 4 to 9, since 0, 1, 2, 3 and 10 are impossible.

1.2   Multinomial Coefficients

We shall now give a natural generalization for binomial coefficients. Imagine we have 12 objects to be placed in 3 boxes. The first box must have 4, the second box must have 6 and the third box must have 2. There are $C(12, 4)$ ways of choosing 4 objects for the first box. Then, since 4 objects have been removed, there are $C(8, 6)$ ways of choosing 6 objects for the second. Finally, there are 2 objects that remain, so these two must be placed in the final box. The product of these three is the number of ways of filling the boxes, which is

$$\frac{12!}{4!8!}\frac{8!}{6!2!}\frac{2!}{2!0!} = \frac{12!}{4!6!2!}.$$

Significant cancelation makes this a simpler problem than one might first have imagined. We generalize this result in the following theorem.

Theorem 1.12 *If n objects are to be placed in k boxes, where k_i of them are placed in the ith box, with $\sum_{i=1}^{t} k_i = n$, then the number of ways of placing the n objects in the t boxes is*

$$\frac{n!}{k_1!k_2!k_3!\cdots k_t!}.$$

Proof The number of ways of placing objects in the boxes is

$$C(n, k_1)C(n - k_1, k_2)C(n - k_1 - k_2, k_3)\cdots C(n - k_1 - k_2 - \cdots - k_{t-1}, k_t)$$

$$= \frac{n!}{k_1!(n - k_1)!}\frac{(n - k_1)!}{k_2!(n - k_1 - k_2)!}\frac{(n - k_1 - k_2)!}{k_3!(n - k_1 - k_2 - k_3)!}\cdots \frac{(n - \sum_{i=1}^{t-1} k_t)!}{k_t!0!}$$

$$= \frac{n!}{k_1!k_2!k_3!\cdots k_t!}.$$

This gives the result.   □

We call this number $\frac{n!}{k_1!k_2!k_3!\cdots k_t!}$ the multinomial coefficient and we denote it by

$$\binom{n}{k_1, k_2, \ldots, k_t}.$$

Example 1.8 Let $n = 9$, $k_1 = 2$, $k_2 = 4$, $k_3 = 3$. Then

$$\binom{9}{2, 4, 3} = \frac{9!}{2!4!3!} = 1260.$$

Theorem 1.13 (The Multinomial Theorem) *Let* $\{x_1, x_2, \ldots, x_t\}$ *be a set of* t *indeterminates. Then*

$$(x_1 + x_2 + \cdots + x_t)^n = \sum \binom{n}{k_1, k_2, k_3, \ldots, k_t} x_1^{k_1} x_2^{k_2} \cdots x_t^{k_t}, \qquad (1.1)$$

where the sum is done over all partitions of n.

Proof We prove this by induction on t. For $t = 2$, it is the binomial theorem proven earlier in the chapter. Then assuming it is true for $t = s - 1$, consider $(x_1 + x_2 + \cdots + x_{t-1} + x_t)$ as $((x_1 + x_2 + \cdots + x_{t-1}) + x_t)$ and apply the binomial theorem again.

The details are left as Exercise 2. An alternate proof is asked for in Exercise 4.   □

Example 1.9 We have the following for $n = 3$.

$$\begin{aligned}
(x + y + z)^3 &= \binom{3}{3, 0, 0} x^3 + \binom{3}{2, 1, 0} x^2 y + \binom{3}{2, 0, 1} x^2 z + \binom{3}{1, 2, 0} xy^2 \\
&\quad + \binom{3}{1, 1, 1} xyz + \binom{3}{1, 0, 2} xz^2 + \binom{3}{0, 3, 0} y^3 + \binom{3}{0, 2, 1} y^2 z \\
&\quad + \binom{3}{0, 1, 2} yz^2 + \binom{3}{0, 0, 3} z^3 \\
&= x^3 + 3x^2 y + 3x^2 z + 3xy^2 + 6xyz + 3xz^2 + y^3 + 3y^2 z + 3yz^2 + z^3.
\end{aligned}$$

Corollary 1.2 *Let* n *be a positive integer. Then*

$$\sum \binom{n}{k_1, k_2, k_3, \ldots, k_t} = t^n,$$

where the sum is done over all possible partitions of n.

Proof Follows from Theorem 1.13, by letting $x_i = 1$ for all i, $1 \le i \le t$.   □

We shall now describe another use of the multinomial coefficient. Consider the English word "syzygy". The word consists of 6 letters but there are not 6! ways of getting distinct arrangements of these letters, since some of the letters repeat. There are 3 letters that are y, and 1 each of s, z, and g. Each of the 6! arrangements of the letters occurs the number of ways they are arranging each of the letters within a set of letters. Hence, for this word there are

$$\binom{6}{3, 1, 1, 1} = \frac{6!}{3!1!1!1!} = 120$$

distinct arrangements of the letters.

In the word "missing", there are 2 letters that are s, 2 that are i, and 1 each of m, n, and g. Therefore, the number of distinct arrangements of the letters of this word is

$$\binom{7}{2, 2, 1, 1, 1} = \frac{7!}{2!2!1!1!1!} = 1260.$$

Exercises

1. Compute $\binom{8}{4,2,2}$.
2. Give the details for the proof of Theorem 1.13.
3. Determine how many distinct ways there are for arranging the letters of Missouri, Pennsylvania, and Mississippi.
4. Give an alternate proof of Theorem 1.13 by counting ways of producing the monomial $x_1^{k_1} x_2^{k_2} \cdots x_t^{k_t}$.

1.3   Pigeonhole Principle

Theorem 1.2 states that if there is an injection from a set A to a set B then $|A| \leq |B|$. The contrapositive of this is that if $|A| > |B|$, then there can be no injections from A to B. This is known as the Pigeonhole Principle, which we state now in its usual form.

Pigeonhole Principle If $n > m$ and n objects are placed in m boxes then at least one box must contain two objects.

This seemingly simple statement has numerous powerful applications in counting. We shall exhibit some of these now. In all applications of this principle, the key is deciding what to call the objects and what to call the boxes.

Example 1.10 We shall show that given any 6 integers, there must be two of them whose difference is a multiple of 5. Let the integers be the objects. Any number has 5 possible remainders when divided by 5, namely, 0, 1, 2, 3, or 4. Place the objects in the box corresponding to its remainder when divided by 5. By the Pigeonhole Principle, some box must have at least 2 elements. Assume they are in the box with remainder r. Then these two numbers are of the form $5k + r$ and $5s + r$ for some integers k and s. Then $(5k + r) - (5s + r) = 5(k - s)$ which is divisible by 5.

Notice that the Pigeonhole Principle does not determine exactly how many objects are in each box. For example, in the previous example, you could pick 6 numbers all of whom have remainder 2 when divided by 5. It does not guarantee that each box is filled but rather that some box must have 2 elements.

Example 1.11 We will show that given any $k + 1$ numbers from 1 to $2k$ that one of them will be a multiple of the other. Write each number from 1 to $2k$ as $2^s m$ where m is an odd number. If two numbers are written like this with the same odd part then one must be a multiple of the other. There are k odd numbers between 1 and $2k$. Let the boxes be the odd parts and place a number $2^s m$ in the box corresponding to m. Therefore, by the Pigeonhole Principle there must be at least two numbers in some box and so one is a multiple of the other.

Example 1.12 Given n people, there must be two people who know the names of exactly the same number of other people in the room, provided that if person A knows person B's name then person B knows person A's name, namely, the relation is symmetric. The number of possible people, other than oneself, that you may know their names, is from 0 to $n - 1$. This means that there are n boxes and n objects so a simplistic application of the Pigeonhole Principle does not apply. First assume, that everyone knows at least one person's name. Then the number of people whose name each of the n may know is $1, 2, \ldots, n - 1$. Hence, there are n people with $n - 1$ possibilities, so some 2 must know the same number of names. Next, assume that at least one person knows no name. Then the number of people whose name each of the n may know is $0, 1, 2, \ldots, n - 2$ since no one knows $n - 1$ people's names (here is where we must have that the relation is symmetric). Again there are n people with $n - 1$ possibilities, so some 2 must know the same number of names.

Example 1.13 Assume we have 21 natural numbers greater than or equal to 1 and less than or equal to 40. We shall show that some 2 of them must sum to 41. Consider the sets $\{1, 40\}, \{2, 39\}, \{3, 38\}, \ldots, \{19, 22\}, \{20, 21\}$. Since there are 20 of these sets and 21 numbers, some 2 of them must be in the same set and so their sum is 41.

We can generalize the Pigeonhole Principle as follows.

Generalized Pigeonhole Principle If $km + 1$ objects are placed in m boxes then at least one box must contain $k + 1$ objects.

Proof Assume no box contains $k + 1$ objects which means that each of the m boxes contains at most k objects. This means that there are at most km objects which is a contradiction. Therefore, some box must have at least $k + 1$ objects. □

We can also use this technique for a proof by contradiction as in the following example.

Example 1.14 Assume that 10 people are handed 40 dollar bills. We shall prove that some two have the same number of dollar bills. Assume that they are all distinct, then the smallest number that can be obtained is $0 + 1 + 2 + 3 + 4 + 5 + 6 + 7 + 8 + 9 = 45$ which is greater than 40.

Of course, we see that the Generalized Pigeonhole Principle applies to all numbers from $km + 1$ to $(k + 1)m$ as well since there are at least $km + 1$ objects.

Example 1.15 We shall show that if you have a set of $37 = 3(12) + 1$ people then at least 4 of them will have birthdays in the same month. Let the twelve months be the boxes and the people the objects. Place each person in the month of their birth. Then, by the Generalized Pigeonhole Principle, one box must have at least 4 objects. Therefore, at least 4 people share the same birth month.

Example 1.16 We shall show that if you have 37 non-zero two-digit numbers then 3 of them will have the same digit sum. There are 18 possible digital sums of two-digit numbers, for example, consider the digit sums of the following: 10, 11, 12, 13, 14, 15, 16, 17, 18, 19, 92, 93, 94, 95, 96, 97, 98, 99. We have $2(18) + 1$ numbers so some three of them will have the same digit sum.

Exercises

1. Prove that if there are 8 colors to paint 27 cars, then at least 4 of them must be painted the same color.
2. Determine how many people you need to have to ensure that at least 8 of them will have birthdays on the same day of the week.
3. Determine how many numbers you need to have to ensure that at least 6 of them will have the same remainder when divided by 6.
4. Prove that if you have 5 points in a square of size 1 then at least two of the points must be within $\frac{1}{\sqrt{2}}$ of each other.
5. Prove that if you place 13 points on a line segment of length 4, then at least 4 of them are within distance 1 of each other.
6. Determine in what range the number of books must be with a page sum of 3585 to ensure that at least one of them has at least 513 pages.
7. Determine how many integers you need to ensure that at least 8 of them have the same remainder when divided by 11.

1.4   Permutations

Permutations are one of the most interesting tools to use to understand a finite combinatorial object. Essentially, you can study the structure of objects by understanding how they can be arranged. We shall develop only the most basic aspects of the theory of permutations, namely, only those parts that we shall use in other parts of the text. The reader is directed to any abstract algebra text for a complete description.

We begin with the definition of a permutation. A permutation on a set A of size n is a bijection $\sigma : A \rightarrow A$. That is, each element of A gets mapped to a unique element of A and each element is the image of a unique element. Recall that a bijection is a map that is both surjective and injective.

The set of all permutations on a set of size n is called the symmetric group on n letters and is denoted by S_n. It is immediate from the results in the basic counting section that $|S_n| = n!$.

We note the following results about the permutations in S_n:

- Composition of functions is associative.
- The identity map is a permutation.
  If σ is the identity than for $a \in A$, $\sigma(a) = a$ so it is surjective. If $\sigma(a) = \sigma(a')$ then $a = a'$ and hence the map is injective.
- If σ and τ are both permutations then the composition $\sigma \circ \tau$ is a permutation. This was proven in Lemma 1.1.
- If σ is a permutation then the inverse map denoted by σ^{-1} is a permutation.
  If $a \in A$ then $\sigma^{-1}(\sigma(a)) = a$ and the map is surjective. If $\sigma^{-1}(a) = \sigma^{-1}(a')$ then $\sigma(\sigma^{-1}(a)) = \sigma(\sigma^{-1}(a'))$ and hence $a = a'$ and therefore the map is injective and it is a permutation.

These items prove that S_n is an object called a group. We shall define groups later in the text in Sect. 10.1, but these properties that we have shown can be used throughout the following sections.

As an example consider the following permutation on $\{1, 2, 3\}$:

$$1 \rightarrow 2$$
$$2 \rightarrow 3$$
$$3 \rightarrow 1$$

We denote this permutation by $(1, 2, 3)$ meaning that 1 is sent to 2, 2 is sent to 3 and 3 is sent to 1. The following permutation on $\{1, 2, 3, 4\}$

$$1 \rightarrow 2$$
$$2 \rightarrow 1$$
$$3 \rightarrow 4$$
$$4 \rightarrow 3$$

would be denoted $(1, 2)(3, 4)$. In general the cycle $(a_1, a_2, \ldots, a_k)$ is the permutation where a_i is sent to a_{i+1}, for i from 1 to $k - 1$ and a_k is sent to a_1.

Since a permutation is a function, then when we multiply permutations we are simply applying the compositions of functions. Therefore when applying the product of cycles the product is read from right to left.

For example, $(1, 2)(2, 3)$ would be $(2, 3, 1)$. If an element is not written in the notation then it is assumed that it is mapped to itself.

A transposition is a permutation of two elements. For example, $(1, 2)$ is the transposition that switches 1 and 2 and leaves the other elements fixed. Notice that every transposition is its own inverse, that is $(a, b)^{-1} = (a, b)$.

Theorem 1.14 *Every permutation can be written as the product of transpositions.*

Proof If $(a_1, a_2, a_3, \ldots, a_n)$ is a permutation then

$$(a_1, a_2, a_3, \ldots, a_n) = (a_1, a_n)(a_1, a_{n-1})(a_1, a_{n-2}) \ldots (a_1, a_3)(a_1, a_2).$$

This gives the result.   □

A permutation is said to be even if it can be written as the product of evenly many transpositions and odd if it can be written as the product of oddly many transpositions. Note that the identity which has 0 transpositions is an even permutation.

Theorem 1.15 *The number of even permutations on the set $\{1, 2, \ldots, n\}$, $n > 1$, is equal to the number of odd permutations on the set $\{1, 2, \ldots, n\}$.*

Proof Let E_n denote the set of even permutations on the set $\{1, 2, \ldots, n\}$ and let O_n denote the set of odd permutations on the set $\{1, 2, \ldots, n\}$. Define the map $f : E_n \to O_n$ by $f(\sigma) = (1, 2)\sigma$. Clearly, this map sends even permutations to odd permutations. Assume $(1, 2)\sigma = (1, 2)\sigma'$ then $(1, 2)(1, 2)\sigma = (1, 2)(1, 2)\sigma'$ which gives $\sigma = \sigma'$. Then f is an injection and so $|E_n| \leq |O_n|$.

Then we let f map O_n to E_n in the same exact manner which gives $|O_n| \leq |E_n|$. This gives that $|E_n| = |O_n|$ and we have the result.   □

It follows immediately that $|E_n| = |O_n| = \frac{n!}{2}$. While the set of all permutations is called the symmetric group, the set of even permutations is called the alternating group.

Theorem 1.16 *If*

$$\sigma = (a_1, a_2, a_3, \ldots, a_n) = (a_1, a_n)(a_1, a_{n-1})(a_1, a_{n-2}) \cdots (a_1, a_3)(a_1, a_2)$$

then

$$\sigma^{-1} = (a_1, a_n, a_{n-1}, \ldots, a_2) = (a_1, a_2)(a_1, a_3)(a_1, a_4) \cdots (a_1, a_{n-1})(a_1, a_n).$$

Proof Simply multiply the permutations:

$$(a_1, a_n)(a_1, a_{n-1})(a_1, a_{n-2}) \cdots (a_1, a_3)(a_1, a_2)(a_1, a_2)$$
$$(a_1, a_3)(a_1, a_4) \cdots (a_1, a_{n-1})(a_1, a_n).$$

Using the fact that each transposition is its own inverse $n - 1$ times gives that this product is the identity. Taking the multiplication in the opposite order gives the same result. □

Multiplication of permutations is, in general, not a commutative operation. For example,

$$(1, 4, 3, 2)(2, 4, 3, 1) = (2, 3, 4)$$

but

$$(2, 4, 3, 1)(1, 4, 3, 2) = (1, 3, 4).$$

However, if the cycles are disjoint then permutations do commute. For example,

$$(1, 5, 3)(2, 4, 6) = (2, 4, 6)(1, 5, 3).$$

1.4.1 Japanese Ladders

We shall give a visual representation for permutations which are known as Japanese ladders. Traditionally, a Japanese ladder (known in Japanese as Amidakuji) was used to distribute objects or tasks to a group of people.

We begin by giving a graphical example before defining them rigorously.

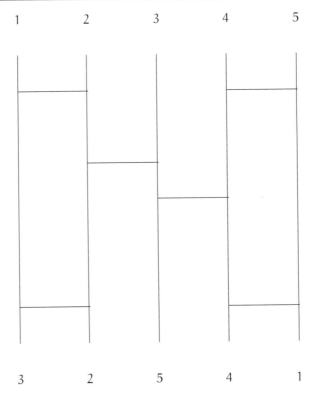

The elements fall down the ladder and cross over any rung it encounters. The element 1 falls down the first post and then moves to the second, then to the third, then to the fourth, and finally to the fifth. The element 2 falls down the second post, moves to the first, and then back to the second. The element 3 falls down the third post, moves to the second, and finally to the first. The element 4 falls down the fourth post, moves to the fifth, and back to the fourth. The element 5 falls down the fifth post then moves to the fourth, and finally to the third.

Since 1 ends in the fifth place we say $1 \to 5$. Likewise we have $2 \to 2$, $3 \to 1$, $4 \to 4$, and $5 \to 3$. As a permutation, this ladder represents $(1, 5, 3)$. Essentially, we have written it as a product of transpositions $(4, 5)(1, 2)(3, 4)(2, 3)(4, 5)(1, 2)$.

We can now define a Japanese ladder rigorously. A Japanese ladder is a representation of a permutation of the form:

$$\prod_{i \in A} (i, i+1), \qquad (1.2)$$

where A is an ordered list of elements of $\{1, 2, \ldots, n\}$.

Here, each transposition $(i, i+1)$ corresponds to a rung of the ladder from the ith post to the $(i+1)$-st post.

There are several benefits to viewing a permutation in this graphical manner. The first benefit of this representation is that the product of two permutations has a very natural representation. Namely, the product of two ladders representing permutations in S_n can be represented by placing one ladder underneath the other. More precisely, the ladder representing $\beta\alpha$ is formed by placing the ladder for β under the ladder for α.

Let α be the permutation: $(1, 4, 3)$. It can be represented as follows:

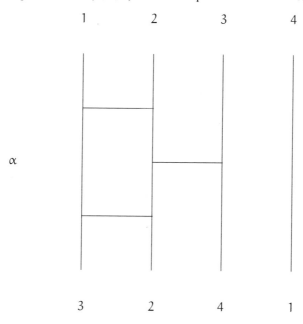

Let β be the permutation: $(1, 2, 4, 3)$. It can be represented as follows:

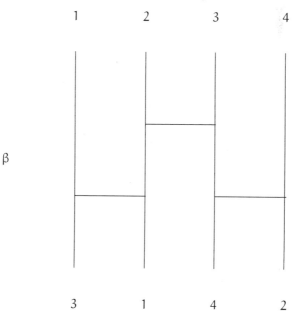

Placing α on the top of β (namely, α is performed first), we obtain the permutation $\beta\alpha = (1, 3, 2, 4)$. We realize it as follows:

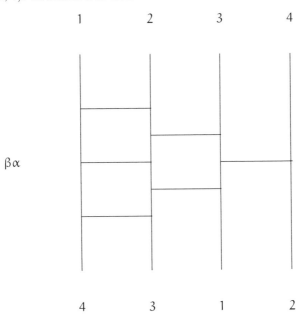

Placing β on the top of α (namely, β is performed first), we obtain the permutation $\alpha\beta = (1, 2, 3, 4)$. We realize it as follows:

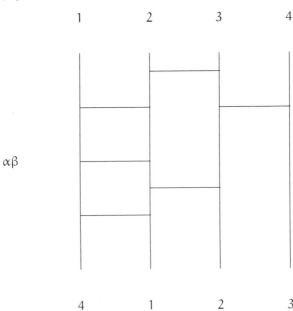

The difference in these two ladders illustrates the non-commutativity of multiplication, that is $(1, 2, 3, 4) = \alpha\beta \neq \beta\alpha = (1, 3, 2, 4)$.

We shall now show that every permutation can be realized as a Japanese ladder. We have already shown that every permutation can be written as a product of transpositions. All that remains is to show that any transposition can be written as a Japanese ladder. Take $a < b$, then the transposition (a, b) can be written as follows:

$$(a, b) = (a, a+1)(a+1, a+2)\cdots(b-1, b)(b-2, b-1)\cdots(a+1, a+2)(a, a+1).$$

As an example we give the ladder for $(1, 5)$:

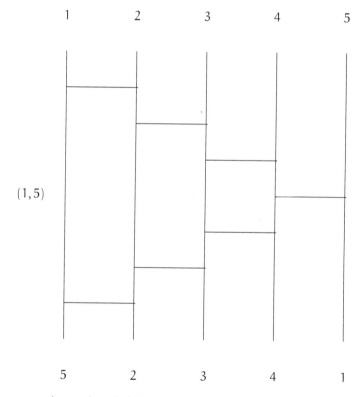

Any permutation can have infinitely many representations as ladders. For example, one can simply add two rungs at the bottom of any two adjacent posts in an existing ladder. For example, if α is a Japanese ladder then $(1, 2)(1, 2)\alpha$ is a Japanese ladder that represents the same permutation. Moreover, two ladders can represent the same permutation in a non-trivial manner as in the following:

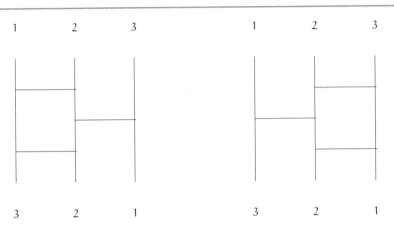

Recall that even permutations can be written as the product of an even number of transpositions, and odd permutations can be written as the product of an odd number of transpositions. When a permutation is realized as a Japanese ladder, even permutations have an even number of rungs and odd permutations have an odd number of rungs. This visualization makes it obvious that the product of two even permutations is even, two odd permutations is even, and the sum of an even permutation and an odd permutation is odd.

Consider the permutation $\alpha = (a_1, a_2)(a_3, a_4)(a_5, a_6) \cdots (a_{k-1}, a_k)$ then $\alpha^{-1} = (a_{k-1}, a_k) \cdots (a_5, a_6)(a_3, a_4)(a_1, a_2)$. When the permutation is realized as a Japanese ladder, the inverse permutation is formed by flipping the ladder upside down.

Recall the permutation $\alpha = (1, 4, 3)$ given above. Then we have that $\alpha^{-1} = (1, 3, 4)$ be represented as follows:

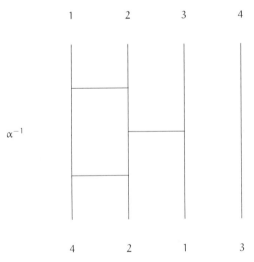

It follows immediately from this visual interpretation that $(\alpha\beta)^{-1} = \beta^{-1}\alpha^{-1}$. Namely, if we have the ladder with β on top of α. Turning it upside down gives α upside down on top of β upside down. Specifically, we place β above α to get $\alpha\beta$ then turning the whole new ladder upside down results in a ladder with the inverse of α on top and the inverse of β on the bottom giving $\beta^{-1}\alpha^{-1}$.

There is an algorithm for constructing a Japanese ladder for a given permutation, see [30] for its relation to a game and [57] for its relation to computer searching. Namely, you can *comb* the permutation into its proper order. Consider a sequence of numbers on the bottom of the ladder: $(a_1, a_2, \ldots, a_n)$. An inversion is any pair a_i, a_j where $i < j$ but $a_i > a_j$. The total number of inversions is the number of minimal number of rungs needed to construct a Japanese ladder for the permutation. Specifically, change each place where $a_i > a_{i+1}$ by placing a rung between the ith and the $(i + 1)$-st rung.

For example, consider the following final state 7 3 4 2 1 5 6. The total number of inversions is $6 + 2 + 2 + 1 = 11$. This corresponds to the permutation $(1, 5, 6, 7)$ $(2, 4, 3)$. We shall comb it in the manner described above to put it in proper order:

$$7\ 3\ 4\ 2\ 1\ 5\ 6$$
$$3\ 7\ 4\ 2\ 1\ 5\ 6$$
$$3\ 4\ 7\ 2\ 1\ 5\ 6$$
$$3\ 4\ 2\ 7\ 1\ 5\ 6$$
$$3\ 4\ 2\ 1\ 7\ 5\ 6$$
$$3\ 4\ 2\ 1\ 5\ 7\ 6$$
$$3\ 4\ 2\ 1\ 5\ 6\ 7$$
$$3\ 2\ 4\ 1\ 5\ 6\ 7$$
$$3\ 2\ 1\ 4\ 5\ 6\ 7$$
$$3\ 1\ 2\ 4\ 5\ 6\ 7$$
$$1\ 3\ 2\ 4\ 5\ 6\ 7$$
$$1\ 2\ 3\ 4\ 5\ 6\ 7$$

This corresponds to the permutation:

$$(1, 2)(2, 3)(3, 4)(4, 5)(5, 6)(6, 7)(2, 3)(3, 4)(2, 3)(1, 2)(2, 3). \qquad (1.3)$$

Notice that these transpositions are in the opposite way than they were given in the algorithm. This is because the algorithm turns 7 3 4 2 1 5 6 into 1 2 3 4 5 6 7 and the ladder is the inverse permutation of this one. Notice that there are precisely 11 rungs in this ladder.

Exercises

1. Multiply the following permutations:

   a. $(1, 3, 4)(2, 4, 5, 6)(1, 6, 5, 2)$
   b. $(1, 2, 3)(1, 5, 2, 3)(3, 2, 4, 5)$
   c. $(4, 5, 1, 3, 2)(2, 3, 4)(1, 6, 5, 3)$
   d. $(1, 2, 3, 4, 5)(5, 4, 3, 2, 1)$
   e. $(1, 2, 3, 4, 5)(1, 5, 4, 3, 2)$
   f. $(2, 5)(1, 3)(4, 6)(7, 8)$

2. Write the following permutations as a product of transpositions and determine whether they are even or odd.

   a. $(1, 6, 3, 4, 5, 2)$
   b. $(1, 4, 3, 2, 5, 6)$
   c. $(2, 4, 6, 1, 3, 5)$
   d. $(6, 2, 3, 1, 5, 4)$
   e. $(3, 5, 7, 9, 1, 2, 4, 6, 8)$
   f. $(4, 5, 3, 6, 2, 7, 8, 1, 9)$

3. Prove that for two permutations σ and τ we have $(\sigma\tau)^{-1} = \tau^{-1}\sigma^{-1}$.

4. Find all 6 permutations in S_3. Then make the 6×6 multiplication table for these permutations.

5. Prove that no permutation can be written both as evenly many transpositions and as oddly many transpositions.

6. Prove that the product of two even permutations is even, the product of two odd permutations is even, and the product of one odd and one even is odd.

7. Let F be a finite field. Let $\sigma_a(x) = ax$. Prove that for each $a \neq 0$ the map σ_a is a permutation of the elements of F. Prove that $\sigma_a(x) \neq \sigma_b(x)$ for all $x \in F - \{0\}$ if $a \neq b$.

8. Construct a Japanese ladder with a minimum number of rungs for each of the following permutations:

   a. $(2, 4, 3, 5, 6, 7, 1)$
   b. $(4, 2, 3, 1, 5, 6, 7)$
   c. $(7, 6, 5, 4, 3, 2, 1)$
   d. $(1, 3, 2, 4)(6, 5, 7)$
   e. $(4, 2, 5)(1, 3, 6, 7)$
   f. $(1, 7, 2, 6, 3, 5, 4)$

9. A Caesar cipher is when there is a permutation on the set of 26 English letters used to send a secret message. (Historically, Caesar simply cycled the letters by 3 positions). Determine how many possible ciphers can be used by this method.

Next, count the number of ciphers that can be created by making a permutation on the set of all 2 letter pairs. Then, count the number of ciphers that can be created by making a permutation on the set of all 3 letter triples.

1.5 Generating Functions

1.5.1 Coin Counting

Generating functions are an excellent example of how to use algebra in counting. Basically, you have a large collection of numbers you want to understand. Instead of considering them as a large (infinite in fact) collection, we can consider them as a single algebraic object, namely, an infinite series. We are not concerned about where the series converges, as in calculus, but rather we consider them as a formal series, meaning they are simply an algebraic object which we use to carry information for us. One of the most popular descriptions of this can be found in Polya's article [68].

We begin by recalling the standard example from calculus of a series, namely, the geometric series:

$$\frac{1}{1-x} = \sum_{n=0}^{\infty} x^n = 1 + x + x^2 + \cdots . \tag{1.4}$$

It is easy to see that the coefficient of x^n in this series is the number of ways of making n cents from pennies, namely, there is 1 way. If we change our pennies to nickels then we can only get multiples of 5 for our sum. If we let $P(x) = \frac{1}{1-x}$, then we can have

$$N(x) = P(x^5) = \sum_{n=0}^{\infty} x^{5n} = 1 + x^5 + x^{10} + x^{15} + \cdots . \tag{1.5}$$

Next, if we want to use both pennies and nickels we can get

$$NP(x) = N(x)P(x) = \frac{1}{1-x}\frac{1}{1-x^5}$$
$$= 1 + x + x^2 + x^3 + x^4 + 2x^5 + 2x^6 + 2x^7 + 2x^8 + 2x^9 + 3x^{10} + \cdots .$$

If we look at the coefficient of x^6, this should tell us how many ways are there in getting 6 cents from pennies and nickels. Namely, the coefficient is 2 which corresponds to 6 pennies and 1 nickel and 1 penny. Similarly, the coefficient of x^{10} is 3 corresponding to 10 pennies, 1 nickel and 5 pennies, and 2 nickels. This series then counts all possible combinations and yet can be considered as a single algebraic object.

We can use this algebraic structure to get other interesting relations. Notice that

$$NP(x) = \frac{1}{1-x}\frac{1}{1-x^5} = \frac{1}{1-x^5}P(x), \tag{1.6}$$

which gives

$$(1 - x^5)NP(x) = P(x)$$
$$NP(x) = x^5 NP(x) + P(x).$$

Let NP_n be the coefficient of x^n in $NP(x)$, and let P_n be the coefficient of x^n in $P(x)$. Then from the previous equation, we have that

$$NP_n = NP_{n-5} + P_n, \qquad (1.7)$$

which gives a recurrence relation to determine the exact value of NP_n. Specifically, the sequence of NP_n, starting with $n = 0$ is

$$1, 1, 1, 1, 1, 2, 2, 2, 2, 2, 3, 3, 3, 3, 3, 4, 4, 4, 4, 4, \ldots \qquad (1.8)$$

A closed form for NP_n is immediate, namely,

$$NP_n = \lfloor \frac{n}{5} \rfloor + 1. \qquad (1.9)$$

We can now add dimes to the collection of coins we are allowed to use. Then

$$D(x) = P(x^{10}) = 1 + x^{10} + x^{20} + x^{30} + \cdots \qquad (1.10)$$

Then

$$DNP(x) = D(x)N(x)P(x) = \frac{1}{1 - x^{10}} \frac{1}{1 - x^5} \frac{1}{1 - x}$$
$$= 1 + x + x^2 + x^3 + x^4 + 2x^5 + 2x^6 + 2x^7 + 2x^8 + 2x^9 + 4x^{10}$$
$$+ 4x^{11} + 4x^{12} + 4x^{13} + 4x^{14} + 6x^{15} + 6x^{16} + 6x^{17} + 6x^{18} + 6x^{19} + 9x^{20} + \cdots$$

Now we have

$$DNP(x) = \frac{1}{1 - x} \frac{1}{1 - x^5} \frac{1}{1 - x^{10}} = \frac{1}{1 - x^{10}} NP(x), \qquad (1.11)$$

which gives

$$(1 - x^{10})DNP(x) = NP(x)$$
$$DNP(x) = x^{10}DNP(x) + NP(x).$$

Then letting DNP_n be the coefficient of x^n in $DNP(x)$ we have

$$DNP_n = DNP_{n-10} + NP_n, \qquad (1.12)$$

which gives a recurrence relation to determine the exact value of NP_n. Specifically, the sequence of NP_n, starting with $n = 0$ is

$$1, 1, 1, 1, 1, 2, 2, 2, 2, 2, 4, 4, 4, 4, 4, 6, 6, 6, 6, 6, 9, 9, 9, 9, 9,$$
$$12, 12, 12, 12, 12, 16, 16, 16, 16, 16 \ldots$$

Finally, we will add quarters to the collection of coins we are allowed to use. Then

$$Q(x) = P(x^{25}) = 1 + x^{25} + x^{50} + x^{75} + \cdots \qquad (1.13)$$

Then

$$QDNP(x) = N(x)P(x)D(x)Q(x) = \frac{1}{1-x}\frac{1}{1-x^5}\frac{1}{1-x^{10}}\frac{1}{1-x^{25}}. \qquad (1.14)$$

Now we have

$$QDNP(x) = \frac{1}{1-x}\frac{1}{1-x^5}\frac{1}{1-x^{10}}\frac{1}{1-x^{25}} = \frac{1}{1-x^{25}}QDNP(x) \qquad (1.15)$$

which gives

$$(1-x^{25})NP(x) = NP(x)$$
$$QDNP(x) = x^{25}QDNP(x) + DNP(x).$$

Then letting $QDNP_n$ be the coefficient of x^n in $QDNP(x)$ we have

$$QDNP_n = QDNP_{n-25} + DNP_n, \qquad (1.16)$$

which gives a recurrence relation to determine the exact value of NP_n. Specifically, the sequence of NP_n, starting with $n = 0$ is

$$1, 1, 1, 1, 1, 2, 2, 2, 2, 2, 4, 4, 4, 4, 4, 6, 6, 6, 6, 6, 9, 9, 9, 9, 9,$$
$$13, 13, 13, 13, 13, 18, 18, 18, 18, 18 \ldots$$

We can use these ideas to solve finite problems. For example, assume we have 3 five dollar bills, 4 ten dollar bills, and 2 twenty dollar bills. We can determine all values that we can obtain from these bills, and how many ways that we can obtain these. For five dollar bills we have $F(x) = 1 + x^5 + x^{10} + x^{15}$. For ten dollar bills we have $T(x) = 1 + x^{10} + x^{20} + x^{30} + x^{40}$. For twenty dollar bills we have $W(x) = 1 + x^{20} + x^{40}$.

Then we can multiply $W(x)T(x)F(x)$ and get

$$1 + x^5 + 2x^{10} + 2x^{15} + 3x^{20} + 3x^{25} + 4x^{30} + 4x^{35} + 5x^{40} + 5x^{45} + 5x^{50}$$
$$+ 5x^{55} + 4x^{60} + 4x^{65} + 3x^{70} + 3x^{75} + 2x^{80} + 2x^{85} + x^{90} + x^{95}.$$

For example, we can see that there are 4 ways of making 35 dollars and 3 ways of making 75 dollars. Hence, we have solved numerous problems all at once using a simple algebraic computation.

We shall now show a technique for determining the coefficient for some frequently appearing generating functions. We begin with a lemma.

Theorem 1.17 *Let* n *be a positive integer. Then*

$$\frac{1}{(1+x)^n} = \sum_{k=0}^{\infty} (-1)^k C(n+k-1, k) x^k \tag{1.17}$$

and

$$\frac{1}{(1-x)^n} = \sum_{k=0}^{\infty} C(n+k-1, k) x^k. \tag{1.18}$$

Proof Let $f(x) = (1+x)^{-n}$. Then the kth derivative of $f(x)$ is $f^{(k)}(x) = (-1)^k n(n+1)\cdots(n+k-1)(1+x)^{-(n+k)}$. Then $f^{(k)}(0) = (-1)^k C(n+k-1, k)$. Then applying Taylor's Theorem we have the first result.

Let $g(x) = (1-x)^{-n}$. Then the kth derivative of $f(x)$ is $f^{(k)}(x) = n(n+1)\cdots(n+k-1)(1+x)^{-(n+k)}$. Then $f^{(k)}(0) = C(n+k-1, k)$. Then applying Taylor's Theorem we have the second result.   □

Example 1.17 Let $n = 2$, then

$$\frac{1}{(1-x)^2} = 1 + 2x + 3x^2 + 4x^3 + 5x^4 + \cdots.$$

Then

$$1 + \frac{x}{(1-x)^2} = 1 + x + 2x^2 + 3x^3 + 4x^4 + 5x^5 + \cdots,$$

which is the generating function counting the number of ways you can choose 1 object out of k objects.

Theorem 1.18 *Let* n *be a positive integer. Then*

$$\frac{1 - x^{n+1}}{1 - x} = 1 + x + x^2 + \cdots + x^n.$$

Proof We have

$$(1-x)(1+x+x^2+\cdots+x^n) = 1+x+x^2+\cdots+x^n - x - x^2 - x^3 - \cdots - x^n - x^{n+1} = 1 - x^{n+1}.$$

This gives the result.   □

We shall show how to use these results to compute coefficients of generating functions.

Example 1.18 We shall determine the coefficient of x^{14} in $(x^2 + x^3 + x^4 + \cdots)^4$. We have

$$(x^2 + x^3 + x^4 + \cdots)^4 = (x^2(1 + x + x^2 + \cdots))^4$$
$$= x^8(1 + x + x^2 + \cdots)^4 = x^8 \frac{1}{(1-x)^4}.$$

Then the coefficient of x^{14} in $(x^2 + x^3 + x^4 + \cdots)^4$ is the coefficient of x^6 in $\frac{1}{(1-x)^4}$ which is $C(4 + 6 - 1, 6) = C(9, 6) = 84$.

1.5.2 Fibonacci Sequence

We shall now investigate one of the most famous and interesting sequences and use a generating function to get a closed form for the sequence. Specifically, we are going to study the Fibonacci sequence. The sequence first appeared in the early thirteenth century text *Liber Abaci* by Leonardo di Pisa, whom we know better as Fibonacci. We define the sequence recursively as follows:

$$\mathcal{F}_0 = 0, \mathcal{F}_1 = 1, \mathcal{F}_n = \mathcal{F}_{n-1} + \mathcal{F}_{n-2}. \tag{1.19}$$

Concretely, the sequence is

$$0, 1, 1, 2, 3, 5, 8, 13, 21, 34, 55, 89, \ldots \tag{1.20}$$

We can then think of the generating function for the series as

$$F(x) = 0 + x + x^2 + 2x^3 + 3x^4 + 5x^5 + 8x^6 + 13x^7 + 21x^8 + 34x^9 + 55x^{10} + 89x^{11} + \cdots \tag{1.21}$$

We would like to find a closed form for this sequence. Consider the following:

$$F(x) = 0 + x + x^2 + 2x^3 + 3x^4 + 5x^5 + 8x^6 + \cdots$$
$$xF(x) = \quad\quad x^2 + x^3 + 2x^4 + 3x^5 + 5x^6 + \cdots$$
$$x^2F(x) = \quad\quad\quad\quad x^3 + x^4 + 2x^5 + 3x^6 + \cdots$$

Using the relation $\mathcal{F}_n = \mathcal{F}_{n-1} + \mathcal{F}_{n-2}$, we see that

$$F(x) - xF(x) - x^2F(x) = x$$
$$F(x) = \frac{x}{1 - x - x^2}.$$

In order to get this into the form we desire, we apply the standard technique of partial fractions on this function. We begin by factoring the denominator. However, we do not want to factor in terms of $(x - \alpha)(x - \beta)$ since this will not help us in terms of series that we know. Rather, we want to factor in terms of $(1 - \alpha x)(1 - \beta x)$.

Therefore, we factor the denominator as

$$1 - x - x^2 = (1 - \frac{1 + \sqrt{5}}{2}x)(1 - \frac{1 - \sqrt{5}}{2}x).$$

We recognize the number $\frac{1+\sqrt{5}}{2}$ as the golden mean. We shall describe this number in geometric terms. Assume a line segment is split into two segments of length L and S, where the ratio of the longer side L to the smaller side S is the same as the ratio of the entire line segment $L + S$ is to the larger side L. Namely, we have

$$\frac{L}{S} = \frac{L + S}{L}. \tag{1.22}$$

This gives

$$L^2 = SL + S^2$$
$$L^2 - SL - S^2 = 0$$
$$(\frac{L}{S})^2 - \frac{L}{S} - 1 = 0.$$

Solving this with the binomial equation gives that the ratio

$$\frac{L}{S} = \frac{1 \pm \sqrt{5}}{2}.$$

The standard notation is to have $\varphi = \frac{1+\sqrt{5}}{2}$ and $\overline{\varphi} = \frac{1-\sqrt{5}}{2}$. Then we can write

$$1 - x - x^2 = (1 - \varphi x)(1 - \overline{\varphi}x).$$

Applying the technique of partial fractions we have

$$\frac{x}{1 - x - x^2} = \frac{\frac{1}{\sqrt{5}}}{1 - \varphi x} - \frac{\frac{1}{\sqrt{5}}}{1 - \overline{\varphi}x}. \tag{1.23}$$

Since we have written these in terms of the standard geometric series, we now have

$$F(x) = \frac{1}{\sqrt{5}}(\sum_{n=0}^{\infty} \varphi^n - \overline{\varphi}^n) \tag{1.24}$$

This gives a closed form for the nth term of the Fibonacci sequence, specifically

$$\mathcal{F}_n = \frac{1}{\sqrt{5}}(\varphi^n - \overline{\varphi}^n). \tag{1.25}$$

This formula was first published by Euler in [34].

Exercises

1. How many ways can you make 15 cents from pennies, nickels, and dimes?
2. How many ways can you make 1 dollar from pennies, nickels, dimes, and quarters?
3. How many ways can you make 25 cents from pennies, nickels, dimes, and quarters?
4. Use generating functions to see how many ways 50 dollars can be made using 5 dollar bills, 10 dollar bills, and 20 dollar bills.
5. Use generating functions to see how many ways you can make 50 cents from dimes, quarters, and half-dollars.
6. Using the same recursion that is used for the Fibonacci sequence but changing the first two values from 0 and 1 to 1 and 2, apply the technique of generating functions to find a closed form for the new sequence.
7. Assuming that the limit exists, show that $\lim_{n \to \infty} \frac{\mathcal{F}_{n+1}}{\mathcal{F}_n} = \varphi$.
8. Lucas numbers are defined to be the sequence defined by $L_0 = 2$, $L_1 = 1$, $L_n = L_{n-1} + L_{n-2}$ for $n > 2$. Write out the first 10 terms of the sequence. Apply the technique of generating functions to find a closed form for the new sequence.

Foundational Algebraic Structures

2.1   Modular Arithmetic

In this section, we shall develop a very small part of the theory of modular arithmetic. The ideas behind it are actually very old, but it was first codified by Gauss in *Disquisitiones Arithmeticae* [39]. In the opinion of the author, Gauss stands with Euler and Archimedes as one of the three greatest mathematicians of all time. Others may dispute this but certainly these three are always included in the top five. In any serious study of discrete and combinatorial mathematics, the names of Euler and Gauss will appear again and again.

We begin with the definition of the relation modulo on the integers. The integers denoted by $\mathbb{Z}$ are the set $\{\ldots, -3, -2, -1, 0, 1, 2, 3, \ldots\}$. A relation on $\mathbb{Z}$ is simply a subset of $\mathbb{Z} \times \mathbb{Z}$.

Definition 2.1 Let a, b be integers and n a positive integer. We say that $a \equiv b$ (mod n) if and only if $a - b = kn$ for some integer k.

As an example, we see that $7 \equiv 1$ (mod 3) since $7 - 1 = 6 = 2(3)$. We also have that $-2 \equiv 1$ (mod 3) since $-2 - 1 = -3 = (-1)3$.

We shall show that the relation modulo n is an equivalence relation. Namely, that it is reflexive, symmetric, and transitive.

Any integer a has $a - a = 0 = 0n$ so $a \equiv a$ (mod n). Therefore the relation is reflexive.

If $a \equiv b$ (mod n) then $a - b = kn$ so $b - a = (-k)n$ and then $b \equiv a$ (mod n) and the relation is symmetric.

If $a \equiv b$ (mod n) and $b \equiv c$ (mod n) then $a - b = kn$ and $b - c = k'n$, then $a - c = a - b + b - c = kn + k'n = (k + k')n$ and so $a \equiv c$ (mod n). This shows that the relation is transitive.

We have shown the following theorem.

© The Editor(s) (if applicable) and The Author(s), under exclusive license   35
to Springer Nature Switzerland AG 2020
S. T. Dougherty, *Combinatorics and Finite Geometry*,
Springer Undergraduate Mathematics Series,
https://doi.org/10.1007/978-3-030-56395-0_2

Theorem 2.1 *The relation* $a \equiv b \pmod{n}$ *is an equivalence relation on the integers.*

The equivalence classes of any equivalence relation form a partition on the ambient space. In other words, you can think of there being n bins where each integer is placed in a unique bin. Instead of doing the arithmetic in the integers we can simply think of doing the arithmetic with bins. We can denote the bins by $0, 1, \ldots, n-1$ as we shall now describe.

We see that if a, b are integers with $a \neq b$ and $0 \leq a, b \leq n-1$ then $a \not\equiv b \pmod{n}$ since $a - b \neq 0$ and $0 < |a - b| < n$ and so $a - b$ is not a multiple of n. Moreover, any integer is equivalent modulo n to some number k with $0 \leq k \leq n-1$. This follows from the division algorithm, namely, an integer $m = qn + r$ with $0 \leq r < n$. That is we simply divide m by n and get remainder r. In this situation $m - r = qn$ and so $m \equiv r \pmod{n}$.

What this implies is that we can always take an integer between 0 and $n-1$ as the representative of the equivalence class. That is, we can always divide a number by n and then the remainder is what the number is equivalent to modulo n.

The ideas behind modular arithmetic are familiar to everyone. One example is the way in which we tell time. If it is 11:00 then in 4 h it will be 3:00. What we have done is simply taken $11 + 4 \equiv 3 \pmod{12}$. Whenever we hit 12:00, we reset to 0. The hours in a day are simply taken modulo 12. If it was 2:00 and someone asked you what time it would be in 42 hours, you would simply calculate $42 + 2 \pmod{12}$ which is $6 + 2 = 8$ and it would be 8:00. Another example is the days of the week. We give a specific name (Sunday, Monday, Tuesday, Wednesday, Thursday, Friday, Saturday) to the seven representatives modulo 7. If it is Tuesday then we know in 6 d it will be Monday. It is a modular computation we are all familiar making. Notice that on a calendar that all numbers in a column are equivalent modulo 7.

2.1.1 Euclidean Algorithm

We shall use modular arithmetic to describe the Euclidean algorithm for finding the greatest common divisor of two integers, since we shall need it in some of the proofs that follow. The greatest common divisor of two integers a and b denoted $\gcd(a, b)$ is the largest integer that divides both a and b, where an integer n divides an integer m if and only if $m = nk$ for some integer k.

Take any two integers a_1 and a_2. Without loss of generality assume that $a_1 > a_2$. We shall say that $a \pmod{b} = c$ if c is the unique number in $\{0, 1, \ldots, b-1\}$ such that $a \equiv c \pmod{b}$. This can be thought of as viewing c as the remainder when a is divided by b. Define a_i for $i > 2$ to be $a_{i-2} \pmod{a_{i-1}}$. The sequence $a_1, a_2, a_3, \ldots$ is a decreasing sequence of non-negative integers. Therefore, there exists an integer s with $a_s \neq 0$ but $a_{s+1} = 0$. That is, we have the following finite sequence:

$$a_1, a_2, a_3, a_4, \ldots, a_{s-1}, a_s, 0.$$

As an example, consider the numbers 47 and 13. The sequence would be

$$47, 13, 8, 5, 3, 2, 1, 0.$$

Theorem 2.2 (Euclidean algorithm) *Let* a_1, a_2 *be two positive integers with* $a_1 > a_2$. *Define* a_i *for* $i > 2$ *to be* a_{i-2} (mod a_{i-1}). *If* $a_{s+1} = 0$, *then the integer* a_s *is the greatest common divisor of* a_1 *and* a_2.

Proof Since a_{s-1} (mod a_s) $= 0$ we have that a_{s-1} is a multiple of a_s. We know a_{i-1} (mod a_i) $= a_{i+1}$. Therefore, there is an integer k with $a_{i-1} - ka_i = a_{i+1}$. This gives that $a_{i-1} = ka_i + a_{i+1}$ and it follows that if b is an integer that divides a_i and a_{i+1} then b divides a_{i-1}. Then, by induction, we have that a_s divides each a_i in the sequence.

Assume d is an integer that divides a_1 and a_2. If d divides a_i and a_{i-1} then we know a_{i-1} (mod a_i) $= a_{i+1}$. Then, there is an integer h with $a_{i-1} - ha_i = a_{i+1}$. This gives that d divides a_{i+1} and so d divides every integer a_i in the sequence by induction. In particular, d divides a_s.

We have shown that a_s divides a_1 and a_2 and that if a number divides both numbers, then it must divide a_s and therefore be less than or equal to a_s. This gives that a_s is the greatest common divisor of a_1 and a_2. ☐

Not only does the Euclidean algorithm give the greatest common divisor, it additionally provides a way to write the greatest common divisor as a linear combination of the two numbers. Specifically, we have the following theorem.

Theorem 2.3 *Let* a_1 *and* a_2 *be integers with* $d = \gcd(a_1, a_2)$. *Then there exists integers* b_1, b_2 *with* $a_1 b_1 + a_2 b_2 = d$.

Proof Apply the Euclidean algorithm to a_1 and a_2 to get the sequence

$$a_1, a_2, a_3, \ldots, a_s, 0.$$

Let $d = a_s$.

We have already seen that $a_{i-1} - k_i(a_i) = a_{i+1}$. Taking one step we get $a_{i-2} - k_{i-1}(a_{i-1}) = a_i$. Then substituting for a_i, we have $a_{i-1} - k_i(a_{i-2} - k_{i-1}(a_{i-1})) = a_{i+1}$. Then, we have $(1 + k_{i-1})(a_{i-1}) - k_i(a_{i-2}) = a_{i+1}$. Thus, if a_{i+1} can be written as a linear combination of a_{i-1} and a_i, then it can be written as a linear combination of a_{i-2} and a_{i-1}. Since a_s is a linear combination of a_{s-1} and a_{s-2} we can apply this repeatedly and write $d = a_s$ as a linear combination of a_1 and a_2. ☐

Example 2.1 We shall give an example based on our first use of the Euclidean algorithm on 47 and 13. Recall that we had $47, 13, 8, 5, 3, 2, 1, 0$ as our sequence.

Then

$$3 - 2 = 1$$
$$3 - (5 - 3) = 1$$
$$2(3) - 5 = 1$$
$$2(8 - 5) - 5 = 1$$
$$2(8) - 3(5) = 1$$
$$2(8) - 3(13 - 8) = 1$$
$$5(8) - 3(13) = 1$$
$$5(47 - 3(13)) - 3(13) = 1$$
$$5(47) - 18(13) = 1.$$

Thus we have the greatest common divisor, which is 1, written as $5(47) - 18(13)$. That is 5 and -18 are the integers guaranteed by the theorem.

It is interesting to note that Euclid described all of this geometrically. Consider two lengths of wood. Use the smaller to measure the larger and when the remaining piece is smaller than the smaller of the original two pieces, cut the wood there. Then with this piece and the smaller do the process again. Continue this until none is left over. The piece you have is the greatest common divisor. This geometric approach has other benefits. For example, Euclid uses this to show when two lengths are incommensurate. Namely, if this process continues forever then there is no length that measures both of them. Euclid applies this to the base and hypotenuse of a right isosceles triangle showing that $\sqrt{2}$ is an irrational number.

2.1.2 Arithmetic Operations

We begin by showing that addition and multiplication are well defined in this setting.

Theorem 2.4 *If $a \equiv a'$ and $b \equiv b'$, then*

$$a + b \equiv a' + b' \pmod{n} \tag{2.1}$$

and

$$a * b \equiv a' * b' \pmod{n}. \tag{2.2}$$

Proof If $a \equiv a'$ and $b \equiv b'$ then $a' = a + kn$ and $b' = b + k'n$ for some $k, k' \in \mathbb{Z}$. Then

$$a' + b' = a + kn + b + k'n = (a + b) + (k + k')n \equiv a + b \pmod{n}$$

and

$$a'b' = (a + kn)(b + k'n) = ab + bkn + ak'n + kk'n^2$$
$$= (ab) + (bk + ak' + kk'n)n \equiv ab \pmod{n}$$

which gives the result.   □

The next theorem follows from the fact that addition and multiplication is commutative on the integers.

Theorem 2.5 *Let n be a positive integer, then*

$$a + b \equiv b + a \pmod{n} \tag{2.3}$$

and

$$a * b \equiv b * a \pmod{n}. \tag{2.4}$$

We can now define $\mathbb{Z}_n$ as the set $\{0, 1, 2, \ldots, n-1\}$ with addition and multiplication of the integers taken modulo n. In terms of abstract algebra, this object is called a ring. It can have properties that are quite unlike the integers or the reals. For example, in $\mathbb{Z}_4$, $2 \cdot 2 = 0$ and $2x = 0$ has two solutions 0 and 2, yet $2x = 1$ has no solutions.

Example 2.2 We give the addition and multiplication tables for $\mathbb{Z}_4$.

$$
\begin{array}{c|cccc}
+ & 0 & 1 & 2 & 3 \\
\hline
0 & 0 & 1 & 2 & 3 \\
1 & 1 & 2 & 3 & 0 \\
2 & 2 & 3 & 0 & 1 \\
3 & 3 & 0 & 1 & 2
\end{array}
\qquad
\begin{array}{c|cccc}
* & 0 & 1 & 2 & 3 \\
\hline
0 & 0 & 0 & 0 & 0 \\
1 & 0 & 1 & 2 & 3 \\
2 & 0 & 2 & 0 & 2 \\
3 & 0 & 3 & 2 & 1
\end{array}
\tag{2.5}
$$

Example 2.3 We give the addition and multiplication tables for $\mathbb{Z}_5$.

$$
\begin{array}{c|ccccc}
+ & 0 & 1 & 2 & 3 & 4 \\
\hline
0 & 0 & 1 & 2 & 3 & 4 \\
1 & 1 & 2 & 3 & 4 & 0 \\
2 & 2 & 3 & 4 & 0 & 1 \\
3 & 3 & 4 & 0 & 1 & 2 \\
4 & 4 & 0 & 1 & 2 & 3
\end{array}
\qquad
\begin{array}{c|ccccc}
* & 0 & 1 & 2 & 3 & 4 \\
\hline
0 & 0 & 0 & 0 & 0 & 0 \\
1 & 0 & 1 & 2 & 3 & 4 \\
2 & 0 & 2 & 4 & 1 & 3 \\
3 & 0 & 3 & 1 & 4 & 2 \\
4 & 0 & 4 & 3 & 2 & 1
\end{array}
\tag{2.6}
$$

The kind of odd occurrences we described for $\mathbb{Z}_4$ simply do not happen for $\mathbb{Z}_5$. Notice that linear equations have a unique solution and if the product of two numbers is 0, then at least one of them must be 0. The reason for this is that p is prime. We shall develop these results in the following theorems.

Theorem 2.6 *The equation* $ax \equiv b \pmod{n}$ *has a solution if and only if* $\gcd(a, n)$ *divides* b.

Proof If $ax \equiv b \pmod{n}$ then $ax - b = kn$ for some k. This gives that $ax - kn = b$. If d divides a and n, with $a = da'$ and $n = dn'$ then $da'x - dkn' = b$ which implies $d(a'x - kn') = b$ and d divides b.

If $d = \gcd(a, n)$ divides b, then write $a = da'$ and $n = dn'$ and $b = db'$. Consider the equation $a'x \equiv b' \pmod{n'}$. We have that $\gcd(a', n') = 1$ so the Euclidean algorithm gives that there exists integers e and f with $a'e + n'f = 1$. This means that $a'e \equiv 1 \pmod{n'}$. Then we have

$$a'x \equiv b' \pmod{n'}$$
$$ea'x \equiv eb' \pmod{n'}$$
$$x \equiv eb' \pmod{n'}.$$

This gives that $a'(eb') - b' = kn'$ for some k and then $da'(eb') - db' = dkn'$ which gives $a(eb') - b = kn$. Hence, eb' is a solution to $ax \equiv b \pmod{n}$. □

Example 2.4 We shall show some examples of solutions to modular equations.

- If $3x \equiv 4 \pmod{6}$ then $3x - 4 = 6k$ for some k in the integers, which implies $3(x - 2k) = 4$ which implies 3 divides 4, which is a contradiction. Therefore no solution to this equation exists.
- If $5x \equiv 2 \pmod{7}$ then using the Euclidean algorithm we get that $3(5) - 2(7) = 1$. This gives that $3(5) \equiv 1 \pmod{7}$. Then multiply both sides by 3 and we have $x \equiv 6 \pmod{7}$.
- If $4x \equiv 8 \pmod{12}$ then $4x - 8 = 12k$ for some k in the integers, then $x - 2 = 3k$ and $x \equiv 2 \pmod{3}$. Hence the solutions are all of the values modulo 12 that are 2 $\pmod{3}$, namely, $2, 5, 8, 11$. The greatest common divisor of 12 and 4 is 4 and hence there are four solutions.

These three examples exhibit the possible situations for modular equations.

Theorem 2.7 *If there do not exist* a, b *such that* $ab \equiv 0 \pmod{n}$, *with neither* a *nor* b *equivalent to* 0 $\pmod{n}$, *then* $ab \equiv cb \pmod{n}$, $b \not\equiv 0 \pmod{n}$ *implies* $a \equiv c \pmod{n}$.

Proof Assume $ab \equiv cb \pmod{n}$. Then $ab - cb \equiv 0 \pmod{n}$ which gives that $(a - c)b \equiv 0 \pmod{n}$. Since there are no two numbers which multiply to 0 (zero divisors), we have that either $a - c$ or b must be equivalent to 0, but we have assumed that b is not. Therefore $a - c \equiv 0 \pmod{n}$ giving that $a \equiv c \pmod{n}$. □

Definition 2.2 Let n be a positive integer.

- An element $a \in \mathbb{Z}_n$ is a unit if there exists an element b such that $ab \equiv 1$ (mod n).
- A non-zero element $c \in \mathbb{Z}_n$ is a zero divisor if there exists a $d \neq 0$ with $cd \equiv 0$ (mod n).

We begin by showing that the set of units is closed under multiplication.

Theorem 2.8 *The product of two units in $\mathbb{Z}_n$ is a unit.*

Proof Let a and b be two units in $\mathbb{Z}_n$. This means there exist c and d with $ac \equiv ca \equiv 1$ (mod n) and $bd \equiv db \equiv 1$ (mod n). Then $(ab)(dc) \equiv a(bd)c \equiv ac \equiv 1$ (mod n) This gives that ab is a unit. $\square$

Theorem 2.9 *The product of a zero divisor and any element is a zero divisor in $\mathbb{Z}_n$.*

Proof Let a be a zero divisor in $\mathbb{Z}_n$ then there exists $b \neq 0$ with $ba \equiv 0$ (mod n). Let c be any element of $\mathbb{Z}_n$. We have $b(ac) \equiv (ba)c \equiv 0c \equiv 0$ (mod n). Therefore ac is a zero divisor in $\mathbb{Z}_n$. $\square$

Lemma 2.1 *If $ac \equiv 0$ (mod n) for $c \not\equiv 0$ (mod n) then a does not have a multiplicative inverse.*

Proof If a has a multiplicative inverse b then $ba \equiv 1$ (mod n). If $ac \equiv 0$ (mod n) then $bac \equiv b0 \equiv 0$ (mod n) which implies $c = 0$ which is a contradiction. $\square$

As an example, 2 has no multiplicative inverse in $\mathbb{Z}_6$ since $2 * 3 = 0$.

Lemma 2.2 *An element $a \in \mathbb{Z}_n$ has a multiplicative inverse if and only if $gcd(a, n) = 1$.*

Proof Assume $gcd(a, n) = k > 1$. Then, let $a = ka'$, $n = n'k$ with $a', n' \in \mathbb{Z}_n$. Consider $an' = a'kn' = a'n = 0$, with $n' \neq 0$. By Lemma 2.1, we know that a does not have a multiplicative inverse. Hence if $gcd(a, n) \neq 1$ then a does not have a multiplicative inverse. This is equivalent to the contrapositive which says that if a has a multiplicative inverse then $gcd(a, n) = 1$.

Assume $gcd(a, n) = 1$. Then by the Euclidean algorithm there exists integers b, c with

$$ab + cn = 1.$$

Reading this equation modulo n we have $ab \equiv 1$ (mod n) and b is the multiplicative inverse of a. $\square$

We let $\mathcal{U}_n$ denote the set of units modulo n. That is $\mathcal{U}_n \subseteq \mathbb{Z}_n - \{0\}$ and $a \in \mathcal{U}_n$ if and only if a is a unit.

Definition 2.3 The Euler ϕ function $\phi : \mathbb{N} \to \mathbb{N}$ is defined to be $\phi(n) = |\mathcal{U}_n|$.

We shall compute this function for a few values.

Theorem 2.10 *Let* p *and* q *be distinct primes. Then* $\phi(p) = p - 1$ *and* $\phi(pq) = (p-1)(q-1)$.

Proof If p is a prime then each number between 1 and $p - 1$ must be relatively prime to p so $\phi(p) = p - 1$.

If p and q are both primes, then the only numbers that are not relatively prime to pq, that are less than pq, are the $(p-1)$ multiples of q and the $(q-1)$ multiples of p. Since p and q are relatively prime then no multiple of p is also a multiple of q. Hence

$$\phi(pq) = pq - 1 - (q-1) - (p-1) = pq - q - p + 1 = (p-1)(q-1).$$

Theorem 2.11 *If* $d = \gcd(a, n)$ *divides* b, *then* $ax \equiv b \pmod{n}$ *has exactly* d *solutions.*

Proof If $ax \equiv b \pmod{n}$ then $ax - b = kn$ for some k. This gives that $ax - kn = b$. If d divides a, b, and n, then let $a = da'$, $b = db'$, and $n = dn'$. This gives that $da'x - db' = dkn'$. Hence $a'x \equiv b' \pmod{n'}$. Since $\gcd(a', n') = 1$ there is a solution by Theorem 2.6. Moreover, it is a unique solution by Lemma 2.2. Moreover, this is easy to see by the Euclidean algorithm. If c is the inverse of a' modulo n' then $ca'x \equiv x \equiv cb' \pmod{n'}$ and there is a unique solution.

Let $e = cb'$. The solutions to $ax \equiv b \pmod{n}$ must all be solutions to $a'x \equiv b' \pmod{n'}$ as well. Therefore, the only possibilities are of the form $e + \alpha n'$. We shall show that $e, e + n', e + 2n', \ldots, e + (d-1)n'$ are all solutions to $ax \equiv b \pmod{n}$. Notice that these are the only possibilities since $e + dn' \equiv e \pmod{n}$.

We have $a(e + \alpha n') \equiv ae + da'\alpha n' \equiv b + a'\alpha n \equiv b \pmod{n}$. Hence they are all solutions and we have exactly d solutions. $\qquad\square$

2.1.3   Fermat's Little Theorem and Euler's Generalization

Exponentiation is also interesting modulo n and is used extensively in public-key encryption. One of the most interesting results in this area is due to Fermat and is known as Fermat's Little Theorem. It is important to denote this theorem by the adjective little since there are three very important Fermat's Theorems. The first is encountered by every calculus student in the beginning study of that subject. The second is the little theorem and the third is Fermat's last theorem which was the most well known and highly sought after conjecture for centuries until it was proven by Andrew Wiles in the late twentieth century.

Theorem 2.12 (Fermat's Little Theorem) *If* $a \not\equiv 0$ (mod p), *with* p *a prime, then* $a^{p-1} \equiv 1$ (mod p).

Proof Consider the elements of $\mathbb{Z}_p - \{0\} = \{1, 2, \ldots, p-1\}$. All of the computations in the proof are done modulo p. Then the set $\{a, 2a, 3a, \ldots, (p-1)a\}$ is exactly this set permuted, since if $ba = ca$ then $b = c$ by Theorem 2.24. Hence we have

$$\prod_{i=1}^{p-1} i \equiv \prod_{i=1}^{p-1} (ai) \quad (\text{mod } p)$$

$$\prod_{i=1}^{p-1} i \equiv a^{p-1} \prod_{i=1}^{p-1} i \quad (\text{mod } p).$$

Then, since we know $\prod_{i=1}^{p-1} i$ is not 0, we can cancel, which gives $a^{p-1} \equiv 1$ (mod p).

$\square$

Notice how useful this is, for example, $237^{113} \equiv 237^{112}237 \equiv (237^4)^{28}237 \equiv 237 \equiv 2$ (mod 5).

The computation was quite simple and did not require multiplying 237 by itself 113 times.

The next corollary follows immediately from the theorem and is often the way Fermat's Little Theorem is stated.

Corollary 2.1 *Let* $a \not\equiv 0$ (mod p), *with* p *a prime, then* $a^p \equiv a$ (mod p).

Proof Simply multiply each side of the congruence $a^{p-1} \equiv 1$ (mod p) by a. $\square$

It is possible to generalize Fermat's Little Theorem using the Euler ϕ function. Namely, Euler's generalization of Fermat's Little Theorem is the following:

Theorem 2.13 (Euler's generalization of Fermat's Little Theorem) *If* a *is a number that is relatively prime to* n *then*

$$a^{\phi(n)} \equiv 1 \quad (\text{mod } n).$$

Proof Consider the set $\mathcal{U}_n$ of units in $\mathbb{Z}_n$. If $a, b, c \in \mathcal{U}_n$ then $ab = ac$ then $b = c$ since a is a unit. Then let $a \in \mathcal{U}_n$ and let $b_1, \ldots, b_{\phi(n)}$ be the elements of $\mathcal{U}_n$.

We have that

$$\prod_{i=1}^{\phi(n)} b_i \equiv \prod_{i=1}^{\phi(n)} ab_i \pmod{n}$$

$$\prod_{i=1}^{\phi(n)} b_i \equiv a^{\phi(n)} \prod_{i=1}^{\phi(n)} b_i \pmod{n}$$

$$1 \equiv a^{\phi(n)} \pmod{n},$$

since $\prod_{i=1}^{\phi(n)} b_i$ is a unit. □

We can use this theorem to find the inverse of a unit in $\mathbb{Z}_n$.

Corollary 2.2 *Let a be a unit in $\mathbb{Z}_n$. Then $a^{-1} = a^{\phi(n)-1}$.*

Proof We have

$$aa^{\phi(n)-1} \equiv a^{\phi(n)} \equiv 1 \pmod{n}$$

which gives the result. □

This gives one of the most useful applications of Euler's generalization.

Corollary 2.3 *Let $\gcd(a, n) = 1$. Then the solution to $ax \equiv b \pmod{n}$ is $x \equiv a^{\phi(n)-1}b$.*

Proof By the previous corollary we have

$$ax \equiv b \pmod{n}$$
$$a^{\phi(n)-1}ax \equiv a^{\phi(n)-1}b \pmod{n}$$
$$x \equiv a^{\phi(n)-1}b \pmod{n}.$$

□

Example 2.5 Consider the modular equation $11x \equiv 7 \pmod{23}$. We need to compute $11^{21} \mod 23$. To illustrate the computational ease, we explicitly show how to computer 11^{21} efficiently. We begin by repeatedly squaring 11 modulo 23, so that we have

$$11^1 \equiv 11 \quad (\text{mod } 23)$$
$$11^2 \equiv 6 \quad (\text{mod } 23)$$
$$11^4 \equiv 13 \quad (\text{mod } 23)$$
$$11^8 \equiv 8 \quad (\text{mod } 23)$$
$$11^{16} \equiv 18 \quad (\text{mod } 23).$$

Then using the base 2 representation of 21 we have

$$11^{21} \equiv 11^{16}11^411^1 \equiv (18)(13)(11) \equiv 21 \quad (\text{mod } 23).$$

Then $x \equiv 7(21) \equiv 9 \pmod{23}$.

Given the importance of Euler's generalization of Fermat's Little Theorem, we want to be able to compute $\phi(n)$ for any positive integer n. To accomplish this, we need to use the Chinese Remainder Theorem. It is an extremely important theorem which goes back to third century China and the mathematician Sunzi. In that form, the theorem was stated in terms of remainders, but we shall state using the modern notation of modular arithmetic.

Theorem 2.14 (Chinese Remainder Theorem) *Let* m, n *be integers with* $\gcd(m, n)$ $= 1$. *There is a unique simultaneous solution* (mod mn) *to the following congruences:*

$$x \equiv a \quad (\text{mod } m)$$
$$x \equiv b \quad (\text{mod } n).$$

Proof If $x \equiv a \pmod{m}$ then $x = a + km$ for some $k \in \mathbb{Z}$. Then consider $a + km \equiv b \pmod{n}$. We get $km \equiv b - a \pmod{n}$. Since $\gcd(m, n) = 1$ there is a unique solution for $k \pmod{n}$. Call this solution c and let $k = c + gn$, for some integer g. Then $x = a + (c + gn)m = a + cm + g(nm)$. This gives $x \equiv a + cm \pmod{mn}$. □

Example 2.6 To solve

$$x \equiv 2 \quad (\text{mod } 5)$$
$$x \equiv 3 \quad (\text{mod } 7),$$

let $x = 2 + 5k$. Then $2 + 5k \equiv 3 \pmod{7}$ which gives $5k \equiv 1 \pmod{7}$. The unique solution to this congruence is $k \equiv 3 \pmod{7}$. Then $k = 3 + 7g$ and $x = 2 + 5k = 2 + 5(3 + 7g) = 17 + 35g$ which gives $x \equiv 17 \pmod{35}$.

Applying Theorem 2.14 repeatedly gives the following corollary.

Corollary 2.4 *Let* $n_1, n_2, \ldots, n_s$ *be positive integers with* $\gcd(n_i, n_j) = 1$ *if* $i \neq j$. *Let* $n = \prod_{i=1}^{s} n_i$. *There is a unique solution modulo* n *to the following congruences:*

$$x \equiv a_1 \quad (\text{mod } n_1)$$
$$x \equiv a_2 \quad (\text{mod } n_2)$$
$$\vdots$$
$$x \equiv a_s \quad (\text{mod } n_s).$$

Example 2.7 To solve the system of congruences:

$$x \equiv 2 \quad (\text{mod } 3)$$
$$x \equiv 1 \quad (\text{mod } 5)$$
$$x \equiv 5 \quad (\text{mod } 7)$$
$$x \equiv 5 \quad (\text{mod } 11),$$

we begin by solving the first two congruences to get $x \equiv 11$ (mod 15). Then we solve this congruence simultaneously with $x \equiv 5$ (mod 7) to get $x \equiv 26$ (mod 105). Solving this congruence with $x \equiv 5$ (mod 11), we have $x \equiv 236$ (mod 1155).

This theorem leads to the next theorem which is very useful in determining $\phi(n)$.

Theorem 2.15 *Let* n *and* m *be relatively prime. Then* $\phi(nm) = \phi(n)\phi(m)$.

Proof If a is a unit in $\mathbb{Z}_{mn}$, then $a = 1 + kmn$ for some $k \in \mathbb{Z}$. Then reading this equation both (mod m) and (mod n), we have that a (mod m) and a (mod n) must both be units. If e is a unit (mod m) and f is a unit (mod n) then there is a unique solution (mod mn) to

$$x \equiv e \quad (\text{mod } m)$$
$$x \equiv f \quad (\text{mod } n).$$

Therefore, the units (mod mn) correspond to the ordered pairs (e, f) where e is a unit (mod m) and f is a unit (mod n). This gives $\phi(nm) = \phi(n)\phi(m)$. $\square$

We shall now determine the value of $\phi(n)$ for prime powers.

Theorem 2.16 *Let* p *be a prime, then* $\phi(p^e) = (p-1)p^{e-1}$.

Proof There are $p^e - 1$ numbers between 1 and $p^e - 1$. The only numbers that are not relatively prime are the multiples of p, of which there are $p^{e-1} - 1$ of them. Then

$$\phi(p^e) = p^e - 1 - (p^{e-1} - 1) = p^e - p^{e-1} = (p-1)p^{e-1}.$$

□

Example 2.8 Let $n = 27 = 3^3$, then $\phi(27) = (3-1)3^2 = 18$.

The next theorem uses our previous results and gives a formula that will determine $\phi(n)$ for any natural number n.

Theorem 2.17 *Let* $n = \prod_{i=1}^{s} p_i^{e_i}$ *where* p_i *is a prime and* $p_i \neq p_j$ *if* $i \neq j$. *Then*

$$\phi(n) = \prod_{i=1}^{s} (p_i - 1)p_i^{e_i - 1}.$$

Proof The theorem follows directly from Theorems 2.15 and 2.16. □

Example 2.9 Let $n = 45 = 3^2(5)$, then $\phi(45) = \phi(3^2)\phi(5) = (3-1)3(5-1) = 24$.

Given the fact that we can easily compute $\phi(n)$ using this theorem and that we can use this result to solve modular equations by Corollary 2.3, it becomes a straightforward matter to solve any linear modular equation.

Exercises

1. Use the Euclidean algorithm to find the greatest common divisor of the following:

    a. 1601 and 543
    b. 3412 and 5464
    c. 234 and 99
    d. 729 and 111
    e. 512 and 248

2. Find the greatest common divisor of the following and write it as a linear combination of these two numbers:

    a. 64 and 14
    b. 93 and 27
    c. 12 and 96
    d. 123 and 18

e. 29 and 37

3. Prove that if p is a prime then $(a + b)^p \equiv a^p + b^p \pmod{p}$.
4. Prove that if n is not a prime then there exists a, b such that $ab \equiv 0 \pmod{n}$, with neither a nor b equal to 0.
5. Prove that the sum of a unit and a unit is not necessarily a unit in $\mathbb{Z}_n$.
6. Prove that in $\mathbb{Z}_n$, if there are no zero divisors, then every non-zero element is a unit.
7. Prove that if n is a prime then there do not exist a, b such that $ab \equiv 0 \pmod{n}$, with neither a nor b equal to 0.
8. Prove that $(n - 1)! \pmod{n}$ is $n - 1$ if n is prime, 2 if $n = 4$ and 0 otherwise.
9. Find all solutions to $x^2 - 1 = 0$ in $\mathbb{Z}_4, \mathbb{Z}_8$, and $\mathbb{Z}_{16}$.
10. Determine $\phi(n)$ for $1 \le n \le 12$.
11. Use Fermat's Little Theorem to find the following:

   a. $573^{112} \pmod 7$
   b. $126^{1432} \pmod{11}$
   c. $146^{1203} \pmod{13}$
   d. $2^{1234} \pmod{17}$
   e. $3^{70} \pmod{23}$

12. Use Euler's generalization to find the following:

   a. $7^{123} \pmod{15}$
   b. $6^{243} \pmod{25}$
   c. $8^{111} \pmod{27}$
   d. $3^{204} \pmod{11}$
   e. $10^{403} \pmod{35}$

13. Solve the following:

   a. $3x \equiv 7 \pmod{19}$
   b. $2x \equiv 16 \pmod{24}$
   c. $13x \equiv 27 \pmod{71}$
   d. $11x \equiv 19 \pmod{23}$
   e. $6x \equiv 9 \pmod{15}$

14. Prove that in $\mathbb{Z}_n$, every non-zero element is either a unit or a zero divisor. By providing a counterexample, show that this is not the case for $\mathbb{Z}$. Determine if this is true in $\mathbb{Q}, \mathbb{R}$, and $\mathbb{C}$.

2.2   Finite Fields

The basic algebraic building block for developing Euclidean geometry is the set of real numbers. Euclid and classical Greek geometers did not take an algebraic approach to geometry. However, even for these geometers the axioms of the real numbers were implicit in their work. For example, they would have taken it for granted that between any two points on a line there is another point. It was not until the early seventeenth century that a clear understanding of the relationship between algebra and geometry would emerge. It is from this vantage point that we can take an algebraic approach to geometry. Namely, we can view the Euclidean space in terms of Cartesian geometry. In this way, we relate n-dimensional Euclidean space with $\mathbb{R}^n$.

The set of real numbers, together with the usual operations of addition and multiplication, is known in algebra as a field. In this section, we shall develop finite fields, namely, structures which satisfy similar algebraic properties as the real numbers, but with only finitely many elements. These objects will serve a similar purpose for finite geometry as the set of real numbers served for infinite geometry. Many of the analytic and topological properties of the real numbers will not be satisfied by these objects, but the vital algebraic properties will still hold.

The study of fields is a very large and interesting branch of algebra. We shall only touch on the most elementary properties of fields which are needed for the construction of geometries. For an excellent introduction to the study of fields see Fraleigh's text [38]. For a classic text on the matter see Bourbaki's text [13]. We begin with the definition of a field.

Definition 2.4 A field $(F, +, *)$ is a set F with two operations $+$ and $*$ such that the following hold

1. (additive closure) If $a, b \in F$ then $a + b \in F$.
2. (additive associativity) For all $a, b, c \in F$ we have $(a + b) + c = a + (b + c)$.
3. (additive identity) There exists an element $0 \in F$ such that $0 + x = x + 0 = x$ for all $x \in F$.
4. (additive inverses) For all $a \in F$ there exists $b \in F$ with $a + b = b + a = 0$.
5. (additive commutativity) For all $a, b \in F$ we have $a + b = b + a$.
6. (multiplicative closure) If $a, b \in F$ then $a * b \in F$.
7. (multiplicative associativity) For all $a, b, c \in F$ we have $(a * b) * c = a * (b * c)$.
8. (multiplicative identity) There exists an element $1 \in F$ such that $1 * x = x * 1 = x$ for all $x \in F$.
9. (multiplicative inverses) For all $a \in F - \{0\}$ there exists $b \in F$ with $a * b = b * a = 1$.
10. (multiplicative commutativity) For all $a, b \in F$ we have $a * b = b * a$.
11. (distributive property) For all $a, b, c \in F$ we have $a * (b + c) = a * b + a * c$ and $(b + c) * a = b * a + c * a$.

We say that the order of the field is $|F|$. That is, it is the cardinality of the underlying set which is the defining parameter of the algebraic object. It is possible for the order to be infinite like the reals, but here we shall only be concerned with fields where the order is a natural number. To avoid trivial cases we assume that the order is at least 2 for any finite field.

As is the usual convention we shall often use juxtaposition to indicate the multiplicative operation, that is, we write ab to mean $a * b$.

Notice that the addition and multiplication tables given in (2.6) show that $\mathbb{Z}_5$ is a field of order 5. Of course, $\mathbb{Z}_n$ is not a field for all n, for example, $\mathbb{Z}_4$ is not a field since 2 has no multiplicative inverse. The real numbers, $\mathbb{R}$, the rational numbers $\mathbb{Q}$, and the complex numbers, $\mathbb{C}$, are all fields with respect to their usual addition and multiplication. The integers are not a field since, for example, the element 2 does not have a multiplicative inverse in the integers.

A structure that satisfies all the axioms except possibly (9) and (10) is known as a ring. Some texts will also allow axiom (8), the existence of an identity, to be violated as well. Sometimes, an object without a multiplicative identity will be known as a rng (that is, ring without the "i") and those with a multiplicative identity will be known as a ring with both words pronounced identically.

It is possible for a structure to satisfy all axioms except for item 10 (multiplicative commutativity). The Quaternions are such an example, see [38] for a complete description. It is not possible for a finite structure to satisfy all axioms except for item 10 (multiplicative commutativity). That is, any finite ring that satisfies the first nine axioms and the eleventh must satisfy all eleven.

Theorem 2.18 *The ring $\mathbb{Z}_n$ is a field if and only if n is a prime.*

Proof By Exercise 3., we have that all of the items except multiplicative inverses are satisfied. By Lemma 2.2, we know that an element has a multiplicative inverse if and only if it is relatively prime to n. If n is prime then every non-zero element of $\mathbb{Z}_n$ is relatively prime to n. If n is not a prime then $n = ab$ for some a, b. Then a is not relatively prime to n and $\mathbb{Z}_n$ is not a field. $\square$

Example 2.10 Consider $n = 6$. Here the elements $2, 3$, and 4 do not have multiplicative inverses so $\mathbb{Z}_6$ cannot be a field.

Not every finite field is $\mathbb{Z}_n$ for some n. For example, the following addition and multiplication tables describe a finite field of order 4.

+	0	1	ω	ω^2
0	0	1	ω	ω^2
1	1	0	ω^2	ω
ω	ω	ω^2	0	1
ω^2	ω^2	ω	1	0

$*$	0	1	ω	ω^2
0	0	0	0	0
1	0	1	ω	ω^2
ω	0	ω	ω^2	1
ω^2	0	ω^2	1	ω

$$(2.7)$$

We shall give a very brief explanation of how fields of other orders are constructed. Consider the following algebraic structure known as a polynomial ring. Let $\mathbb{Z}_p[x] = \{a_0 + a_1x + a_2x^2 + \cdots + a_{e-1}x^{e-1} \mid a_i \in \mathbb{Z}_p, e \in \mathbb{N}\}$. It can be easily verified that this object is a ring. It is not, however, a field, since, for example, the element x has no multiplicative inverse.

For any element $p(x) \in \mathbb{Z}_p[x]$, let $\langle p(x) \rangle = \{a(x)p(x) \mid a(x) \in \mathbb{Z}_p[x]\}$. In terms of ring theory, this object is known as an ideal. We can mod out by such an ideal in a manner similar to the technique we used to perform modular arithmetic. Likewise, it was the fact that we were moding out by a prime (namely, a number that was not the non-trivial product of smaller numbers) that made $\mathbb{Z}_p$ a field. Without giving a detailed proof, we can state the following theorem which shows how to construct fields of orders that are not prime.

Theorem 2.19 *Let* $p(x)$ *be an irreducible polynomial of degree e in* $\mathbb{Z}_p[x]$. *Then* $\mathbb{Z}_p[x]/\langle p(x) \rangle$ *is a field of order* p^e.

Example 2.11 Consider the polynomial $x^2 + 1$ over $\mathbb{Z}_3$. If it were to factor, it would have to factor into linear terms which would mean that either $0, 1,$ or 2 would have to be a root of the polynomial in $\mathbb{Z}_3$. It is easy to verify that none of these is a root. Therefore, $x^2 + 1$ is an irreducible polynomial of degree 2 over $\mathbb{Z}_3$. In practical terms, this means that in $\mathbb{Z}_3[x]/\langle x^2 + 1 \rangle$, we have that $x^2 + 1 = 0$. This means that every time we encounter x^2 in a computation we can replace it with -1 which is 2 in $\mathbb{Z}_3$. Therefore, the elements of this object are the set:

$$\{0, 1, 2, x, x+1, x+2, 2x, 2x+1, 2x+2\}.$$

The addition and multiplication are given as follows.

+	0	1	2	x	x + 1	x + 2	2x	2x + 1	2x + 2
0	0	1	2	x	x + 1	x + 2	2x	2x + 1	2x + 2
1	1	2	0	x + 1	x + 2	x	2x + 1	2x + 2	2x
2	2	0	1	x + 2	x	x + 1	2x + 2	2x	2x + 1
x	x	x + 1	x + 2	2x	2x + 1	2x + 2	0	1	2
x + 1	x + 1	x + 2	x	2x + 1	2x + 2	2x	1	2	0
x + 2	x + 2	x	x + 1	2x + 2	2x	2x + 1	2	0	1
2x	2x	2x + 1	2x + 2	0	1	2	x	x + 1	x + 2
2x + 1	2x + 1	2x + 2	2x	1	2	0	x + 1	x + 2	x
2x + 2	2x + 2	2x	2x + 1	2	0	1	x + 2	x	x + 1

$$(2.8)$$

*	0	1	2	x	x + 1	x + 2	2x	2x + 1	2x + 2
0	0	0	0	0	0	0	0	0	0
1	0	1	2	x	x + 1	x + 2	2x	2x + 1	2x + 2
2	0	2	1	2x	2x + 2	2x + 1	x	x + 2	x + 1
x	0	x	2x	2	x + 2	2x + 2	1	x + 1	2x + 1
x + 1	0	x + 1	2x + 2	x + 2	2x	1	2x + 2	2	x
x + 2	0	x + 2	2x + 1	2x + 2	1	x	x + 1	2x	2
2x	0	2x	x	1	2x + 2	x + 1	2	2x + 1	x + 2
2x + 1	0	2x + 1	x + 2	x + 1	2	2x	2x + 1	x	1
2x + 2	0	2x + 2	x + 1	2x + 1	x	2	x + 2	1	2x

$$(2.9)$$

This object is a finite field of order 9.

We seek to not only construct finite fields but to classify them as well. We begin with the necessary definitions. Let F and K be fields. A field homomorphism is a map $\Phi : F \to K$ such that $\Phi(ab) = \Phi(a)\Phi(b)$ and $\Phi(a + b) = \Phi(a) + \Phi(b)$, for all $a, b \in F$. If a field homomorphism is also a bijection then it is said to be an isomorphism.

We say that two fields F and K are isomorphic if there exists an isomorphism $\Phi : F \to K$. Isomorphic fields are essentially identical in structure. For example, we have the following easy results.

Theorem 2.20 *Let F be a field with additive identity 0 and multiplicative identity 1 and let K be a field with additive identity $0'$ and multiplicative identity $1'$. Let $\Phi : F \to K$ be a field isomorphism. Then $\Phi(0) = 0'$ and $\Phi(1) = 1'$.*

Proof For all $a \in F$ we have $a + 0 = 0 + a = a$. Then $\Phi(a) = \Phi(a + 0) = \Phi(a) + \Phi(0)$ and $\Phi(a) = \Phi(0 + a) = \Phi(0) + \Phi(a)$. Since Φ is a bijection then every element of K is of the form $\Phi(a)$ for some $a \in F$. Therefore, for all $a' \in K$ we have $a' + \Phi(0) = \Phi(0) + a' = a'$. Then $\Phi(0)$ is an additive identity in K. Since the additive identity in a field is unique we have $\Phi(0) = 0'$.

For all $a \in F$ we have $a1 = 1a = a$. Then $\Phi(a) = \Phi(a1) = \Phi(a)\Phi(1)$ and $\Phi(a) = \Phi(1a) = \Phi(1)\Phi(a)$. Since Φ is a bijection then every element of K is of the form $\Phi(a)$ for some $a \in F$. Therefore, for all $a' \in K$ we have $a'\Phi(1) = \Phi(1)a' = a'$. Then $\Phi(1)$ is an additive identity in K. Since the additive identity in a field is unique we have $\Phi(1) = 1'$. □

This leads naturally to the following theorem.

Theorem 2.21 *Let Φ be a field isomorphism between F and K. If $a \in F$ then $\Phi(-a) = -\Phi(a)$ and if $a \neq 0$ then $\Phi(a^{-1}) = \Phi(a)^{-1}$.*

Proof Let $a \in F$ then $a + (-a) = 0$ which gives $\Phi(a + (-a)) = \Phi(0)$. Then, we have $\Phi(a) + \Phi(-a) = 0'$ where $0'$ is the additive identity of K. Therefore $\Phi(-a) = -\Phi(a)$.

Let $a \in F \setminus \{0\}$ then $aa^{-1} = 1$ which gives $\Phi(aa^{-1}) = \Phi(1)$. Then, we have $\Phi(a)\Phi(a^{-1}) = 1'$ where $1'$ is the additive identity of K. Therefore $\Phi(a^{-1}) = \Phi(a)^{-1}$. □

Example 2.12 Consider the isomorphism $\Phi : \mathbb{F}_4 \to \mathbb{F}_4$, $\Phi(a) = a^2$, where $\mathbb{F}_4$ is the finite field given in Eq. 2.7. The isomorphism is realized as follows:

$$0 \to 0$$
$$1 \to 1$$
$$\omega \to \omega^2$$
$$\omega^2 \to \omega$$

We note that the additive identity is sent to the additive identity and the multiplicative identity is sent to the multiplicative identity. The multiplicative inverse of ω is ω^2. Then Φ sends ω to ω^2 and its inverse ω^2 to ω. Hence inverses go to inverses.

The following is a well-known theorem of algebra whose proof is beyond the scope of this text. The proof can be found in [38].

Theorem 2.22 *Finite fields exist if and only if their order is of the form* p^e *where* p *is a prime and* $e > 0$. *Any two finite fields of the same order are isomorphic.*

With this theorem in mind, we can denote *the* finite field of order q, when q is a prime power, by $\mathbb{F}_q$ since we know such a field exists and up to isomorphism there is only one such field. This is not true for infinite fields. For example, the real and the complex numbers are not isomorphic. This can be easily seen by the fact that if there were an isomorphism $\Phi : \mathbb{C} \to \mathbb{R}$, then $\Phi(i)$ would need to satisfy $x^2 = -1$ by the previous theorems. Since there is no solution to this equation in the reals, then there cannot be an isomorphism.

We shall give a few more theorems to indicate the structure of a field.

Theorem 2.23 *If* $ab = 0$ *in a field then either* $a = 0$ *or* $b = 0$.

Proof If a is not 0 then there exists a multiplicative inverse of a denoted by a^{-1}. Then we have $a^{-1}ab = a^{-1}0$ which implies $b = 0$. Hence either a or b must be 0. □

This theorem says that there are no zero divisors in a field. For infinite structures there are rings that have no zero divisors, such as the integers, that are not fields. It will not occur for finite fields as we shall see in Theorem 2.25.

Theorem 2.24 *In a field, if* $ab = cb$, $b \neq 0$ *then* $a = c$ *and if* $ba = bc$, $b \neq 0$ *then* $a = c$.

Proof Assume $ab = cb$ where $b \neq 0$. Then there exists an element b^{-1} with $bb^{-1} = 1$. Then

$$ab = cb$$
$$abb^{-1} = cbb^{-1}$$
$$a(1) = b(1)$$
$$a = b.$$

The second result follows from the first proof and the fact that multiplication is commutative. $\qquad\square$

It is often a good idea, given Theorem 2.23, to show a ring has zero divisors when showing that it is not a field. This indicates that not every non-zero element has a multiplicative inverse.

An integral domain satisfies all field axioms except that it need not satisfy axiom (9). It must also have no zero divisors, that is, if $ab = 0$ then either $a = 0$ or $b = 0$. An infinite example would be the integers. The next theorem shows that there are no finite examples.

Theorem 2.25 *A finite integral domain is a field.*

Proof Let D be a finite integral domain. All that we need to do is to show that every non-zero element has a multiplicative inverse.

If a is non-zero and $ac = ab$ then $ac - ab = 0$ and $a(c - b) = 0$. Then since there are no zero divisors c must equal b. Consider the non-zero elements of $D' = D - \{0\}$. From the previous statement we know that each element of aD' must be distinct and so must coincide with the elements of D'. Hence, there must be a $b \in D'$ with $ab = 1$ since $1 \in D'$. $\qquad\square$

The characteristic of a finite field is a if $\sum_{i=1}^{a} x = 0$ for all $x \in F$, and a is the smallest such number satisfying this property. That is, adding any number to itself a times will result in 0. For example, the characteristic of $\mathbb{Z}_2$ and the field of order 4 already exhibited is 2 and the characteristic of the field of order 3 is 3.

Theorem 2.26 *The characteristic of a finite field must be a prime.*

Proof Assume the characteristic is a and $a = bc$ with neither b nor c equal to 1. Then $0 = \sum_{i=1}^{a} 1 = (\sum_{i=1}^{b} 1)(\sum_{i=1}^{c} 1)$. This implies the existence of zero divisors and we have a contradiction. $\qquad\square$

Example 2.13 The characteristic of the field of order 4 given in Eq. 2.7 is 2. The characteristic of the field of order 9 given in Example 2.11 is 3. We note that both characteristics are prime.

Often we shall use the space F^n which is shorthand for the cross product of F with itself n times, $F \times F \times \cdots \times F$. Concretely

$$F^n = \{(v_1, v_2, \ldots, v_n) \mid v_i \in F\}.$$

For example,

$$\mathbb{F}_2^3 = \{(0,0,0), (0,0,1), (0,1,0), (0,1,1), (1,0,0), (1,0,1), (1,1,0), (1,1,1)\}.$$

The generalized multiplicative counting principle gives that there are $|F|^n$ elements in this space. We refer to it as a space rather than a set because it is naturally a vector space. We shall describe vector spaces completely in Chap. 6. We shall use these vector spaces extensively to construct combinatorial objects known as finite geometries.

Exercises

1. Verify that the object with addition and multiplication in Eq. 2.7 is a field.
2. Find two solutions to $x^2 + x + 1 = 0$ in the field of order 4.
3. Show that $\mathbb{Z}_n$ satisfies all axioms of a field except for item 9 (multiplicative inverses) for all n. That is, show that $\mathbb{Z}_n$ is a commutative ring with identity.
4. Prove that the real numbers and the complex numbers are not isomorphic fields. Hint: assume there is such an isomorphism from the complex numbers to the real numbers and examine the image of the element i under that isomorphism.
5. Prove that in a field F of characteristic p that $\Phi(a) = a^p$ is a homomorphism from F to F. This map is not necessarily the identity as a novice might assume from Fermat's Little Theorem. For example, in the field of order 4 described earlier, which has characteristic 2, we have $w^2 \neq w$.
6. Determine if the set of n by n matrices with entries from a field is a field. If not what axiom(s) does it violate?
7. Prove that for a field F, $F \times F$, where $(a, b) + (c, d) = (a + c, b + d)$ and $(a, b) * (c, d) = (a * c, b * d)$, is not a field. What axiom does it violate?
8. Determine if

$$\left\{ \begin{pmatrix} a & -b \\ b & a \end{pmatrix} \mid a, b \in \mathbb{R} \right\}$$

   is a field with the usual matrix addition and multiplication.
9. Find the multiplicative inverses for the elements in $\mathbb{Z}_{11}$.
10. Prove that a field isomorphism sends the additive identity to the additive identity and the multiplicative identity to the multiplicative identity, that is, if $f : F \to E$ with f an isomorphism then $f(0_F) = 0_E$ and $f(1_F) = 1_E$.

11. Prove that in a field there is a unique solution to the equation $ax + b = c$ for all non-zero a.

12. Prove that in any finite field there must be an a satisfying the definition of the characteristic.

13. Prove that no element can be the multiplicative inverse for two distinct elements in a field.

14. Prove that in a field the equation $x^2 = a$ can have at most two solutions. Use this to prove that in the finite field $\mathbb{Z}_p$, p a prime, precisely half of the non-zero elements are squares. Determine which elements are squares for $\mathbb{Z}_7$ and $\mathbb{Z}_{11}$.

15. Prove that if a is non-zero in a field F, and b is any element in F, then there exists $c \in F$ with $ac = b$.

16. Prove that the inverse of a field isomorphism is a field isomorphism.

17. Determine if $x^3 + x + 1$ is irreducible over $\mathbb{Z}_2$. If it is irreducible, then use it to construct a field of order 8.

18. Let Φ be an isomorphism between two fields F and K. Assume α is a solution to $a_0 + a_1x + a_2x^2 + \ldots a_{n-1}x^{n-1}$ in $F[x]$. Prove that $\Phi(\alpha)$ is a solution to $\Phi(a_0) + \Phi(a_1)x + \Phi(a_2)x^2 + \ldots \Phi(a_{n-1})x^{n-1}$ in $K[x]$.

19. Consider the set $S = \{0, 1, u, 1 + u\}$ where all operations are done modulo 2. There are four distinct possibilities, namely, $u^2 = 0$, $u^2 = 1$, $u^2 = u$, and $u^2 = 1 + u$ which give algebraic structures. Determine which of these are a field and which are not a field.

2.3 Combinatorial Numbers

2.3.1 Geometric Numbers

The first collection of combinatorial numbers correspond to geometric objects. We shall examine these numbers as a sequence. A sequence is a function $f : \mathbb{N} \to \mathbb{R}$. More efficiently, we can often denote $f(n)$ by a_n and write the sequence as $a_0, a_1, a_2, \ldots$.

The square numbers are those numbers that can be described as a collection of items placed in a square.

$$n = 1$$

•

$$(2.10)$$

$$n = 2$$

$$\begin{matrix} \bullet & \bullet \\ \bullet & \bullet \end{matrix}$$

$$(2.11)$$

$$n = 3$$

• • •
• • •
• • •

(2.12)

$$n = 4$$

• • • •
• • • •
• • • •
• • • •

(2.13)

This sequence can be described explicitly as $f(n) = n^2$. Recursively, we can describe it as $a_0 = 0$, $a_{n+1} = a_n + 2n + 1$. We now prove that these two give the same sequence.

Theorem 2.27 *The sequence* $f(n) = n^2$ *matches the sequence given recursively by* $a_0 = 0$, $a_{n+1} = a_n + 2n + 1$.

Proof We shall prove the result by induction. If $n = 0$, $f(0) = 0$ and $a_0 = 0$.

Next assume that $f(n) = a_n$. Then $f(n+1) = (n+1)^2 = n^2 + 2n + 1 = f(n) + 2n + 1 = a_{n+1}$. Therefore, by the principle of mathematical induction we have the result. □

Concretely, the square numbers are

$$0, 1, 4, 9, 16, 25, 36, 49, 64, 81, 100, 121, 144, 169, \ldots$$

Theorem 2.28 *If* p *is prime that divides* n^2, *then* p^2 *divides* n^2.

Proof If p divides $n^2 = n \cdot n$, then either p divides n or p divides n by the definition of a prime. Either case gives p divides n, giving $n = pk$ for some integer k. Then $n^2 = p^2 k^2$ and p^2 divides n^2. □

As a consequence of this theorem, we note that any even square number must be 0 (mod 4). Also if a square number ends in 0 (divisible by 5 and 2) it must end in 00 (divisible by $5^2 2^2$). More generally, we have that any square number that is 0 (mod p) must be 0 (mod p^s).

We note also that a square number written in base 10 must end in a square (mod 10), namely, 0, 1, 4, 5, 6, or 9.

We can also get a formula for the sum of the first k positive square numbers.

Theorem 2.29 *Let k be a positive integer. Then*

$$\sum_{n=1}^{k} n^2 = \frac{k(k+1)(2k+1)}{6}. \tag{2.14}$$

Proof We shall prove the result by induction. If $k = 1$ the left side of the equation is $1^2 = 1$ and the right side is $\frac{1(2)(3)}{6} = 1$ giving that the formula is correct for $k = 1$.
Assume the formula is correct for k and consider the sum of the first $k + 1$ squares:

$$\sum_{n=1}^{k+1} n^2 = (\sum_{n=1}^{k} n^2) + (k+1)^2$$

$$= \frac{k(k+1)(2k+1)}{6} + \frac{6(k+1)^2}{6}$$

$$= (k+1)\frac{k(2k+1) + 6(k+1)}{6}$$

$$= (k+1)\frac{2k^2 + 7k + 6}{6}$$

$$= \frac{(k+1)(k+2)(2k+3)}{6},$$

which is the desired result. Therefore, by the principle of mathematical induction, the formula is correct for all positive k. □

The cubic numbers are those numbers that can be described as a collection of items placed in a cube.

$$n = 1$$

$$\bullet$$

$$\tag{2.15}$$

$n = 2$

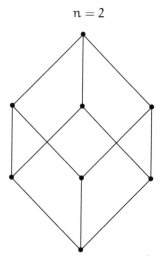

This sequence can be described explicitly as $f(n) = n^3$. Recursively, we can describe it as $a_0 = 0$, $a_{n+1} = a_n + 3n^2 + 3n + 1$. We now prove that these two give the same sequence.

Theorem 2.30 *The sequence* $f(n) = n^3$ *matches the sequence given recursively by* $a_0 = 0$, $a_{n+1} = a_n + 3n^2 + 3n + 1$.

Proof We shall prove the result by induction. If $n = 0$, $f(0) = 0$, and $a_0 = 0$.

Next assume that $f(n) = a_n$. Then $f(n+1) = (n+1)^3 = n^3 + 3n^2 + 3n + 1 = f(n) + 3n^2 + 3n + 1 = a_{n+1}$. Therefore, by the principle of mathematical induction, we have the result. $\square$

Concretely, the cubic numbers are

$$0, 1, 8, 27, 64, 125, 216, 343, 512, \ldots$$

Theorem 2.31 *If* p *is prime that divides* n^3, *then* p^3 *divides* n^3.

Proof If p divides $n^3 = n \cdot n \cdot n$, then either p divide n by the definition of a prime. This gives p divides n, giving $n = pk$ for some integer k. Then $n^3 = p^3 k^3$ and p^3 divides n^3. $\square$

This theorem says that if a cubic number is $0 \pmod{p}$ then it is $0 \pmod{p^3}$ as well. However, unlike square numbers, this does not reduce the possible final digit in a cubic number in its base 10 representation since all numbers $\pmod{10}$ are cubes.

We can determine the sum of the first k cubic numbers.

Theorem 2.32 *Let* k *be a positive integer. Then*

$$\sum_{n=1}^{k} n^3 = \frac{k^2(k+1)^2}{4}. \tag{2.16}$$

Proof We shall prove the result by induction. If $k = 1$ the left side of the equation is $1^3 = 1$ and the right side is $\frac{1(4)}{4} = 1$ giving that the formula is correct for $k = 1$.
   Assume the formula is correct for k and consider the sum of the first $k + 1$ squares:

$$\begin{aligned}
\sum_{n=1}^{k+1} n^3 &= (\sum_{n=1}^{k} n^3) + (k+1)^3 \\
&= \frac{k^2(k+1)^2}{4} + (k+1)^3 \\
&= (k+1)^2 \frac{k^2 + 4k + 4}{4} \\
&= \frac{(k+1)^2(k+2)^2}{4},
\end{aligned}$$

which is the desired result. Therefore, by the principle of mathematical induction, the formula is correct for all positive k. □

   The triangular numbers are those numbers that can be described as a collection of items placed in a triangle.

$$n = 1$$

•

$$\tag{2.17}$$

$$n = 2$$

•
• •

$$\tag{2.18}$$

$$n = 3$$

•
• •
• • •

$$\tag{2.19}$$

$$n = 4$$

•
• •
• • •
• • • •

$$\tag{2.20}$$

We note that the triangular numbers are the sum of the first n numbers. We can determine a formula for this sum.

Theorem 2.33 *Let n be a positive integer. Then $\sum_{j=1}^{n} j = \frac{n(n+1)}{2}$.*

Proof If $n = 0$, the left side of the equation is 1 and the right side is $\frac{1(2)}{2} = 1$. Therefore, the formula is correct for $n = 1$.

Assume the formula is correct for n and consider

$$\sum_{j=1}^{n+1} j = (\sum_{j=1}^{n} n) + (n+1)$$
$$= \frac{n(n+1)}{2} + \frac{2(n+1)}{2}$$
$$= (n+1)\frac{n+2}{2} = \frac{(n+1)(n+2)}{2},$$

which is the desired result. Therefore, by the principle of mathematical induction, the formula is correct for all positive n. □

We note then that $\sum_{j=1}^{n} j = C(n+1, 2)$. Additionally, we have the following corollary.

Corollary 2.5 *For any positive integer k, we have*

$$\sum_{n=1}^{k} n^3 = (\sum_{n=1}^{k} n)^2. \tag{2.21}$$

Proof Follows directly from Theorem 2.32 and Theorem 2.33. □

This sequence can be described explicitly as $f(n) = \frac{n(n+1)}{2}$. Recursively, we can describe it as $a_0 = 0$, $a_{n+1} = a_n + n + 1$. We now prove that these two give the same sequence.

Theorem 2.34 *The sequence $f(n) = \frac{n(n+1)}{2}$ matches the sequence given recursively by $a_0 = 0$, $a_{n+1} = a_n + n + 1$.*

Proof We shall prove the result by induction. If $n = 0$, $f(0) = 0$, and $a_0 = 0$.

Next assume that $f(n) = a_n$. Then $f(n+1) = \frac{(n+1)(n+2)}{2} = \frac{n(n+1)}{2} + \frac{2(n+1)}{2} = a_n + (n+1)$. Therefore, by the principle of mathematical induction, we have the result. □

Concretely, the triangular numbers are

$$0, 1, 3, 6, 10, 15, 21, 28, 36, 45, 55, \ldots$$

Notice if we add each number to its successor in the sequence we get the following sequence:

$$1, 4, 9, 16, 25, 36, 49, 64, 81, 100, \ldots$$

which are the squares. We state this explicitly in the following theorem.

Theorem 2.35 *The sum of two consecutive triangular numbers is a square.*

Proof Let n be a positive integer. Then

$$\frac{n(n+1)}{2} + \frac{(n+1)(n+2)}{2} = \frac{(n^2+n) + (n^2+3n+2)}{2}$$

$$= \frac{2n^2 + 4n + 2}{2} = n^2 + 2n + 1 = (n+1)^2.$$

This gives the result.   □

We note also that this theorem says that any square number is the sum of two triangular numbers.

We can also find the sum of the first k triangular numbers.

Theorem 2.36 *Let k be a positive integer. Then*

$$\sum_{n=1}^{k} \frac{n(n+1)}{2} = \frac{n(n+1)(n+2)}{6}. \tag{2.22}$$

Proof We shall prove the result by induction. If $k = 1$ the left side of the equation is $\frac{1(2)}{2} = 1$ and the right side is $\frac{1(2)(3)}{6} = 1$ giving that the formula is correct for $k = 1$.

Assume the formula is correct for k and consider the sum of the first $k + 1$ triangular numbers:

$$\sum_{n=1}^{k+1} \frac{n(n+1)}{2} = \left(\sum_{n=1}^{k} \frac{n(n+1)}{2} \right) + \frac{(n+1)(n+2)}{2}$$

$$= \frac{n(n+1)(n+2)}{6} + \frac{(n+1)(n+2)}{2}$$

$$= (n+1) \frac{n(n+2) + 3(n+2)}{6}$$

$$= \frac{(n+1)(n+2)(n+3)}{6},$$

which is the desired result. Therefore, by the principle of mathematical induction, the formula is correct for all positive k. □

A perfect number is a number that is equal to the sum of its positive divisors that are less than the number. For example, $6 = 1 + 2 + 3$ and $28 = 1 + 2 + 4 + 7 + 14$. Therefore 6 and 28 are perfect numbers. It is not known if there exist odd perfect numbers.

Even perfect numbers are related in a natural way to Mersenne primes. A Mersenne prime is a prime number of the form $2^p - 1$ where p is a prime. We illustrate in the following lemma why p must be prime.

Lemma 2.3 *Let n be an integers with $n > 1$. If n is composite then $2^n - 1$ is composite.*

Proof Recall that $x^k - 1 = (x - 1)(1 + x + x^2 + \cdots + x^{k-1})$. If $n = ab$ with $1 < a, b < n$, then

$$2^n - 1 = 2^{ab} - 1 = (2^a)^b - 1.$$

Applying the factorization above we have

$$2^n - 1 = (2^a - 1)(1 + 2^a + (2^a)^2 + \cdots + (2^a)^{b-1}).$$

Since $1 < a, b < n$, $1 < (2^a - 1) < 2^n - 1$ and $1 < (1 + 2^a + (2^a)^2 + \cdots + (2^a)^{b-1}) < 2^n - 1$, giving that $2^n - 1$ is composite. □

The number $2^p - 1$ is not guaranteed to be prime when p is prime, but when it is we have that $(2^p - 1)2^{p-1}$ is a perfect number. For example, if $p = 2, 3, 5, 7, 13, 17,$ s19, 31, then $2^p - 1$ is a prime. Generally, the largest known prime is a Mersenne prime. This is because there is a much better algorithm for determining if $2^p - 1$ is prime than there is for arbitrary n. This algorithm is known as the Lucas-Lehmer test. As an example, the number $2^{82,589,933} - 1$ is known to be a prime number. This number has millions of digits.

Lemma 2.4 *If $2^p - 1$ is prime then $(2^p - 1)2^{p-1}$ is a perfect number.*

Proof If $2^p - 1$ is prime then the divisors of $(2^p - 1)2^{p-1}$ are numbers of the form $(2^p - 1)2^i, 0 \leq i \leq p - 2$ and $2^j, 0 \leq j \leq p - 1$. The sum of the first type is $(2^p - 1)(2^{p-1} - 1)$ and the sum of the second type is $2^p - 1$. Then $(2^p - 1)(2^{p-1} - 1) + (2^p - 1) = (2^p - 1)2^{p-1}$ and we have the result. □

Example 2.14 Let $p = 3$, then $2^3 - 1 = 7$ which is prime. Then $(2^p - 1)(2^{p-1} - 1) = 7(3) = 21$ which is $7 + 14$ and $(2^p - 1) = 7$ which is $1 + 2 + 4$. Therefore, $(2^p - 1)2^{p-1} = 7(4) = 28$ is a perfect number.

Theorem 2.37 *The number* $(2^p - 1)2^{p-1}$ *is a triangular number.*

Proof Let $n = 2^p - 1$. Then $\frac{n(n+1)}{2} = \frac{(2^p-1)2^p}{2} = (2^p - 1)2^{p-1}$. Therefore, $(2^p - 1)2^{p-1}$ is a triangular number. $\qquad\square$

Exercises

1. Determine a recursive formula for $f(n) = n^4$. Prove your formula is correct.
2. Prove that if p is a prime dividing n^s, then p^s divides n^s.
3. Determine the possible final digit for a square number written in base 8.
4. Determine the possible final digit for a cubic number written in base 8.

2.3.2   Catalan Numbers

Catalan numbers were first described by Eugène Catalan in [22]. They have numerous connections to a variety of combinatorial problems. We begin by defining the Catalan numbers to be the numbers satisfying the following recursion:

$$C_{n+1} = \sum_{k=0}^{n} C_k C_{n-k}, \quad C_0 = 1. \tag{2.23}$$

It follows that the first 8 Catalan numbers are the following.

$$\begin{array}{c|ccccccccc} n & 0 & 1 & 2 & 3 & 4 & 5 & 6 & 7 & 8 \\ \hline C_n & 1 & 1 & 2 & 5 & 14 & 42 & 132 & 429 & 1430 \end{array} \tag{2.24}$$

We shall determine a closed formula for C_n as is given in Stanley's text [81]. We shall use generating functions to find the closed formula. First, we require a lemma due to Newton.

Lemma 2.5 *Let* $k \in \mathbb{R}$, $n \in \mathbb{N}$. *Then, for* $-1 < x < 1$,

$$(1+x)^k = \sum_{n=0}^{\infty} C(k,n)x^n, \tag{2.25}$$

where $C(k,n) = \frac{k(k-1)(k-2)\cdots(k-n+1)}{n!}$, *and* $C(k,0) = 1$.

Proof The result is a direct consequence of Taylor's Theorem. $\qquad\square$

We notice that this definition of $C(k,n)$ matches the previous definition if k and n are both natural numbers.

Let $p(x) = \sum_{n \geq 0} C_n x^n$. Multiply the recurrence by x^n and sum to obtain

$$\sum_{n \geq 0} C_{n+1} x^n = \sum_{n \geq 0} \left(\sum_{k=0}^{n} C_k C_{n-k} \right) x^n. \tag{2.26}$$

We note that

$$x \sum_{n \geq 0} C_{n+1} x^n = \sum_{n \geq 1} c_n x^n = p(x) - 1. \tag{2.27}$$

Consider the series $p(x)^2 = (\sum_{n \geq 0} C_n x^n)^2$. The coefficient of x^n in this series is

$$\sum_{k=0}^{n} C_k C_{n-k}.$$

This follows simply from the definition of multiplication of series. Using this and applying Eqs. 2.26 and 2.27 we have

$$\frac{p(x) - 1}{x} = p(x)^2. \tag{2.28}$$

Writing this as a binomial equation we have $xp(x)^2 - p(x) + 1 = 0$.

Apply the quadratic equation to this equation with $p(x)$ as the variable and we have

$$p(x) = \frac{1 \pm (1 - 4x)^{\frac{1}{2}}}{2x}.$$

Now we can apply Lemma 2.5 to $(1 - 4x)^{\frac{1}{2}}$ (replacing x with $4x$) for both the choice of the plus sign and the minus sign. Using the plus sign gives $p(x) = \frac{1}{x} - 1 - x + \ldots$ which is not the series we are seeking. So we use the minus sign.

This gives that

$$p(x) = \sum_{n \geq 0} C_n x^n = \frac{1}{2x}(1 - (1 - 4x)^{\frac{1}{2}})$$

$$= \frac{1}{2x}(1 - \sum_{n \geq 0} C(\frac{1}{2}, n)(-4x)^n)$$

$$= \frac{1}{2x}(\sum_{n \geq 1} C(\frac{1}{2}, n)(-1)^n 4^n x^n)$$

$$= \sum_{n \geq 1} C(\frac{1}{2}, n)(-1)^n \frac{1}{2} 4^n x^{n-1}$$

$$= \sum_{n \geq 1} \frac{(1(-1)(-3)(-5) \cdots (-(2n-3)))}{n! 2^n} (-1)^n \frac{1}{2} 4^n x^{n-1}$$

$$= \sum_{n \geq 1} \frac{(1(1)(3)(5) \cdots (2n-3))}{n!} 2^{n-1} x^{n-1}$$

$$= \sum_{n \geq 0} \frac{((1)(3)(5) \cdots (2n-1))}{(n+1)!} 2^n x^n$$

$$= \sum_{n \geq 0} \frac{(2n)!}{n!(n+1)!} x^n.$$

This gives the following result.

Theorem 2.38 *Let n be a non-negative integer and let C_n satisfy the recursion $C_{n+1} = \sum_{k=0}^{n} C_k C_{n-k}$. Then*

$$C_n = \frac{1}{n+1} C(2n, n). \tag{2.29}$$

Example 2.15 By the recursion in Eq. 2.23, we have

$$C_8 = C_0 C_7 + C_1 C_6 + C_2 C_5 + C_3 C_4 + C_4 C_3 + C_5 C_2 + C_6 C_1 + C_7 C_0.$$

Using the numbers in Eq. 2.24, we have

$$\begin{aligned}
C_8 &= C_0 C_7 + C_1 C_6 + C_2 C_5 + C_3 C_4 + C_4 C_3 + C_5 C_2 + C_6 C_1 + C_7 C_0 \\
&= 1(429) + 1(132) + 2(42) + 5(14) + 14(5) + 42(2) + 132(1) + 429(1) \\
&= 429 + 132 + 84 + 70 + 70 + 84 + 132 + 429 \\
&= 1430.
\end{aligned}$$

If $n = 8$, then $\frac{1}{n+1} C(2n, n) = \frac{1}{9} C(16, 8) = \frac{1}{9}(12870) = 1430$.

Notice that we did not begin the discussion of Catalan numbers with a combinatorial motivation, but rather we defined them via a recursive formula and from this we obtained a closed formula for the Catalan numbers. Therefore, all we need to do to show that these numbers solve a given combinatorial problem is to show that, that problem can be solved with that recursion. In [81], Stanley gives 214 combinatorial applications of the Catalan numbers. Here we shall restrict ourselves to a much smaller number of applications, pointing the interested reader to [81].

The first combinatorial application of the Catalan numbers is sometimes called a ballot sequence. Consider a sequence a_i of length $2n$ consisting of entries from the set $\{1, -1\}$ where each appears n times, with the property that $\sum_{i=1}^{k} a_i \geq 0$ for all k with $1 \leq k \leq 2n$.

For $n = 1$ there is one such sequence, $1, -1$. For $n = 2$, there are two possible sequences:

$$1, 1, -1, -1$$
$$1, -1, 1, -1.$$

For $n = 3$ there are five possible sequences:

$$1,1,1,-1,-1,-1$$
$$1,1,-1,-1,1,-1$$
$$1,1,-1,1,-1,-1$$
$$1,-1,1,1,-1,-1$$
$$1,-1,1,-1,1,-1$$

We see that these first few terms are the Catalan numbers. We prove this result.

Theorem 2.39 *The number of sequences a_i of length $2n$ consisting of entries from the set $\{1, -1\}$ where each appears n times, with the property that $\sum_{i=1}^{k} a_i \geq 0$ for all k, with $1 \leq k \leq 2n$, is C_n.*

Proof We begin by counting the number of sequences of length $2n$ with n elements that are 1 and n elements that are -1. This number is the number of ways of choosing n objects out of $2n$ objects which is $C(2n, n) = \frac{(2n)!}{n!n!}$.

Next, we need to count the number of such sequences that do not have non-negative partial sums and subtract this number from $\frac{(2n)!}{n!n!}$.

Consider a sequence that has a partial sum that is negative. That is, $\sum_{i=1}^{s} a_i < 0$. Let t be the smallest such s such that the partial sum is negative. Since t is minimal and $\sum_{i=1}^{t} a_i < 0$, there must be an equal number of occurrences of 1 and -1 preceding a_t which gives that t is odd. This means that there are $\frac{t-1}{2}$ occurrences of 1 before t and $n - \frac{t-1}{2}$ and $\frac{t-1}{2}$ occurrences of -1 before t and $n - \frac{t+1}{2}$ since a_t must be -1. If you form a new sequence b_i of length $2n$ where $b_i = a_i$ if $i > t$ and $b_i = -a_i$ if $i \leq t$, then the sum of this sequence must be 2 and have $n + 1$ occurrences of 1 and $n - 1$ occurrences of -1. The number of such sequences is $\frac{(2n)!}{(n+1)!(n-1)!}$.

Then we have

$$\frac{(2n)!}{n!n!} - \frac{(2n)!}{(n+1)!(n-1)!} = \frac{(2n)!(n+1)}{(n+1)n!n!} - \frac{(2n)!n}{(n+1)!(n-1)!n}$$

$$= \frac{(2n!)}{(n+1)!n!}$$

$$= \frac{1}{n+1} C(2n, n) = C_n.$$

$\square$

The next application involves triangulations. Consider the number of triangulations of a convex $(n+2)$-gon which was first studied by Euler. Equivalently, this number is the number of $n - 1$ non-crossing diagonals of a convex polygon with $n + 2$ sides. Consider the cases for $n = 1, 2, 3$ given in Fig. 2.1.

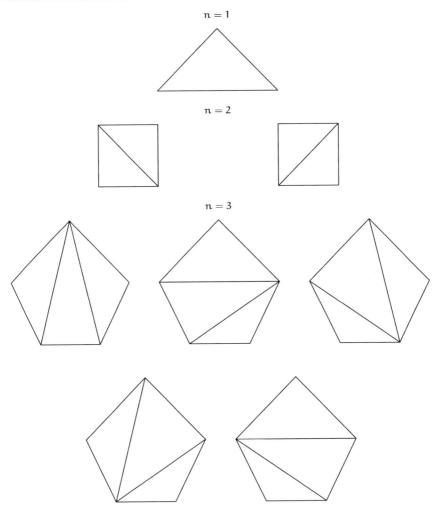

Fig. 2.1 Catalan numbers for $n = 1, 2$, and 3

This gives the number of such triangulations is 1 for $n = 1$, 2 for $n = 2$, and 5 for $n = 3$.

To find the five triangulations for $n = 3$, consider the one with $n = 1$. Opening it and adjoining two new sides, each triangulated with the possibilities for $n = 2$, and give the second and fifth triangulations for $n = 3$. The other three can be found by opening the two for $n = 2$ and adjoining one new side with the triangulation for $n = 1$. A straightforward generalization of this argument gives that the number of triangulations of a convex $(n + 2)$-gon satisfies the desired recursion, which gives the following.

Theorem 2.40 *The number of triangulations of a convex $(n + 2)$-gon is C_n.*

1. Construct all 14 triangularization of hexagons.
2. Construct all ballot sequences for $n = 4$.

2.3.3 Stirling Numbers

Stirling numbers are named for James Stirling who wrote about them in [87].

Stirling numbers of the first kind counts the number of ways to partition a set of n things into k cycles. We denote this number by $\begin{bmatrix} n \\ k \end{bmatrix}$.

We can write a cycle as $[1, 2, 3, 4, 5]$. We notice that

$$[1, 2, 3, 4, 5] = [2, 3, 4, 5, 1] = [3, 4, 5, 1, 2] = [4, 5, 1, 2, 3] = [5, 1, 2, 3, 4].$$

We can always specify a single element from this equivalence class by distinguishing a certain element to be the first in the representation of the cycle.

Example 2.16 Let $n = 3$ and $k = 1$. Then we shall count the number of ways that the set $\{1, 2, 3\}$ can be written in terms of one cycle. There are two ways, namely, $[1, 2, 3]$ and $[1, 3, 2]$. This gives that $\begin{bmatrix} 3 \\ 1 \end{bmatrix} = 2$.

Lemma 2.6 *Let n be a positive integer, then*

$$\begin{bmatrix} n \\ 1 \end{bmatrix} = (n - 1)!$$

and

$$\begin{bmatrix} n \\ n \end{bmatrix} = 1.$$

Proof Without loss of generality we can assume the symbols are $1, 2, 3, \ldots, n$.

Let $k = 1$. Since the elements are written in one cycle, every element is written in that cycle. We can assume that we take the representation where 1 is the first element in the cycle representation. Then every cycle is of the form $[1, \sigma]$ where σ is a permutation of the remaining $n - 1$ elements. Therefore $\begin{bmatrix} n \\ 1 \end{bmatrix} = (n - 1)!$, since there are $(n - 1)!$ permutations on a set with $n - 1$ elements.

If $k = n$ then they are written as n singleton cycles and there is only 1 way to do this. $\qquad \square$

Theorem 2.41 *Let* n *and* k *be positive integers. Then*

$$\begin{bmatrix} n \\ k \end{bmatrix} = (n-1)\begin{bmatrix} n-1 \\ k \end{bmatrix} + \begin{bmatrix} n-1 \\ k-1 \end{bmatrix}.$$

Proof Consider a set A with cardinality $n+1$. Choose an element from the set, call it a_{n+1}. Splitting into cycles puts a_{n+1} in a cycle by itself, there $\begin{bmatrix} n-1 \\ k-1 \end{bmatrix}$ such cycles. Otherwise a_{n+1} is put into one of the k cycles. There are $n-1$ different places to insert a_{n+1}. Hence, there are $(n-1)\begin{bmatrix} n-1 \\ k \end{bmatrix}$ such cycles. This gives the result. ☐

Lemma 2.6 and Theorem 2.41 allow us to make a Pascal like triangle of the values of the Stirling numbers of the first kind, where the rows are indexed by n and the columns by k. We write it this way to make the recursion evident.

$$
\begin{array}{ccccccccccccccccc}
&&&&&&&& 1 &&&&&&&& \\
&&&&&&& 1 && 1 &&&&&&& \\
&&&&&& 2 && 3 && 1 &&&&&& \\
&&&&& 6 && 11 && 6 && 1 &&&&& \\
&&&& 24 && 50 && 35 && 10 && 1 &&&& \\
&&& 120 && 274 && 225 && 85 && 15 && 1 &&& \\
&& 720 && 1764 && 1624 && 735 && 175 && 21 && 1 && \\
& 5040 && 13068 && 13132 && 6769 && 1960 && 322 && 28 && 1 & \\
40320 && 109584 && 118124 && 67284 && 22449 && 4536 && 546 && 36 && 1 \\
\end{array}
$$

$$\vdots$$

Stirling numbers of the second kind counts the number of ways to partition a set of n things into k non-empty subsets. We denote this number by $\begin{Bmatrix} n \\ k \end{Bmatrix}$.

Example 2.17 Consider the set of cardinality 3, $\{1, 2, 3\}$. There is one partition into 3 sets, namely, $\{1\}, \{2\}, \{3\}$. There is one partition into 1 set, namely, $\{1, 2, 3\}$. There are three partitions into 2 sets, namely,

$$\{1, 2\}, \{3\}$$
$$\{1, 3\}, \{2\}$$
$$\{2, 3\}, \{1\}.$$

This gives that $\begin{Bmatrix} 3 \\ 1 \end{Bmatrix} = \begin{Bmatrix} 3 \\ 3 \end{Bmatrix} = 1$ and $\begin{Bmatrix} 3 \\ 2 \end{Bmatrix} = 3$.

Lemma 2.7 *Let* n *and* k *be non-negative integers, then*

$$\begin{Bmatrix} n \\ 1 \end{Bmatrix} = \begin{Bmatrix} n \\ n \end{Bmatrix} = 1. \tag{2.30}$$

Proof There is only one way of partitioning a set of cardinality n into sets of cardinality n, namely, the set itself, therefore $\begin{Bmatrix} n \\ n \end{Bmatrix} = 1$.

There is only one way of partitioning a set of cardinality n into sets of cardinality 1, namely each element of the set is in its own set of the partition, therefore $\begin{Bmatrix} n \\ 1 \end{Bmatrix} = 1$.

□

We now give the recursion for Stirling numbers of the second kind.

Theorem 2.42 *Let n and k be non-negative integers, then*

$$\begin{Bmatrix} n \\ k \end{Bmatrix} = k \begin{Bmatrix} n-1 \\ k \end{Bmatrix} + \begin{Bmatrix} n-1 \\ k-1 \end{Bmatrix}. \tag{2.31}$$

Proof Consider a set A with cardinality $n + 1$. Chose an element from the set, call it a_{n+1}. Any partition of the set of cardinality n, $A - \{a_{n+1}\}$, into $k - 1$ sets can be made a partition of A into k sets by adjoining $\{a_{n+1}\}$ to the partition. There are $\begin{Bmatrix} n-1 \\ k-1 \end{Bmatrix}$ such partitions. Any partition of $A - \{a_{n+1}\}$ into k sets can be made a partition of A into k sets by placing a_{n+1} in any of the k sets of the partition. There are $k \begin{Bmatrix} n-1 \\ k \end{Bmatrix}$ such ways of accomplishing this. This gives the result.   □

Lemma 2.7 and Theorem 2.42 allow us to make a Pascal like triangle of the values of the Stirling numbers of the second kind, where the rows are indexed by n and the columns by k. We write it this way to make the recursion evident.

$$
\begin{array}{ccccccccccc}
 & & & & & 1 & & & & & \\
 & & & & 1 & & 1 & & & & \\
 & & & 1 & & 3 & & 1 & & & \\
 & & 1 & & 7 & & 6 & & 1 & & \\
 & 1 & & 15 & & 25 & & 10 & & 1 & \\
1 & & 31 & & 90 & & 65 & & 15 & & 1 \\
1 & 63 & & 301 & & 350 & & 140 & & 21 & 1 \\
1 & 127 & 966 & & 1701 & & 1050 & & 266 & 28 & 1 \\
1 & 255 & 3025 & 7770 & & 6951 & & 2646 & 462 & 36 & 1
\end{array}
$$

.
.
.

Exercises

1. Prove that $\begin{bmatrix} n \\ k \end{bmatrix} = \begin{Bmatrix} n \\ k \end{Bmatrix}$ where $k = n$ or $k = n - 1$. Determine the values in these cases.

2. Prove that triangular numbers are Stirling numbers of the second kind.

2.3.4   Towers of Hanoi

The problem of the Towers of Hanoi was first presented by Édouard Lucas in 1883 [60]. The problem involves a collection of disks of different sizes and three poles. The disks are placed on the pole.

   Objective: Move the entire stack of disks from one pole to another.
   Rules:

1. A move consists of taking the top disk from one of the poles and putting it on the top of another pole.
2. Only one disk can be moved at a time.
3. A larger disk can never be placed on a smaller disk.

Example 2.18 Consider the case with two disks, which we shall call 1 and 2, where a larger number indicates a larger disk. We shall move the disks to the second pole.

$$
\begin{array}{c}
1 \\
\underline{2} \ \_ \ \_
\end{array}
\ \longrightarrow \
\underline{2} \ \_ \ \underline{1}
\ \longrightarrow \
\_ \ \underline{2} \ \underline{1}
\ \longrightarrow \
\begin{array}{c}
1 \\
\_ \ \underline{2} \ \_
\end{array}
$$

We see that the it takes $2^2 - 1 = 3$ moves to complete the transfer.

Example 2.19 Consider the case with three disks, which we shall call 1, 2, and 3, where a larger number indicates a larger disk. We shall move the disks to the second pole.

$$
\begin{array}{c}
1 \\
2 \\
\underline{3} \ \_ \ \_
\end{array}
\ \longrightarrow \
\begin{array}{c}
2 \\
\underline{3} \ \underline{1} \ \_
\end{array}
\ \longrightarrow \
\underline{3} \ \underline{1} \ \underline{2}
\ \longrightarrow \
\begin{array}{c}
1 \\
\underline{3} \ \_ \ \underline{2}
\end{array}
\ \longrightarrow
$$

$$
\begin{array}{c}
1 \\
\_ \ \underline{3} \ \underline{2}
\end{array}
\ \longrightarrow \
\underline{1} \ \underline{3} \ \underline{2}
\ \longrightarrow \
\begin{array}{c}
2 \\
\underline{1} \ \underline{3} \ \_
\end{array}
\ \longrightarrow \
\begin{array}{c}
1 \\
2 \\
\_ \ \underline{3} \ \_
\end{array}
$$

We see that it takes $2^3 - 1 = 7$ moves to complete the transfer.

   These two examples lead us to the following theorem.

Theorem 2.43 *Let a_n be the number of moves required to move n disks from one pole to another. Then, recursively*

$$a_{n+1} = 2a_n + 1, a_1 = 1 \tag{2.32}$$

and $a_n = 2^n - 1$.

Proof First, we notice that $a_1 = 1$ since it requires 1 move to move a single disk from one pole to another.

Next, consider the disks labeled $1, 2, \ldots, n$. If they are on the first pole, it takes a_{n-1} moves to move $1, 2, \ldots, n-1$ to the third pole. Then n can be moved to the second pole. Then it takes a_{n-1} moves $1, 2, \ldots, n-1$ to the second pole. This gives the recurrence $a_{n+1} = 2a_n + 1$.

Finally, we show that this recurrence gives the closed formula. We note that if $n = 1$, $a_1 = 1$, and $2^1 - 1 = 1$.

Then if $a_{n-1} = 2^{n-1} - 1$, then $a_n = 2a_{n-1} + 1 = 2(2^{n-1} - 1) + 1 = 2^n - 2 + 1 = 2^n - 1$. Therefore, by the principal of mathematical induction $a_n = 2^n - 1$ for all n. $\qquad\square$

Exercises

1. Assume you have n disks and k poles, with $k > n$. Prove that the number of moves to the disks is $2n - 1$.
2. Write an algorithm to implement the solution to the Towers of Hanoi problem.

Mutually Orthogonal Latin Squares

3

3.1   36 Officers and Latin Squares

In 1782, the great mathematician Leonhard Euler asked the following question: Can you arrange 36 officers of 6 ranks and 6 regiments in a 6 by 6 square so that each row and column contains each rank and regiment exactly once?

Euler asked the question in his paper "Recherches sur une nouvelle espace de quarré magique" [35]. This paper began the study of Latin squares and was one of the most important papers in the history of combinatorics. It is fairly long by mathematical standards but it is written in a style which has completely vanished from mathematical publications. He wrote as if he were explaining the material to a good friend who was sitting at his side. There is no intimidation here, no hopelessly complicated language or notation. Euler's writing, like his mathematics, is a delight. While Euler's work was the one that prompted the study of Latin squares, they had been defined earlier by the Korean mathematician Choi Seok-Jeong and the French mathematician Jacques Ozanam. The work of Choi Seok-Jeong was particularly interesting and inventive, defining orthogonality and constructing magic squares. He even gave practical applications for the ideas, but this work was largely forgotten. The work of Jacques Ozanam related Latin squares to a problem about playing cards. See [74], for a complete description of who deserves recognition for introducing Latin squares.

This question seems remarkably easy to answer. At first glance most people seem to think that they could be arranged in such a manner. Unfortunately, it proved to be quite a difficult problem. It was not solved until 1901 when Col. Tarry, a colonel in the French army in Algeria, and an amateur mathematician, showed that such an arrangement was, in fact, impossible [90]. His solution was simply to go through all possible arrangements by hand and see that none would work. While there are some shortcuts which can be better understood after reading the explanation of the problem which will follow, his proof was essentially an exhaustive search. In modern

© The Editor(s) (if applicable) and The Author(s), under exclusive license
to Springer Nature Switzerland AG 2020
S. T. Dougherty, *Combinatorics and Finite Geometry*,
Springer Undergraduate Mathematics Series,
https://doi.org/10.1007/978-3-030-56395-0_3

parlance, it would be known as a computer proof, although computers were not yet in existence. To many in the pursuit of mathematical truth such proofs are highly undesirable. This kind of inelegant proof does not explain *why* something is true. A refinement of this proof was made by Fisher and Yates in 1934 [37], in which they significantly reduced the number of Latin squares to consider by subdividing the set of all Latin squares into classes. So as not to leave the reader in suspense, there are proofs which solve the problem without any resort to an exhaustive search. The interested reader can find them in [28,84]. Additionally, there is a letter from the astronomer Heinrich Schumacher to Carl Gauss earlier than Tarry's proof, that his assistant Thomas Clausen had a solution to the problem, but this solution was never published or made known. See [74] for a complete description.

In an attempt to solve the problem, Euler decided to denote the ranks and regiments by Greek and Latin letters. Specifically, the ranks can be A, B, C, D, E, F and the regiments can be $\alpha, \beta, \gamma, \delta, \epsilon, \zeta$. Listing just the ranks in a 6 by 6 square with each row and column containing each rank exactly once produced a Latin square. For example, the following is a Latin square:

$$\begin{pmatrix} A & B & C & D & E & F \\ B & A & D & C & F & E \\ C & D & E & F & A & B \\ D & C & F & E & B & A \\ E & F & A & B & C & D \\ F & E & B & A & D & C \end{pmatrix}.$$

To solve the problem you would need to find a corresponding Greek square such that when overlapped with this square all possible pairs would appear.

Let us look at a smaller example. Can you arrange 9 officers of 3 ranks and regiments in a 3 by 3 square so that each row and column contains each rank and regiment exactly once? Now the ranks are A, B, C and the regiments are α, β, γ.

The following is a Latin square of order 3. Saying it is of order 3 means it is 3 by 3, which would mean that the above square was of order 6.

$$\begin{pmatrix} A & B & C \\ C & A & B \\ B & C & A \end{pmatrix}. \tag{3.1}$$

The following is a Greek square of order 3.

$$\begin{pmatrix} \alpha & \beta & \gamma \\ \beta & \gamma & \alpha \\ \gamma & \alpha & \beta \end{pmatrix}. \tag{3.2}$$

Overlaying them gives the following Graeco-Latin square.

$$\begin{pmatrix} A\alpha & B\beta & C\gamma \\ C\beta & A\gamma & B\alpha \\ B\gamma & C\alpha & A\beta \end{pmatrix}. \tag{3.3}$$

Notice that each pair appears exactly once, so the answer for order 3 is that they can be arranged. It may seem that the solution to order 3 was quite simple. The Latin square was formed by listing the ranks as the first row and then cycling them to the right. The Greek square was formed by listing the regiments as the first row and then cycling them to the left.

We can try the same with 4. The Latin square would be

$$\begin{pmatrix} A & B & C & D \\ D & A & B & C \\ C & D & A & B \\ B & C & D & A \end{pmatrix} \tag{3.4}$$

and the Greek square would be

$$\begin{pmatrix} \alpha & \beta & \gamma & \delta \\ \beta & \gamma & \delta & \alpha \\ \gamma & \delta & \alpha & \beta \\ \delta & \alpha & \beta & \gamma \end{pmatrix}. \tag{3.5}$$

The Graeco-Latin square would be

$$\begin{pmatrix} A\alpha & B\beta & C\gamma & D\delta \\ D\beta & A\gamma & B\delta & C\alpha \\ C\gamma & D\delta & A\alpha & B\beta \\ B\delta & C\alpha & D\beta & A\gamma \end{pmatrix}. \tag{3.6}$$

Notice that $A\alpha$ appears twice but $A\beta$ never appears. Similar problems occur throughout this square. This technique of cycling in opposite directions works if the order is an odd number but it does not work if the order is an even number. In particular, it does not help us to answer the 36 officer problem.

There is a way to arrange 16 officers. In fact you can do more. The three squares that follow can be put together any two at a time to make a Graeco-Latin square (with the provision of changing the numbers to either Latin or Greek letters). In general, we simply use the digits between 0 and 1 less than the order of the square as elements (or the numbers between 1 and the order) to simplify matters, but it is aesthetically pleasing to use different letters.

$$\begin{pmatrix} A & B & C & D \\ B & A & D & C \\ C & D & A & B \\ D & C & B & A \end{pmatrix} \begin{pmatrix} \alpha & \beta & \gamma & \delta \\ \gamma & \delta & \alpha & \beta \\ \delta & \gamma & \beta & \alpha \\ \beta & \alpha & \delta & \gamma \end{pmatrix} \begin{pmatrix} 0 & 1 & 2 & 3 \\ 3 & 2 & 1 & 0 \\ 1 & 0 & 3 & 2 \\ 2 & 3 & 0 & 1 \end{pmatrix}$$

$$\begin{pmatrix} A\alpha 0 & B\beta 1 & C\gamma 2 & D\delta 3 \\ B\gamma 3 & A\delta 2 & D\alpha 1 & C\beta 0 \\ C\delta 1 & D\gamma 0 & A\beta 3 & B\alpha 2 \\ D\beta 2 & C\alpha 3 & B\delta 0 & A\gamma 1 \end{pmatrix}.$$

When two Latin squares can be overlapped so that each pair appears exactly once then we call the squares orthogonal. If we have a collection of squares so that any two of them are orthogonal then we say that we have mutually orthogonal squares or MOLS for short.

We shall now make rigorous definitions of these ideas.

Definition 3.1 Let A be an alphabet of size n. Then an n by n array L is a Latin square of order n if each element of A appears exactly once in every row and every column.

Example 3.1 Let A be the alphabet $\{a, b\}$ then the following is a Latin square of order 2:

$$\begin{pmatrix} a & b \\ b & a \end{pmatrix}.$$

Definition 3.2 If $L = (\ell_{ij})$ is a Latin square of order n over an alphabet A and $M = (m_{ij})$ is a Latin square of order n over an alphabet B then L and M are said to be orthogonal if every pair of $A \times B$ occurs exactly once in $\{(\ell_{ij}, m_{ij})\}$. If $L_1, L_2, \ldots, L_k$ are Latin squares of order n then they are a set of k Mutually Orthogonal Latin Squares (MOLS) if L_i and L_j are orthogonal for all $i, j, i \neq j$.

Hence, a Graeco-Latin square can be realized as 2 MOLS.

Latin squares where each row is determined by cycling each element one space to the right (left), as was seen in the squares denoted given in (3.1), (3.2), (3.4), and (3.5), are known as right (left) circulant Latin squares.

We shall show that cycling does not produce orthogonal squares when n is even.

Theorem 3.1 *A left circulant Latin square and a right circulant Latin square are not orthogonal when n is even.*

Proof Without loss of generality let $A = \mathbb{Z}_n = \{0, 1, 2, \ldots, n-1\}$ and use as the first row $0 \ 1 \ \ldots \ (n-1)$. Using this alphabet allows us to use the arithmetic we have already defined for $\mathbb{Z}_n$. Let L be the left circulant square and R be the right circulant square. It is easy to see that $L_{i,j} = i + j$ and $R_{i,j} = j - i$. Let $n = 2k$. Consider the row $L_{k,j} = k + j$ and $R_{k,j} = j - k$. Notice that in $\mathbb{Z}_{2k}$, we have that $k = -k$. This gives that $j + k = j - k$ and so $L_{k,j} = R_{k,j}$. Hence (a, a) appears when overlapping the k-th row for each a and in the top row when $i = 0$. Since these pairs appear twice, the two squares are not orthogonal. $\square$

Example 3.2 For $n = 6$ the left and right circulant Latin squares are as follows:

$$
\begin{pmatrix}
0\ 1\ 2\ 3\ 4\ 5 \\
1\ 2\ 3\ 4\ 5\ 0 \\
2\ 3\ 4\ 5\ 0\ 1 \\
3\ 4\ 5\ 0\ 1\ 2 \\
4\ 5\ 0\ 1\ 2\ 3 \\
5\ 0\ 1\ 2\ 3\ 4
\end{pmatrix}
\qquad
\begin{pmatrix}
0\ 1\ 2\ 3\ 4\ 5 \\
5\ 0\ 1\ 2\ 3\ 4 \\
4\ 5\ 0\ 1\ 2\ 3 \\
3\ 4\ 5\ 0\ 1\ 2 \\
2\ 3\ 4\ 5\ 0\ 1 \\
1\ 2\ 3\ 4\ 5\ 0
\end{pmatrix}.
$$

Overlapping the squares gives us the following:

$$
\begin{pmatrix}
00\ 11\ 22\ 33\ 44\ 55 \\
15\ 20\ 31\ 42\ 53\ 04 \\
24\ 35\ 40\ 51\ 02\ 13 \\
33\ 44\ 55\ 00\ 11\ 22 \\
42\ 53\ 04\ 15\ 20\ 31 \\
51\ 02\ 13\ 24\ 35\ 40
\end{pmatrix}.
$$

Notice that this is not a Graeco-Latin square as 00 appears twice but 01 does not appear at all. In fact, every pair that appears, does so twice, and half of the possible pairs do not appear at all.

Notice that when $n = 2$ the two squares are actually identical, that is,

$$
L = R = \begin{pmatrix} 0\ 1 \\ 1\ 0 \end{pmatrix}.
$$

Theorem 3.2 *A left circulant Latin square and a right circulant Latin square with the same first row are orthogonal if the order of the square is odd.*

Proof Again, as in the previous theorem, without loss of generality let $A = \mathbb{Z}_n = \{0, 1, 2, \ldots, n - 1\}$, and we let L be the left circulant square and R be the right circulant square. We have $L_{i,j} = i + j$ and $R_{i,j} = j - i$. If $(i + j, j - i) = (i' + j', j' - i')$ then $i + j = i' + j'$ and $j - i = j' - i'$. Adding these equations we get $2j = 2j'$. Since n is odd we have that 2 is a unit in $\mathbb{Z}_n$, that is, 2 has a multiplicative inverse. This gives that $j = j'$. Using $i + j = i' + j'$ we have that $i = i'$. This gives that L and R are orthogonal. $\square$

Example 3.3 For $n = 5$, the left and right circulant Latin squares are as follows:

$$
\begin{pmatrix}
0\ 1\ 2\ 3\ 4 \\
1\ 2\ 3\ 4\ 0 \\
2\ 3\ 4\ 0\ 1 \\
3\ 4\ 0\ 1\ 2 \\
4\ 0\ 1\ 2\ 3
\end{pmatrix}
\qquad
\begin{pmatrix}
0\ 1\ 2\ 3\ 4 \\
4\ 0\ 1\ 2\ 3 \\
3\ 4\ 0\ 1\ 2 \\
2\ 3\ 4\ 0\ 1 \\
1\ 2\ 3\ 4\ 0
\end{pmatrix}.
$$

Overlapping the squares gives us the following:

$$\begin{pmatrix} 00 \ 11 \ 22 \ 33 \ 44 \\ 14 \ 20 \ 31 \ 42 \ 03 \\ 23 \ 34 \ 40 \ 01 \ 12 \\ 32 \ 43 \ 04 \ 10 \ 21 \\ 41 \ 02 \ 13 \ 24 \ 30 \end{pmatrix}.$$

Notice that this is a Graeco-Latin square as each possible pair appears exactly once.

This shows that there is always at least 2 MOLS of order n where n is odd.

Theorem 3.3 *The maximum number of MOLS of order n is $n - 1$.*

Proof Assume that the symbols are $0, 1, \ldots, n - 1$ for each square. Without loss of generality, we assume that the first row is

$$0 \ 1 \ 2 \ldots n - 1.$$

(If it is not, we simply rename the elements.) There are $n - 1$ choices for the first element of the second row, since it cannot be 0. In each square, these must be distinct since each element appears already with itself in the first row. This means we can have at most $n - 1$ MOLS of order n.

For example, if the order is 5 then the first row of the squares can be made to be

$$0 \ 1 \ 2 \ 3 \ 4.$$

There can not be a 0 in the first coordinate of the second row for any square since it already appears in that column. If any number, say 2, were to appear twice in that place then $(2, 2)$ would appear both in the first row and in the second when those two squares were overlapped and the squares would not be orthogonal. $\qquad\square$

When there are $n - 1$ MOLS of order n we say that there is a complete set of MOLS. In other words, we have the largest number of MOLS we can possibly have.

Example 3.4 Consider $n = 3$.

$$\begin{pmatrix} 0 \ 1 \ 2 \\ 1 \ 2 \ 0 \\ 2 \ 0 \ 1 \end{pmatrix} \quad \begin{pmatrix} 0 \ 1 \ 2 \\ 2 \ 0 \ 1 \\ 1 \ 2 \ 0 \end{pmatrix} \quad \begin{pmatrix} 0 \ 1 \ 2 \\ \square \end{pmatrix}.$$

The first two squares are orthogonal. In the third square, there is no possible entry to place in the box since 1 and 2 have already been used in the first two squares.

The fundamental question of Latin squares is the following.

Fundamental Question: What is the maximum number of Mutually Orthogonal Latin Squares of order n, for any n?

Let us return to the original problem. The way Col. Tarry solved the problem was to write out all possible Latin squares using $1, 2, 3, 4, 5, 6$ realizing that the first row could be

$$1\ 2\ 3\ 4\ 5\ 6$$

Then he took each one and overlapped it with every other one until he had tested them all in this manner. He then reported that none of the squares were orthogonal. Today, of course, this could be done fairly easily with a computer. The first open order is 10, meaning the maximum number of MOLS of order 10 is unknown although we do know of a pair of MOLS. The reader may ask why not just do it with a computer. The answer is that there are far too many for any computer.

Euler had conjectured that there would be no Graeco-Latin square of order 6 and, as usual, he was right. He also conjectured that there would be no Graeco-Latin square of order for any even number that was not divisible by 4, that is,

$$2, 6, 10, 14, 18, 22, 26, \ldots.$$

It is obviously true for 2 but Euler was uncharacteristically wrong for $10, 14, \ldots$. This was not known until 1960 [11,12] when Bose, Shrikhande, and Parker showed how to construct 2 MOLS of any desired order of this type greater than 6. In fact, other than 2 and 6 there is a Graeco-Latin square of any order. We shall prove this later.

Here is the Graeco-Latin square of order 10 that they found.

$$\begin{pmatrix} \alpha A & \theta E & \iota B & \eta H & \kappa C & \delta J & \zeta I & \epsilon D & \beta G & \gamma F \\ \eta I & \beta B & \theta F & \iota C & \alpha H & \kappa D & \epsilon J & \zeta E & \gamma A & \delta G \\ \zeta J & \alpha I & \gamma C & \theta G & \iota D & \beta H & \kappa E & \eta F & \delta B & \epsilon A \\ \kappa F & \eta J & \beta I & \delta D & \theta A & \iota E & \gamma H & \alpha G & \epsilon C & \zeta B \\ \delta H & \kappa G & \alpha J & \gamma I & \epsilon E & \theta B & \iota F & \beta A & \zeta D & \eta C \\ \iota G & \epsilon H & \kappa A & \beta J & \delta I & \zeta F & \theta C & \gamma B & \eta E & \alpha D \\ \theta D & \iota A & \zeta H & \kappa B & \gamma J & \epsilon I & \eta G & \delta C & \alpha F & \beta E \\ \epsilon B & \zeta C & \eta D & \alpha E & \beta F & \gamma G & \delta A & \theta H & \iota I & \kappa J \\ \beta C & \gamma D & \delta E & \epsilon F & \zeta G & \eta A & \alpha B & \kappa I & \theta J & \iota H \\ \gamma E & \delta F & \epsilon G & \zeta A & \eta B & \alpha C & \beta D & \iota J & \kappa H & \theta I \end{pmatrix}. \qquad (3.7)$$

At present, it is unknown if there exists a complete set of MOLS of order n where n is not a prime power.

Theorem 3.4 *A complete set of MOLS exists of order n if there is a field of order n.*

Proof Let L be the addition table for a field $F = \{a_0, a_1, \ldots, a_{n-1}\}$, i.e. $L_{i,j} = a_i + a_j$. Then let $L^1 = L$ and let L^k be formed by replacing the h-th row of L with the $a_k a_h$ row of L. We know that L is a Latin square so permuting the rows does not affect that it is a Latin square, specifically each row and column still has every element exactly once.

Assume $(L^b_{i,j}, L^c_{i,j}) = (L^b_{i',j'}, L^c_{i',j'})$ where $b \neq c$. Then we have

$$a_b a_i + a_j = a_b a_{i'} + a_{j'}$$
$$a_c a_i + a_j = a_c a_{i'} + a_{j'}.$$

Subtracting these equations we get

$$(a_b - a_c)a_i = (a_b - a_c)a_{i'}. \tag{3.8}$$

Since $b \neq c$ then $a_b - a_c$ has a multiplicative inverse since we are in a field, so $a_i = a_{i'}$ giving $i = i'$. It follows immediately that $a_j = a_{j'}$ and hence the two squares are orthogonal. This gives that $L_1, L_2, \ldots, L_{n-1}$ are a set of $n - 1$ MOLS of order n and hence are complete. $\qquad\square$

Example 3.5 We exhibit the six MOLS of order 7 using Theorem 3.4.

$$L^1 = \begin{pmatrix} 0\,1\,2\,3\,4\,5\,6 \\ 1\,2\,3\,4\,5\,6\,0 \\ 2\,3\,4\,5\,6\,0\,1 \\ 3\,4\,5\,6\,0\,1\,2 \\ 4\,5\,6\,0\,1\,2\,3 \\ 5\,6\,0\,1\,2\,3\,4 \\ 6\,0\,1\,2\,3\,4\,5 \end{pmatrix} \quad L^2 = \begin{pmatrix} 0\,1\,2\,3\,4\,5\,6 \\ 2\,3\,4\,5\,6\,0\,1 \\ 4\,5\,6\,0\,1\,2\,3 \\ 6\,0\,1\,2\,3\,4\,5 \\ 1\,2\,3\,4\,5\,6\,0 \\ 3\,4\,5\,6\,0\,1\,2 \\ 5\,6\,0\,1\,2\,3\,4 \end{pmatrix}$$

$$L^3 = \begin{pmatrix} 0\,1\,2\,3\,4\,5\,6 \\ 3\,4\,5\,6\,0\,1\,2 \\ 6\,0\,1\,2\,3\,4\,5 \\ 2\,3\,4\,5\,6\,0\,1 \\ 5\,6\,0\,1\,2\,3\,4 \\ 1\,2\,3\,4\,5\,6\,0 \\ 4\,5\,6\,0\,1\,2\,3 \end{pmatrix} \quad L^4 = \begin{pmatrix} 0\,1\,2\,3\,4\,5\,6 \\ 4\,5\,6\,0\,1\,2\,3 \\ 1\,2\,3\,4\,5\,6\,0 \\ 5\,6\,0\,1\,2\,3\,4 \\ 2\,3\,4\,5\,6\,0\,1 \\ 6\,0\,1\,2\,3\,4\,5 \\ 3\,4\,5\,6\,0\,1\,2 \end{pmatrix}$$

$$L^5 = \begin{pmatrix} 0\,1\,2\,3\,4\,5\,6 \\ 5\,6\,0\,1\,2\,3\,4 \\ 3\,4\,5\,6\,0\,1\,2 \\ 1\,2\,3\,4\,5\,6\,0 \\ 6\,0\,1\,2\,3\,4\,5 \\ 4\,5\,6\,0\,1\,2\,3 \\ 2\,3\,4\,5\,6\,0\,1 \end{pmatrix} \quad L^6 = \begin{pmatrix} 0\,1\,2\,3\,4\,5\,6 \\ 6\,0\,1\,2\,3\,4\,5 \\ 5\,6\,0\,1\,2\,3\,4 \\ 4\,5\,6\,0\,1\,2\,3 \\ 3\,4\,5\,6\,0\,1\,2 \\ 2\,3\,4\,5\,6\,0\,1 \\ 1\,2\,3\,4\,5\,6\,0 \end{pmatrix}.$$

These six Latin squares form a complete set of MOLS of order 7.

Definition 3.3 A magic square of order n is an n by n matrix where each element of the set $\{0, 1, 2, \ldots, n^2 - 1\}$ appears once and the sum of each row and column is the same.

Example 3.6 The following is an example of a magic square of order 3:

$$\begin{pmatrix} 0 & 4 & 8 \\ 5 & 6 & 1 \\ 7 & 2 & 3 \end{pmatrix}.$$

Here the elements $0, 1, 2, \ldots, 8$ each appear once and the sum of every row and column is 12.

Theorem 3.5 *Let* L *and* K *be orthogonal Latin squares of order* n *over the alphabet* $\mathbb{Z}_n$. *Then the square matrix* M *formed by*

$$M_{i,j} = L_{i,j} + nK_{i,j}, \tag{3.9}$$

where the operations are done in the integers, is a magic square of order n *and the sum of the elements of any row or column of* M *is* $\frac{(n-1)n(n+1)}{2}$.

Proof In every row or column each element of $\mathbb{Z}_n$ appears exactly once as a and once as b in the set $\{a + bn\}$. Hence the sum of the row is

$$\left(\sum_{i=1}^{n-1} i\right)n + \left(\sum_{i=1}^{n-1} i\right) = \frac{(n-1)n}{2}n + \frac{(n-1)n}{2} = \frac{(n-1)n}{2}(n+1).$$

Since the squares are orthogonal each pair (a, b) appears exactly once in their overlap, hence each number from 0 to $n^2 - 1$ appears once in the set $\{a + bn\}$. $\qquad\square$

Example 3.7 As an example consider the following orthogonal Latin squares and the magic square they form.

$$\begin{pmatrix} 0 & 1 & 2 & 3 & 4 \\ 1 & 2 & 3 & 4 & 0 \\ 2 & 3 & 4 & 0 & 1 \\ 3 & 4 & 0 & 1 & 2 \\ 4 & 0 & 1 & 2 & 3 \end{pmatrix} + 5 \begin{pmatrix} 0 & 1 & 2 & 3 & 4 \\ 4 & 0 & 1 & 2 & 3 \\ 3 & 4 & 0 & 1 & 2 \\ 2 & 3 & 4 & 0 & 1 \\ 1 & 2 & 3 & 4 & 0 \end{pmatrix} = \begin{pmatrix} 0 & 6 & 12 & 18 & 24 \\ 21 & 2 & 8 & 14 & 15 \\ 17 & 23 & 4 & 5 & 11 \\ 13 & 19 & 20 & 1 & 7 \\ 9 & 10 & 16 & 22 & 3 \end{pmatrix}. \tag{3.10}$$

Each element from 0 to 24 appears once and the sum of each row and column is $\frac{4(5)(6)}{2} = 60$.

We shall now define how to take the direct product of two Latin squares. Let $A = (A_{ij})$ and $B = (B_{ij})$ be Latin squares of order n and m, respectively. Define

$$A \times B = \begin{pmatrix} A_{0,0}B & A_{0,1}B & \cdots & A_{0,n-1}B \\ A_{1,0}B & A_{1,1}B & \cdots & A_{1,n-1}B \\ \vdots & & & \\ A_{n-1,0}B & A_{n-1,1}B & \cdots & A_{n-1,n-1}B \end{pmatrix}, \tag{3.11}$$

where (i, j) is rewritten as $im + j$.

As an example, consider the following cross product of a Latin square of order 2 with a Latin square of order 3.

$$\begin{pmatrix} 0 & 1 \\ 1 & 0 \end{pmatrix} \times \begin{pmatrix} 0 & 1 & 2 \\ 2 & 0 & 1 \\ 1 & 2 & 0 \end{pmatrix} = \begin{pmatrix} 0 & 1 & 2 & 3 & 4 & 5 \\ 2 & 0 & 1 & 5 & 3 & 4 \\ 1 & 2 & 0 & 4 & 5 & 3 \\ 3 & 4 & 5 & 0 & 1 & 2 \\ 5 & 3 & 4 & 2 & 0 & 1 \\ 4 & 5 & 3 & 1 & 2 & 0 \end{pmatrix}. \tag{3.12}$$

Theorem 3.6 *If L and L' are orthogonal Latin squares of order n and M and M' are orthogonal Latin squares of order m then $L \times M$ and $L' \times M'$ are orthogonal Latin squares of order nm.*

Proof The proof is straightforward and is left as an exercise. See Exercise 10.   □

This theorem is enough to tell us a great deal about the existence of a pair of MOLS.

The following is a direct consequence of Theorem 2.19.

Lemma 3.1 *There exist finite fields of every prime power order.*

This lemma implies, by Theorem 3.4, that there is always a complete set of MOLS for p^e where p is a prime.

Theorem 3.7 *If $n \not\equiv 2 \pmod 4$ then there is a pair of MOLS of order n.*

Proof If $n = \prod p_i^{e_i}$ where p_i is a prime and $p_i \neq p_j$ for $i \neq j$ then we know there exists MOLS of every order as long as the exponent of 2 is not 1, i.e., as long as $n \not\equiv 2 \pmod 4$. Then we take the direct product of these MOLS to produce 2 MOLS of order n by Theorem 3.6.   □

Given this theorem it is easy to see why Euler conjectured that there would be no set of MOLS when $n \equiv 2 \pmod 4$.

Example 3.8 Let $n = 588 = 2^2 \cdot 3 \cdot 7$. There exists a pair of MOLS of order 4, 3, and 7^2 by Theorem 3.4. Then by applying Theorem 3.6 twice, we have that there is a pair of MOLS of order 588.

We shall examine some ideas used in studying the construction of MOLS. Given a Latin square L a transversal is a set of n coordinates in the square such that each symbol appears once in the set and the set intersects each row and column exactly once. The bold elements form a transversal in the following Latin square.

$$
\begin{pmatrix}
\mathbf{1} & 2 & 0 & 4 & 3 \\
2 & \mathbf{1} & 3 & 0 & 4 \\
0 & \mathbf{4} & 1 & 3 & 2 \\
4 & 3 & 2 & 1 & \mathbf{0} \\
3 & 0 & 4 & \mathbf{2} & 1
\end{pmatrix} .
\tag{3.13}
$$

An orthogonal mate is a set of n mutually disjoint transversals. To show a square has no orthogonal mates, it is sufficient to show it has no transversals or even that it has no transversals through a particular coordinate.

The following Latin square has no transversals.

$$
\begin{pmatrix}
0 & 1 & 2 & 3 \\
1 & 2 & 3 & 0 \\
2 & 3 & 0 & 1 \\
3 & 0 & 1 & 2
\end{pmatrix} .
\tag{3.14}
$$

Theorem 3.8 *Given a set of* k-*MOLS* $L^1, L^2, \ldots, L^k$, *if there is a Latin square* L *such that* L *is orthogonal to each* L^i, *then each set of coordinates corresponding to a symbol in* L *is a transversal to* L^i *for each* i, $1 \le i \le k$.

Proof The set of coordinates of each symbol must be a transversal to L^i, since L^i is orthogonal to L.   □

Given a Latin square, a k-transversal is a set of kn coordinates in the square such that each symbol appears k times in the set and the set intersects each row and column k times. (These objects can also be called a k-plex.) The union of k transversals will form a k-transversal but there exist k transversals that are not the union of transversals. For example, the previous Latin square has the following 2-transversal.

$$
\begin{pmatrix}
\mathbf{0} & 1 & 2 & \mathbf{3} \\
\mathbf{1} & 2 & 3 & \mathbf{0} \\
2 & \mathbf{3} & \mathbf{0} & 1 \\
3 & \mathbf{0} & \mathbf{1} & 2
\end{pmatrix} .
\tag{3.15}
$$

There are two important conjectures concerning transversals. They are as follows.

Conjecture 3.1 *Every Latin square of odd order has a transversal.*

Conjecture 3.2 *Every Latin square of order n has $\lfloor \frac{n}{2} \rfloor$ disjoint 2-transversals.*

An extremely large literature exists about Latin squares and their generalizations. A good first reference would be the text by Denes and Keedwell, Latin Squares and Applications [25] published first in 1974 (a second edition appeared in 2015). A glance at the text indicates how much was known about Latin squares and how many applications there were for Latin squares. Since the publication of [25], the number of papers studying Latin squares has grown immensely. For a more recent text, see Discrete Mathematics using Latin Squares, by Laywine and Mullen, [59].

Exercises

1. Prove that the addition table of a finite field is a Latin square.
2. Determine if the following Latin square has any 1-transversals or 2-transversals:

$$\begin{pmatrix} 1\;2\;3\;4\;0 \\ 2\;1\;0\;3\;4 \\ 3\;4\;2\;0\;1 \\ 4\;0\;1\;2\;3 \\ 0\;3\;4\;1\;2 \end{pmatrix}. \tag{3.16}$$

3. Prove that the multiplication table of non-zero elements in a finite field is a Latin square.
4. Prove that if there exists $n-2$ MOLS of order n then there exists a unique Latin square that completes the set to $n-1$ MOLS.
5. Prove that any Latin square of order 3 has an orthogonal mate.
6. Construct a magic square from the orthogonal Latin squares of order 10.
7. Find four MOLS of order 5. The first is given below (3.17) and the other 3 can be found by permuting the order of the rows of the first.

$$\begin{pmatrix} 0\;1\;2\;3\;4 \\ 1\;2\;3\;4\;0 \\ 2\;3\;4\;0\;1 \\ 3\;4\;0\;1\;2 \\ 4\;0\;1\;2\;3 \end{pmatrix}. \tag{3.17}$$

8. Find all possible Latin squares of order 4 where the first row is $1, 2, 3, 4$.
9. Show that over $\mathbb{Z}_n$, the square $L_{ij} = ai + j$ is a Latin square if and only if $\gcd(a, n) = 1$.
10. Prove that the cross product of Latin squares is a Latin square and prove Theorem 3.6.
11. Let L and M be two orthogonal Latin squares of order 3. Compute $L \times L$ and $M \times M$ and verify that the two Latin squares of order 9 are orthogonal.

12. Construct 2 MOLS of order 12 from the MOLS of order 3 and 4.
13. Determine the minimum number of MOLS that can exist of order $p_1^{e_1}p_2^{e_2}$, where p_1, p_2 are distinct primes and $e_i > 0$.

3.2   Forming Latin Squares

Our next task will be to find ways to construct Latin squares from existing Latin squares and how to group them together.

Let $L = (L_{ij})$ be a Latin square of order n. Then the matrix formed by reversing the roles of rows and columns is again a Latin square. This matrix is called the transpose of the matrix and is denoted by L^t, namely, $L^t = (L_{ji})$.

Lemma 3.2 *If* $L = (L_{ij})$ *is a Latin square of order* n, *then* $L^t = (L_{ji})$ *is a Latin square of order* n.

Proof Each row and each column of the transpose contains each element of the alphabet exactly once since each column and each row contains each element exactly once. □

As an example, consider the following Latin square and its transpose:

$$L = \begin{pmatrix} 0\,1\,2\,3 \\ 2\,0\,3\,1 \\ 3\,2\,1\,0 \\ 1\,3\,0\,2 \end{pmatrix}, \quad L^t = \begin{pmatrix} 0\,2\,3\,1 \\ 1\,0\,2\,3 \\ 2\,3\,1\,0 \\ 3\,1\,0\,2 \end{pmatrix}. \tag{3.18}$$

Notice that these Latin squares may not be distinct. The following is equal to its transpose.

$$L = \begin{pmatrix} 0\,1\,2\,3 \\ 1\,0\,3\,2 \\ 2\,3\,0\,1 \\ 3\,2\,1\,0 \end{pmatrix}. \tag{3.19}$$

Any Latin square that is equal to its transpose, $L = L^t$, is called a symmetric Latin square.

Theorem 3.9 *If* L *is a Latin square of order* n *with* M *orthogonal to* L, *then* M^t *is orthogonal to* L^t.

Proof If $\{(L_{i,j}, M_{i,j}) \mid 1 \leq i, j \leq n\}$ gives all possible ordered pairs then it is immediate that $\{(L_{j,i}, M_{j,i}) \mid 1 \leq i, j \leq n\}$ gives all possible ordered pairs since it is the same set. This gives the result. □

Example 3.9 The following two Latin squares are orthogonal:

$$L = \begin{pmatrix} 0\ 1\ 2\ 3\ 4 \\ 2\ 3\ 4\ 0\ 1 \\ 4\ 0\ 1\ 2\ 3 \\ 1\ 2\ 3\ 4\ 0 \\ 3\ 4\ 0\ 1\ 2 \end{pmatrix} \quad M = \begin{pmatrix} 0\ 1\ 2\ 3\ 4 \\ 3\ 4\ 0\ 1\ 2 \\ 1\ 2\ 3\ 4\ 0 \\ 4\ 0\ 1\ 2\ 3 \\ 2\ 3\ 4\ 0\ 1 \end{pmatrix}.$$

Their transposes are as follows and they are also orthogonal:

$$L^t = \begin{pmatrix} 0\ 2\ 4\ 1\ 3 \\ 1\ 3\ 0\ 2\ 4 \\ 2\ 4\ 1\ 3\ 0 \\ 3\ 0\ 2\ 4\ 1 \\ 4\ 1\ 3\ 0\ 2 \end{pmatrix} \quad M^t = \begin{pmatrix} 0\ 3\ 1\ 4\ 2 \\ 1\ 4\ 2\ 0\ 3 \\ 2\ 0\ 3\ 1\ 4 \\ 3\ 1\ 4\ 2\ 0 \\ 4\ 2\ 0\ 3\ 1 \end{pmatrix}.$$

This exercise shows that if we are looking for squares that have an orthogonal mate we need not consider the transpose after we have considered the original square. This significantly reduces the number we have to consider (almost by half, remembering that some squares are equal to their transpose). This is the type of reasoning that Fisher and Yates used so well to reduce the computation in showing that there are not two MOLS of order 6. We shall examine other ways of grouping squares together.

First, we shall show an alternate representation of a Latin square. Suppose we have the first Latin square given in (3.18) where the rows and columns are indexed by $\mathbb{Z}_4 = \{0, 1, 2, 3\}$. Then we can simply label the coordinates and the entry in those coordinates. We use the first coordinate to label the row, the second to label the column, and the third to label entry. For this square we would have

$$\begin{pmatrix} R\ C\ S \\ 0\ 0\ 0 \\ 0\ 1\ 1 \\ 0\ 2\ 2 \\ 0\ 3\ 3 \\ 1\ 0\ 2 \\ 1\ 1\ 0 \\ 1\ 2\ 3 \\ 1\ 3\ 1 \\ 2\ 0\ 3 \\ 2\ 1\ 2 \\ 2\ 2\ 1 \\ 2\ 3\ 0 \\ 3\ 0\ 1 \\ 3\ 1\ 3 \\ 3\ 2\ 0 \\ 3\ 3\ 2 \end{pmatrix}. \qquad\qquad (3.20)$$

This is known as an orthogonal array. For a detailed survey of combinatorial structures, which are equivalent to MOLS, see [53]. It is easy to see that we can add additional columns for Latin squares orthogonal to the first and get a larger orthogonal array but for right now we simply want to look at the representation of one Latin square.

If we reverse the first and second columns we simply get the transpose of the matrix. It should be clear that we can switch any two columns of this array and still have a Latin square. If we switch the first and the third, that is, change the roles of rows and symbols, then we call the resulting Latin square the row adjugate of the original and denote it by L^r. If we switch the second and the third, that is, change the roles of columns and symbols, then we call the resulting Latin square the column adjugate of the original and denote it by L^c.

Lemma 3.3 *If L is a Latin square of order n then L^c and L^r are Latin squares of order n.*

Proof See Exercise 5. □

We have three operations described here which can be done in any order which correspond to the 6 different ways of arranging the three columns of the orthogonal array. Basically, we are permuting the set $\{R, C, S\}$. If we rename this set as $\{1, 2, 3\}$ we can use the notation previously given for permutations. Namely, we have the following correspondence.

- The identity corresponds to the original Latin square.
- $(1, 2)$ corresponds to the transpose.
- $(1, 3)$ corresponds to the row adjugate.
- $(2, 3)$ corresponds to the column adjugate.
- $(1, 2, 3)$ corresponds to the transpose of the column adjugate.
- $(1, 3, 2)$ corresponds to the transpose of the row adjugate.

Example 3.10 We shall examine all 6 squares formed by these operations on the Latin square of order 4 given in (3.18).

$$L = \begin{pmatrix} 0\,1\,2\,3 \\ 2\,0\,3\,1 \\ 3\,2\,1\,0 \\ 1\,3\,0\,2 \end{pmatrix} \quad L^t = \begin{pmatrix} 0\,2\,3\,1 \\ 1\,0\,2\,3 \\ 2\,3\,1\,0 \\ 3\,1\,0\,2 \end{pmatrix} \quad L^r = \begin{pmatrix} 0\,1\,3\,2 \\ 3\,0\,2\,1 \\ 1\,2\,0\,3 \\ 2\,3\,1\,0 \end{pmatrix}. \tag{3.21}$$

$$L^c = \begin{pmatrix} 0\,1\,2\,3 \\ 1\,3\,0\,2 \\ 3\,2\,1\,0 \\ 2\,0\,3\,1 \end{pmatrix} \quad (L^r)^t = \begin{pmatrix} 0\,3\,1\,2 \\ 1\,0\,2\,3 \\ 3\,2\,0\,1 \\ 2\,1\,3\,0 \end{pmatrix} \quad (L^c)^t = \begin{pmatrix} 0\,1\,3\,2 \\ 1\,3\,2\,0 \\ 2\,0\,1\,3 \\ 3\,2\,0\,1 \end{pmatrix}. \tag{3.22}$$

Theorem 3.10 *Let* L *be a Latin square. Then*

$$(L^t)^r = (L^c)^t$$

and

$$(L^t)^c = (L^r)^t.$$

Proof If we start with RCS, then taking the transpose gives CRS. Then the row adjugate is SRC. If we take that column adjugate of RCS we have RSC, then taking the transpose gives SRC. This gives the first equation.

If we start with RCS, then taking the transpose gives CRS. Then the column adjugate is CSR. If we take that row adjugate of RCS we have SCR, then taking the transpose gives CSR. This gives the second equation.   □

Given a pair of mutually orthogonal Latin squares we can construct an orthogonal array with 4 columns by using the same first two columns and writing the symbols of the second square in the fourth column. For example, given the two MOLS of order 3 given in (3.3) we construct the following orthogonal array:

$$\begin{pmatrix} R & C & S_1 & S_2 \\ 0 & 0 & 0 & 0 \\ 0 & 1 & 1 & 1 \\ 0 & 2 & 2 & 2 \\ 1 & 0 & 2 & 1 \\ 1 & 1 & 0 & 2 \\ 1 & 2 & 1 & 0 \\ 2 & 0 & 1 & 2 \\ 2 & 1 & 2 & 0 \\ 2 & 2 & 0 & 1 \end{pmatrix}. \tag{3.23}$$

Lemma 3.4 *If* L *is a Latin square of order* n *with an orthogonal mate* M *then* L^r *and* L^c *have orthogonal mates.*

Proof Consider the orthogonal array with 4 columns described above. If we take the row adjugate of each we have $S_1 CR$ and $S_2 CR$ as the latin squares. Since each pair appears in the original two squares when overlapped, this means that each pair appears when overlapping the row adjugates. This is because no pair appears twice in any two columns. This is the same proof for the column adjugates.   □

There are $4! = 24$ ways of permuting the columns in (3.23). Any permutation of the columns will produce a pair of MOLS.

There are more ways to produce Latin squares from an existing square. Consider a Latin square L of order n. Any of the $n!$ ways of permuting the rows results in a Latin square and any of the $n!$ ways of permuting the columns results in a Latin square. The fact there are $n!$ ways of doing this follows from Theorem 1.5.

Lemma 3.5 *If* L *and* M *are orthogonal and* L' *is formed by permuting the rows and columns of* L *then the square* M' *formed by performing the same permutations on* M *is orthogonal to* L'.

Proof If every pair occurs exactly once when L and M are overlaid, then exactly every pair occurs exactly once when L' and M' are overlaid. They are simply in a different position. □

For example, the following Latin squares are orthogonal.

$$L = \begin{pmatrix} 0 & 1 & 2 \\ 2 & 0 & 1 \\ 1 & 2 & 0 \end{pmatrix} \quad M = \begin{pmatrix} 0 & 1 & 2 \\ 1 & 2 & 0 \\ 2 & 0 & 1 \end{pmatrix}. \tag{3.24}$$

Now apply the following permutation to the rows:

$$R_0 \rightarrow R_1,$$
$$R_1 \rightarrow R_2,$$
$$R_2 \rightarrow R_0.$$

and the following permutations to the columns:

$$C_0 \rightarrow C_0,$$
$$C_1 \rightarrow C_2,$$
$$C_2 \rightarrow C_1.$$

The resulting squares are

$$L' = \begin{pmatrix} 1 & 0 & 2 \\ 0 & 2 & 1 \\ 2 & 1 & 0 \end{pmatrix}, \quad \text{and } M' = \begin{pmatrix} 2 & 1 & 0 \\ 0 & 2 & 1 \\ 1 & 0 & 2 \end{pmatrix}, \tag{3.25}$$

which are orthogonal.

Definition 3.4 If L is a Latin square, then any Latin square that can be obtained by any combination of permuting the rows, permuting the columns, permuting the symbols, taking the row adjugate, taking the column adjugate, and taking the transpose is said to be in the main class of L.

Theorem 3.11 *If* L *and* L′ *are Latin squares in the same main class then* L *has an orthogonal mate if and only if* L′ *has an orthogonal mate.*

Proof Follows from Theorem 3.9, Lemmas 3.4 and 3.5.   □

The importance of this theorem is that if you are looking for squares that have orthogonal mates you need only look through the collection of main classes rather than the entire set of Latin squares which can be quite large. This is essentially the proof of Fisher and Yates for the 36 officer problem, given that they actually find representatives of the main classes and check them for orthogonality.

To give an idea of just how large the set of Latin squares can be. The number of Latin squares of order 9 (given in [6,66]) is

$$(9!)(8!)(377,597,570,964,258,816) \approx 10^{27}.$$

This should give the reader an idea of how hard it would be to search through them all to find orthogonal mates, let alone trying to find 3 MOLS.

The reason the number is given in this manner is that they find all Latin squares whose first row and first column is $1, 2, \ldots, 9$. Then by Theorem 1.5, there are 9! ways of permuting the symbols in the first row and 8! ways of permuting the first column, given that the first element was changed by the permutation of the rows and reduces the number of possible permutations.

Similarly, in [64], it was shown that the number of Latin squares of order 10 is

$$7580721483160132811489280(10!)(9!) \approx 9(10^{36}),$$

and in [50], it was shown that the number of Latin squares of order 11 is

$$122161773153692292614825540(11!)(10!) \approx 10^{39}.$$

It is easily seen that given the size of these numbers, searching for orthogonal mates with an exhaustive search is not computationally feasible.

Exercises

1. Prove there exists a symmetric Latin square for each order $n > 0$.
2. Prove that the addition table of a finite field is a symmetric Latin square.
3. Prove that the multiplication table of the non-zero elements of a finite field is a symmetric Latin square.
4. Produce the MOLS formed by allowing the permutation $(1, 3, 2, 4)$ on the columns of (3.23).
5. Prove Lemma 3.3.

6. Find all 6 Latin squares formed from these operations applied to the Latin square

$$\begin{pmatrix} 0\ 1\ 2\ 3 \\ 3\ 2\ 1\ 0 \\ 2\ 3\ 0\ 1 \\ 1\ 0\ 3\ 2 \end{pmatrix}. \tag{3.26}$$

7. Use Lemma 3.4 to find Latin squares that are orthogonal to L, L^t, L^r, L^c, $(L^c)^t$ and $(L^r)^t$ if L is the Latin square

$$\begin{pmatrix} 0\ 1\ 2 \\ 2\ 0\ 1 \\ 1\ 2\ 0 \end{pmatrix}. \tag{3.27}$$

3.3 Structure of Latin Squares

We shall now describe some aspects of Latin squares which helps to understand their structure.

Definition 3.5 Let L be a Latin square of order n, indexed by the elements of $\mathbb{Z}_n$. If there are subsets $R, C \subset \mathbb{Z}_n$ with $|R| = |C| = m$, where L_{ij}, $i \in R$, $j \in C$, is a Latin square of order m, then this is said to be a subsquare of order m.

As an example, consider the following Latin square. The bold elements form a subsquare of order 2.

$$\begin{pmatrix} 0\ 1\ 2\ 3 \\ 2\ 3\ 0\ 1 \\ 3\ 2\ 1\ 0 \\ 1\ 0\ 3\ 2 \end{pmatrix}. \tag{3.28}$$

Theorem 3.12 *The circulant Latin square of order n has a subsquare of order 2 if and only if n is even.*

Proof Let $L_{ij} = i + j$ where the matrix is indexed by $\mathbb{Z}_n$. If i_1, i_2 are the rows and j_1, j_2 are the columns then we have that

$$\begin{pmatrix} L_{i_1,j_1} & L_{i_1,j_2} \\ L_{i_2,j_1} & L_{i_2,j_2} \end{pmatrix}$$

is a Latin square. This gives that

$$L_{i_1,j_1} = L_{i_2,j_2}$$
$$L_{i_1,j_2} = L_{i_2,j_1}.$$

That is,

$$i_1 + j_1 = i_2 + j_2$$
$$i_1 + j_2 = i_2 + j_1.$$

Then we have $i_1 - i_2 = j_2 - j_1 = i_2 - i_1$ which gives that $2i_1 = 2i_2$. If n is odd, then 2 is a unit so $i = 1 = i_2$ and there is no subsquare. If n is $2k$ then let $i_1 = 0$ and $i_2 = k$. Then solving $j_2 - j_1 = k$ gives a Latin subsquare of size 2.   □

Notice that in the proof we actually can find many subsquares for different values of i_1, i_2, j_1, and j_2.

While circulant Latin squares of odd order do not have a subsquare of order 2 a circulant Latin square of odd order can have a subsquare. Consider the following, the bold elements form a subsquare of order 3.

$$\begin{pmatrix} \mathbf{0}\ 1\ 2\ 3\ \mathbf{4}\ 5\ 6\ \mathbf{7}\ 8 \\ 8\ 0\ 1\ 2\ 3\ 4\ 5\ 6\ 7 \\ 7\ 8\ 0\ 1\ 2\ 3\ 4\ 5\ 6 \\ 6\ 7\ \mathbf{8}\ 0\ 1\ \mathbf{2}\ 3\ 4\ \mathbf{5} \\ 5\ 6\ 7\ 8\ 0\ 1\ 2\ 3\ 4 \\ 4\ 5\ 6\ 7\ 8\ 0\ 1\ 2\ 3 \\ \mathbf{3}\ 4\ 5\ \mathbf{6}\ 7\ 8\ \mathbf{0}\ 1\ 2 \\ 2\ 3\ 4\ 5\ 6\ 7\ 8\ 0\ 1 \\ 1\ 2\ 3\ 4\ 5\ 6\ 7\ 8\ 0 \end{pmatrix}. \tag{3.29}$$

We can generalize this idea in the following theorem.

Theorem 3.13 *If s and n are relatively prime then there is no subsquare of order s in the circulant Latin square of order n*

Proof Let $L_{i,j} = i + j$ as before. Consider two rows in the subsquare:

$$L_{i_1,j_1}\ L_{i_1,j_2}\ L_{i_1,j_3}\ \cdots\ L_{i_1,j_s}$$
$$L_{i_2,j_1}\ L_{i_2,j_2}\ L_{i_2,j_3}\ \cdots\ L_{i_2,j_s}$$

These rows are permutations of the same sets of elements. This gives that their sums must be identical, that is,

$$si_1 + \sum_{a=1}^{s} j_a = si_2 + \sum_{a=1}^{s} j_a.$$

This gives $si_1 = si_2$. If $\gcd(s,n) = 1$ then s is a unit and so $i_1 = i_2$ and there is no subsquare.   □

Definition 3.6 A partial Latin square of order n is an n by n array, where cells may be empty or from a symbol set A of size n, such that each symbol occurs at most once in any row or column.

Consider the extremely popular game of Sudoku. This game gives a partial Latin square of order 9 with the property that each of 9 different 3 by 3 subsquares must contain each of the elements $1, 2, \ldots, 9$. The aim of the game is to complete the square to its unique completion.

As an example, the following is a partial Latin square of order 5.

$$
\begin{pmatrix}
1 & 2 & 3 & 4 & 5 \\
4 & & 2 & 1 & \\
3 & 1 & & 2 & 4 \\
& 4 & 1 & 3 & \\
2 & & 4 & & 3
\end{pmatrix}.
\tag{3.30}
$$

Theorem 3.14 *There exists a non-completable partial Latin square for all orders* $n \geq 2$.

Proof Let L be any Latin square of order n using symbols $0, 1, 2, \ldots, n-1$. Let M be the partial Latin square:

$$
M = \begin{pmatrix} L & \\ & n \end{pmatrix}.
$$

In the open spaces of M no element from $0, 1, 2, \ldots, n-1$ can be used since these are represented in every row and column already since L is a Latin square. The open spaces also cannot be filled with n since this already appears in the last row and last column. Therefore, this partial Latin square is non-completable. $\qquad\square$

Example 3.11 The following partial Latin squares are not completable:

$$
\begin{pmatrix} 0 & \\ & 1 \end{pmatrix}, \quad
\begin{pmatrix} 0 & 1 & \\ 1 & 0 & \\ & & 2 \end{pmatrix}, \quad
\begin{pmatrix} 0 & 1 & 2 & \\ 1 & 2 & 0 & \\ 2 & 0 & 1 & \\ & & & 3 \end{pmatrix}.
$$

It is also possible for a partial Latin square to complete in two different ways. For example, consider the partial Latin square:

$$
\begin{pmatrix} 0 & & \\ & 0 & \\ & & 0 \end{pmatrix}.
$$

This square can complete to either of the following:

$$\begin{pmatrix} 0\ 2\ 1 \\ 1\ 0\ 2 \\ 2\ 1\ 0 \end{pmatrix} \qquad \begin{pmatrix} 0\ 1\ 2 \\ 2\ 0\ 1 \\ 1\ 2\ 0 \end{pmatrix}.$$

Exercises

1. Complete the partial Latin square in Eq. 3.30 to a Latin square. Determine if the completion is unique.
2. Find all other subsquares in the Latin square in Eq. 3.28.
3. Prove that if $\gcd(s, n) \neq 1$ then there is a subsquare of order $\gcd(n, s)$ in the circulant Latin square of order n. Hint: We need to have $si_1 = si_2 = si_3 = \cdots = si_s$. Then how many solutions are there for $sx \equiv si_1 \pmod{n}$. Use these to construct the subsquares.
4. Show that there are partial Latin squares that are not completable for all $n > 2$ using a different proof than the one given in Theorem 3.14.

Affine and Projective Planes

4

4.1 Introduction

The origins of all modern mathematics are in ancient Greek mathematics. For this culture, mathematics was geometry and the most important textbook was Euclid's Elements [91]. This textbook would have been a necessary prerequisite for any study of mathematics. It was the model for not only the content of mathematics but also how mathematics should be done. The importance of this textbook to the development of mathematics and logic cannot be overstated. It served as the model of mathematics and of reasoning in general for a variety of cultures over the past 2300 years. When modern mathematics was being developed this text not only provided the model for the techniques of mathematics but also furnished some extremely interesting open questions which fueled mathematical inquiry for millennia. Until very recently this text was an integral part of most people's education. In fact, in Plato's academy, over the door was written that no one destitute of mathematics should enter through the portal. This means that education in any subject required knowledge of geometry and the reasoning skills developed studying it.

The text began by making some basic definitions and then stating the basic axioms of Euclidean geometry. An axiom is simply a mathematical statement that we accept without proof. Euclid was willing to accept these statements as obvious and necessary for the development of geometry. These axioms are the foundation of the entire discipline. From these axioms, theorems are developed which are then used to construct further theorems.

These axioms, in modern terminology, were as follows:

1. Any two points can be connected by a line.
2. Any line segment can be produced continuously in a straight line.
3. Given a center and a radius, a circle can be produced with that center and radius.

© The Editor(s) (if applicable) and The Author(s), under exclusive license to Springer Nature Switzerland AG 2020
S. T. Dougherty, *Combinatorics and Finite Geometry*, Springer Undergraduate Mathematics Series,

Fig. 4.1 Axiom 5

4. All right angles are equal to one another. (Note that no mention is made of degrees
   or radians. A right angle is simply defined as one half of the angle of a straight
   line.)
5. If a straight line intersecting two lines made two interior angles on the same side
   less than two right angles, then the two lines will intersect on that side.

From the perspective of finite geometry, we are not really concerned with Axioms
2, 3, and 4 since they largely pertain to infinite planes.

The fifth axiom was one of the most important and controversial questions in all
of intellectual history. The axiom can be restated into the following form: Given a
line ℓ and a point p not on that line there exists a unique line through p that is parallel
to ℓ. This is shown in Fig. 4.1.

In an infinite plane, it can be shown that there is at least one line parallel to ℓ. It was
widely believed that this axiom could be proven from the first four. It turns out, thanks
to some fantastic nineteenth-century mathematics done by Lobachevsky and Bolyai
that the fifth axiom is independent of the first four. This means that there are planes
where the first four axioms are true but the fifth axiom is false. In particular, there are
planes where there are infinitely many lines through p that are parallel to ℓ. These
geometries are called hyperbolic planes. Their discovery sent shock waves through
the mathematical and philosophical worlds and sparked a tremendous amount of
further study and discussion. In the final analysis, Euclid is vindicated since it was
absolutely necessary to state the fifth axiom.

If Axiom 2 is eliminated, a geometry can be produced that has no parallel lines.
These planes are called projective planes. A projective or hyperbolic plane is called
a non-Euclidean plane, since in these planes Euclid's fifth axiom is false. A plane
where the fifth axiom holds is called an affine or Euclidean plane.

Many standard theorems in geometry require the fifth axiom. For example, one
of the first theorems of Euclidean geometry that students are exposed to is that the
sum of the angles of a triangle is a straight line. This follows from the following two
diagrams:

Namely, in Fig. 4.2, it is shown that alternate interior angles are equal if they are
formed by a transversal to two parallel lines. In Fig. 4.3, a line is drawn through the
top point of the triangle parallel to the base of the triangle. Notice that this requires
Euclid's fifth axiom. Then using Fig. 4.2, we see that the angles marked α are equal
and the angles marked β are equal. Then since at the top of the triangle α, β, γ make
a straight line, then the sum of the angles in the triangle make a straight line.

The theorem that the sum of the angles in a triangle is a straight line is not true
for hyperbolic nor projective planes. In a hyperbolic plane the sum of the angles is
less than a straight line and for a projective plane the sum of the angles is more than
a straight line.

Fig. 4.2 Alternate interior angles

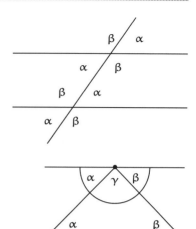

Fig. 4.3 The sum of the angles in a triangle

It can be shown that a hyperbolic plane, that is, a plane that has more than one line parallel to a given line through a point off that line, must be an infinite plane so we restrict ourselves to affine and projective planes.

4.2   Definitions

We shall now apply these geometric ideas in a combinatorial setting. Namely, we shall see what combinatorial objects we can construct that satisfy the geometric axioms. We leave the terms points and lines undefined but recognize that a point can be understood as the lines that are incident with it and a line can be understood as the points that are incident with it. Euclid gave definitions for points and lines that are not completely rigorous. He defined points as that which had no part and lines as breadthless length. Of course, our intuitive idea of a point and line is very much fixed. These intuitive ideas must be modified to study non-Euclidean geometry and finite geometry. As an example, consider the surface of the earth. The lines on the surface of the earth are the great circle routes, meaning those circles that lie on the surface of the earth whose center coincides with the center of the earth. These are the canonical lines on the earth as any airline traveler can tell you, but they may not match our intuitive idea of a line. We shall show later that, in fact, these are lines in a projective geometry.

We shall consider incidence structures which are a set of points $\mathcal{P}$, a set of lines $\mathcal{L}$, and an incidence relation $I \subseteq \mathcal{P} \times \mathcal{L}$, where $(p, \ell) \in I$ means that p is incident with ℓ. We say two lines are parallel if they have no point that is incident with both of them or if they are identical.

We make the following definitions.

Definition 4.1 (*Affine Plane*) An affine plane is a set of points $\mathcal{A}$, a set of lines $\mathcal{M}$, and an incidence relation $\mathcal{J} \subseteq \mathcal{A} \times \mathcal{M}$ such that

Fig. 4.4 Trivial geometry

1. Through any two points there is a unique line incident with them. More precisely, given any two points p, q ∈ $\mathcal{A}$, there exists a unique line ℓ ∈ $\mathcal{M}$ with (p, ℓ) ∈ $\mathcal{J}$ and (q, ℓ) ∈ $\mathcal{J}$.
2. If p is a point not incident with ℓ then there exists a unique line through p parallel to ℓ. More precisely, if p ∈ $\mathcal{A}$, ℓ ∈ $\mathcal{M}$ with (p, ℓ) ∉ $\mathcal{J}$, then there exists a unique line m ∈ $\mathcal{M}$ with (p, m) ∈ $\mathcal{J}$ and ℓ parallel to m.
3. There exists at least three non-collinear points.

Definition 4.2 (*Projective Plane*) A projective plane is a set of points $\mathcal{P}$ and a set of lines $\mathcal{L}$ and an incidence relation $\mathcal{I} \subseteq \mathcal{P} \times \mathcal{L}$ such that

1. Through any two points there is a unique line incident with them. More precisely, given any two points p, q ∈ $\mathcal{P}$, there exists a unique line ℓ ∈ $\mathcal{L}$ with (p, ℓ) ∈ $\mathcal{I}$ and (q, ℓ) ∈ $\mathcal{I}$.
2. Any two lines meet in a unique point. More precisely, given any two lines ℓ, m ∈ $\mathcal{L}$, there exists a unique point p ∈ $\mathcal{P}$ with (p, ℓ) ∈ $\mathcal{I}$ and (p, m) ∈ $\mathcal{I}$.
3. There exists at least 4 points, no three of which are collinear.

Notice that in a projective plane there are no parallel lines. In particular, Axiom 2 of affine planes is false for these planes.

The third axiom of the affine plane requires the existence of a triangle and the third axiom of the projective plane requires the existence of a quadrangle. These axioms are there simply to eliminate trivial cases which could satisfy the axioms. For example, the diagram in Fig. 4.4 represents a system that would satisfy the axioms trivially but is not an object we wish to study:

If $\Pi = (\mathcal{P}, \mathcal{L}, \mathcal{I})$ is a projective plane, then the plane $(\mathcal{L}, \mathcal{P}, \mathcal{I})$ is called the dual plane. It follows from this that there is a duality in the axioms of the planes, see Exercise 1. This means that for any theorem we prove about lines, there is a corresponding theorem about points; and any theorem we prove about points, there is a corresponding theorem about lines.

Let $\Pi = (\mathcal{P}, \mathcal{L}, \mathcal{I})$ be a projective plane. A projective plane is said to have order n when there are n + 1 points on a line. It will become apparent after we build a projective plane why the order is n while there are n + 1 points on a line.

Lemma 4.1 *On a projective plane of order* n *there are* n + 1 *points on each line and* n + 1 *lines through each point.*

Proof Let L be a line and p a point off L. Any line through p must meet L as in the diagram below since any two lines must meet. Moreover, through any point on L there exists a unique line through that point and p. Then since any line must meet L we have the result.

We are now in a position to count the total number of points and lines (Fig. 4.5).

Fig. 4.5 Number of points
on a line and lines through a
point

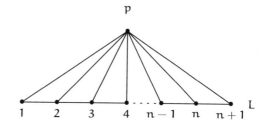

Theorem 4.1 *Let* $\Pi = (\mathcal{P}, \mathcal{L}, \mathcal{I})$ *be a projective plane of order* n *then*

$$|\mathcal{P}| = |\mathcal{L}| = n^2 + n + 1. \tag{4.1}$$

Proof Consider a line L and the points $p_1, p_2, \ldots, p_{n+1}$ on L. Through each p_i there are n lines distinct from L. If M is such a line through p_i and M' is such a line through p_j then $M \neq M'$ since the only line through p_i and p_j is L. Now since any line in the plane intersects L there are $n(n + 1)$ lines distinct from L and so $n(n + 1) + 1 = n^2 + n + 1$ lines in the plane.

Of course, by the duality of the definitions, we have that the number of points and lines are the same. However, we ask for a direct proof of the fact that $|\mathcal{P}| = n^2 + n + 1$ in Exercise 3. $\qquad\square$

We shall consider the best-known example, namely, the projective plane of order 2. It has 7 points each with 3 lines through them and 7 lines each containing 3 points. What follows is simply a representation of the plane, and it is important to remember that each line consists only of three points. The novice may err and think that there are points everywhere on the line as if it were a line in the Cartesian plane.

If the lines are listed $L_1, \ldots, L_7$ then we have the following correspondence between points and lines:

$$L_1 \leftrightarrow \{A, B, C\},$$
$$L_2 \leftrightarrow \{C, D, G\},$$
$$L_3 \leftrightarrow \{A, F, G\},$$
$$L_4 \leftrightarrow \{C, E, F\},$$
$$L_5 \leftrightarrow \{A, D, E\},$$
$$L_6 \leftrightarrow \{B, E, G\},$$
$$L_7 \leftrightarrow \{B, D, F\}.$$

Another way of looking at a representation of the plane is to look at its incidence matrix. The incidence matrix is simply a matrix indexed by lines and points where there is a 1 at location (L, p) if p is incident with L and a 0 otherwise. For this plane the incidence matrix in Table 4.1.

Table 4.1 Incidence matrix for the projective plane of order 2

	A	B	C	D	E	F	G
L_1	1	1	1	0	0	0	0
L_2	0	0	1	1	0	0	1
L_3	1	0	0	0	0	1	1
L_4	0	0	1	0	1	1	0
L_5	1	0	0	1	1	0	0
L_6	0	1	0	0	1	0	1
L_7	0	1	0	1	0	1	0

Fig. 4.6 The projective
plane of order 2

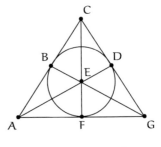

Fig. 4.7 The projective
plane of order 1

Notice that each row and column of the matrix has exactly 3 entries with a 1 in them. Later in this text, we shall show how this representation of the plane can actually be used to communicate with deep space probes or along telephone lines (Fig. 4.6).

The third axiom gives that this is actually the smallest case of a projective plane, since there must be at least 4 points and the smallest number of the form $n^2 + n + 1$ greater than 4 is 7, making $n = 2$. However, some people are willing to allow the case for $n = 1$, that is a projective plane of order 1. It would consist of 3 points and 3 lines and would be represented as follows (Fig. 4.7).

If we had begun with the assumption that the number of lines and points were equal to v and that there are $n + 1$ points on a line, then we could have counted the number of points and lines as follows. We know that through any two points there exists a unique line and there are $C(v, 2)$ ways of picking two points. Also we know that there are $C(n + 1, 2)$ ways of picking two points on a line, meaning that each line is counted $C(n + 1, 2)$ times in the $C(v, 2)$ ways of counting lines. This gives

$$\frac{C(v, 2)}{C(n + 1, 2)} = v$$

$$\frac{v(v - 1)}{2} \Big/ \frac{(n + 1)(n)}{2} = v$$

$$v(v - 1) = v(n + 1)(n)$$

$$v - 1 = n^2 + n$$

$$v = n^2 + n + 1.$$

We shall now give an example of a model of an infinite projective plane. Consider a sphere S with center at the origin in three-dimensional Euclidean space. We will only consider the top half of that sphere and only the points on the equator that fall in the western hemisphere. The lines on this plane are the great circle routes. Now it is easy to see that through any two points there is a line (simply slice the sphere with the plane formed by the two points and the center of the sphere). Any two great circle routes will in fact meet. It is easy to see now why we only considered the top hemisphere since two great circle routes will meet twice on a sphere, once in each hemisphere. Two lines that meet on the equator would meet on the opposite side on the equator as well, which is why we eliminated half of the equator. The second axiom of Euclid is, of course, not true since a line does not extend infinitely far in both directions. These lines end at the equator.

We now turn our attention to affine planes. The first thing to consider in an affine plane is the notion of parallelism. As a matter of definition we say that a line is parallel to itself and two distinct lines are parallel if they have no points in common.

Theorem 4.2 *Parallelism is an equivalence relation on the lines of an affine plane.*

Proof Every line is parallel to itself by definition and hence the relation is reflexive.

If ℓ is parallel to m then either they are equal or disjoint. In either case m is parallel to ℓ and the relation is symmetric.

Assume ℓ is parallel to m and m is parallel to k. If k and ℓ had a point of intersection p then there would be two lines through p parallel to m contradicting Axiom 2. Hence the relation is transitive and therefore an equivalence relation. $\square$

We shall do similar counts for the affine plane that we did earlier for the projective plane. Let us consider an affine plane. Let $\pi = (\mathcal{A}, \mathcal{M}, \mathcal{J})$ be an affine plane. Assume that each line has n points incident with it. We say that such a plane has order n. Let ℓ be a line and let m be a line that intersects ℓ at the point p. There are $n - 1$ points on m that are not on ℓ. Through each of these there must be a unique line parallel to ℓ. Of course none of these could be the same since then through the point of intersection there would be two lines parallel to ℓ. Since ℓ is parallel to itself, we see that there are at least n lines in a parallel class. If there was another line parallel to ℓ it would have to intersect m, but we have considered all such lines and hence we have the following.

Lemma 4.2 *In an affine plane of order* n *each parallel class has exactly* n *lines.*

This lemma allows us to determine the cardinality of the point set.

Lemma 4.3 *Let* $\pi = (\mathcal{A}, \mathcal{M}, \mathcal{J})$, *then* $|A| = n^2$.

Proof Let ℓ be a line. There are n lines in the parallel class of ℓ. Each of these has n points incident with it and no two lines have any points in common. Hence, there are n^2 points on these lines. If there were another point in the plane not yet counted there would be a line through it parallel to ℓ which is a contradiction. Therefore, there are n^2 points on the plane.   □

We can now count the number of parallel classes.

Lemma 4.4 *There are* $n + 1$ *parallel classes in an affine plane of order* n.

Proof Let p be a point in the plane. There are $n + 1$ lines through p. For any parallel class, there must be a line from that class through p since the lines in any parallel class cover all points in the plane. Hence each line comes from a different parallel class since if two were in the same class then they would intersect and not be parallel.   □

We can use these lemmas to determine the number of lines in an affine plane.

Theorem 4.3 *In an affine plane of order* n *there are* $n^2 + n$ *lines.*

Proof By Lemma 4.4, there are $n + 1$ parallel classes and by Lemma 4.2, each class has n lines in it. Therefore, there are $(n + 1)n = n^2 + n$ lines in the plane.   □

We summarize all the counting results in the next theorem.

Theorem 4.4 *Let* $\pi = (\mathcal{A}, \mathcal{M}, \mathcal{J})$ *be an affine plane of order* n. *There are* n *points on a line,* $n + 1$ *lines through a point,* $|\mathcal{A}| = n^2$, $|\mathcal{M}| = n^2 + n$, *and parallelism is an equivalence relation on the set of lines with* n *lines in each of* $n + 1$ *parallel classes. Let* $\Pi = (\mathcal{P}, \mathcal{L}, \mathcal{I})$ *be a projective plane of order* n. *There are* $n + 1$ *points on each line,* $n + 1$ *lines through each point, and* $|\mathcal{P}| = |\mathcal{L}| = n^2 + n + 1$.

We give an example with the affine plane of order 2 in Fig. 4.8.

If the lines are listed as $L_1, \ldots, L_6$, then we have the following correspondence between points and lines:

Fig. 4.8 Affine plane of
order 2

$$L_1 \leftrightarrow \{A, B\},$$
$$L_2 \leftrightarrow \{C, D\},$$
$$L_3 \leftrightarrow \{A, C\},$$
$$L_4 \leftrightarrow \{B, D\},$$
$$L_5 \leftrightarrow \{A, D\},$$
$$L_6 \leftrightarrow \{B, C\}.$$

There are three parallel classes here, namely, $\{L_1, L_2\}$, $\{L_3, L_4\}$, and $\{L_5, L_6\}$.
Note that each parallel class partitions the set of points.

Affine and projective planes are canonically connected via the next two theorems.

Theorem 4.5 *Let $\pi = (\mathcal{A}, \mathcal{M}, \mathcal{J})$ be an affine plane of order n. Let q_i be a new point corresponding to the i-th parallel class of π. Consider the incidence structure with $\mathcal{P} = \mathcal{A} \cup \{q_i\}$, $\mathcal{L} = \{m \cup \{q_i\} \mid m \in \mathcal{M}, m$ in the i-th parallel class $\} \cup \{L_\infty\}$ where L_∞ is incident with a point p in $\mathcal{P}$ if and only if $p = q_i$, and incidence is induced by the incidence of π and q_i is incident with each line in the i-th parallel class. Then $\Pi = (\mathcal{P}, \mathcal{L}, \mathcal{I})$ is a projective plane of order n.*

Proof Let p and q be two distinct points. If they were both points in the affine plane incident with the line ℓ in the i-th parallel class then the line $L = \ell \cup \{q_i\}$ is a line incident with p and q. Since the only other possible line is L_∞ which has only points not in π on it we see that the line is unique. If neither p nor q are in π then L_∞ is the unique line through the two points. If p is a point of π and $q = q_i$ not a point of π then the unique line through these two points is the unique line of the i-th parallel class through p.

Consider two lines L and M. If $L = \ell \cup \{q_i\}$ and $M = m \cup \{q_j\}$ then if $i = j$, the lines meet at q_i and if $i \neq j$ then the lines meet at the unique point of intersection of ℓ and m. If $L = \ell \cup \{q_i\}$ and $M = L_\infty$ then the unique point of intersection is q_i.   □

Basically, what we have done in this theorem is to make all lines meet at a point that were parallel in the affine case. The language that is often used in this is that we make parallel lines meet at a point at infinity and then all of these points at infinity make a line at infinity. It is often said that the affine plane completes to a projective plane.

Theorem 4.6 *If* $\Pi = (\mathcal{P}, \mathcal{L}, \mathcal{I})$ *is a projective plane of order* n, *then the incidence structure formed by removing any line* L *and the points incident with* L *is an affine plane of order* n. *Precisely, if* $\mathcal{A} = \mathcal{P} - \{p \mid (p, L) \in \mathcal{I}\}$, $\mathcal{M} = \mathcal{L} - \{L\}$, *and incidence* $\mathcal{J}$ *is induced by* $\mathcal{I}$ *then* $\pi = (\mathcal{A}, \mathcal{M}, \mathcal{J})$ *is an affine plane.*

Proof See Exercise 9.   □

The line that is removed is usually called the line at infinity. Of course, any line can be removed and the construction works. Intuitively, this means that in the transition from affine to projective and back, any line can be thought of as the line at infinity.

Consider the projective plane of order 2 given in Fig. 4.6. We shall take L_7 to be the line at infinity. This is the standard choice for this plane because somehow it *looks* like a line at infinity. Remove L_7 and the points B, D, F and the plane that remains is given in the Fig. 4.9.

Notice that this plane is essentially the same as the previous description of the affine plane of order 2. In fact, later we shall see that they are exactly the same.

This plane has the incidence matrix given in Table 4.2.

While the projective plane of order 1 seems perfectly natural, the same cannot be said for what would be the affine plane of order 1. By removing a line and the points on it from the projective plane of order 1 we would have 1 point with 2 lines through it, which seems to be not at all what we wanted. From this point on when discussing planes we always assume we have the third axiom, that is, the order of the plane is always at least 2.

Let us consider our infinite example. Consider the usual Euclidean plane given by $z = 0$ in standard three-dimensional Euclidean space $\mathbb{R}^3$. Place the top half of the sphere with half of the equator as described before so that the center of the sphere is at $(0, 0, 1)$. For any point (x, y) on the plane draw the unique line from that

Fig. 4.9 Affine plane formed from the projective plane of order 2

Table 4.2 Incidence matrix for the affine plane of order 2

	A	C	E	G
L_1	1	1	0	0
L_2	0	1	0	1
L_3	1	0	0	1
L_4	0	1	1	0
L_5	1	0	1	0
L_6	0	0	1	1

point to the point $(0,0,1)$. This line intersects the sphere at a unique point. Hence, there is a unique point on the half-sphere associated with each point on the plane, except for those points on the equator. For each parallel class with slope α where $\alpha \in \mathbb{R}$ or $\alpha = \infty$ for lines with no slope, associate the point on the equator that is the intersection of the plane that is perpendicular to the plane $z = 0$ and intersects it at the line $y = \alpha x$ or $x = 0$ if $\alpha = \infty$. Then notice that each line on the plane traces a great circle route on the half-sphere. The points on the plane correspond to the non-equatorial points on the half-sphere, the equatorial points are the points at infinity, the equator is the line at infinity, and the lines of the plane have a corresponding line in the half-sphere, with the equator serving as L_∞. We see that this is precisely the projective completion of the affine plane. Notice also that this finite half-sphere contains all of the geometric information of the infinite plane, plus a bit more!

Exercises

1. Let $\Pi = (\mathcal{P}, \mathcal{L}, \mathcal{I})$ be a projective plane. Show that $(\mathcal{L}, \mathcal{P}, \mathcal{I})$ is also a projective plane. That is, each line is now called a point and each point is now called a line. Show also that reversing the roles of lines and points in an affine plane will not produce another affine plane.
2. Assume there are $n + 1$ lines through each point in a projective plane and prove that there must be exactly $n + 1$ points on each line.
3. Prove that in a projective plane of order n, we have that $|\mathcal{P}| = n^2 + n + 1$.
4. Determine the lines that are incident with each point in the projective plane of order 2 given in Fig. 4.6.
5. Prove that in an affine plane there are $n + 1$ lines through a point.
6. Prove Theorem 4.3.
7. List the lines in the affine plane of order 2 through each point given in Fig. 4.8.
8. Show that the construction in Theorem 4.5 corresponds to the counting information in Theorem 4.4.
9. Prove Theorem 4.6.
10. Show that the construction in Theorem 4.6 corresponds to the counting information in Theorem 4.4.
11. Construct an affine and projective plane of order 3.

4.3 Planes from Fields

The first affine plane that students encounter is the real affine plane. It is usually described in terms of the Cartesian coordinate system with an x- and y-axis. The points are given by (x, y) where x and y are real numbers and the lines are all of the form $y = mx + b$ or $x = c$. We can replace the real numbers with any field, including finite fields, and we can still construct an affine plane. Throughout the remainder of the section, we shall let F denote a field.

Let $\mathcal{A} = \{(x, y) \mid x, y \in F\}$, $\mathcal{M} = \{y = mx + b \mid m, b \in F\} \cup \{x = c \mid c \in F\}$. We say that the line $y = mx + b$ contains all points (x, y) that satisfy the equation and $x = c$ contains all points (c, y) for any y. This is, of course, the natural way to regard these structures.

First we shall show that it does, in fact, form a plane. Let $(a, d), (a', d')$ be two distinct points. If $a = a'$ then the two points are on the line $x = a$. If they were both on the same line of the form $y = mx + b$ then we would have $d = ma + b = d'$ and the two points would not be distinct. If $a \neq a'$ then there is no line of the form $x = c$ that both points are on. If both points are on $y = mx + b$ then we have

$$d = ma + b \ \text{ and } \ d' = ma' + b$$
$$\Rightarrow d - ma = d' - ma'$$
$$\Rightarrow d - d' = m(a - a')$$
$$\Rightarrow m = \frac{d - d'}{a - a'}.$$

Then m is uniquely determined and $b = d - ma$ is uniquely determined as well. Therefore, through any two points there exists a unique line. Notice how necessary it was that we had a field. Namely, we needed to be able to divide by $a - a'$ when it is non-zero, i.e., when $a \neq a'$. If the algebraic structure were not a field, for example, $\mathbb{Z}_4$, we would not be able to guarantee that there is a line through any two points. For example, if we were looking over $\mathbb{Z}_4$ there is not a line incident with the points $(3, 1)$ and $(1, 2)$.

Any two lines of the form $x = c$ are parallel. Given two lines $y = mx + b$ and $x = c$ then their unique point of intersection is $(c, cm + b)$. Given two lines $y = mx + b$ and $y = m'x + b'$. If $m \neq m'$ and (x, y) satisfies both equations then

$$y = mx + b \ \text{ and } \ y = m'x + b'$$
$$\Rightarrow mx + b = m'x + b'$$
$$\Rightarrow b - b' = (m' - m)x$$
$$\Rightarrow x = \frac{b - b'}{m' - m}$$

and x is uniquely determined. Then $y = m\frac{b-b'}{m'-m} + b$ and there is a unique point of intersection. Notice that again it was necessary to be able to divide by a non-zero element.

Given two lines $y = mx + b$ and $y = mx + b'$, if there were a point of intersection, we would have $b = y - mx = b'$ and the lines would have to be identical. Therefore, any two lines with identical m (slope) are parallel.

Let $|F| = n$. Since there are n choice for m there are n parallel lines in each class. Also we note that there are n^2 points in the plane. It should be evident now why the order of the plane is given by n, namely, that the plane that comes from a field of size n has order n. We have proven the following.

Theorem 4.7 *Let F be a field of order n. Let $\mathcal{A} = \{(x, y) \mid x, y \in F\}$ and $\mathcal{M} = \{y = mx + b \mid m, b \in F\} \cup \{x = c \mid c \in F\}$ with the natural incidence relation $\mathcal{I}$ then $\pi = (\mathcal{A}, \mathcal{M}, \mathcal{I})$ is an affine plane of order n.*

Fig. 4.10 Affine plane of order 2

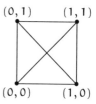

We shall do a simple example.

Let $F = \mathbb{F}_2 = \{0, 1\}$ the field of order 2. The points are

$$(0, 0), (0, 1), (1, 0), (1, 1).$$

The lines are

$$x = 0, x = 1, y = 0, y = 1, y = x, y = x + 1.$$

We note that there are $4 = 2^2$ points and $6 = 2^2 + 2$ lines.

The following diagram describes this plane (Fig. 4.10).

The bottom horizontal line is $y = 0$ and consists of the points $(0, 0)$ and $(1, 0)$. The top horizontal line is $y = 1$ and consists of the points $(0, 1)$ and $(1, 1)$. The left vertical line is $x = 0$ and consists of the points $(0, 0)$ and $(0, 1)$. The right vertical line is $x = 1$ and consists of the points $(1, 0)$ and $(1, 1)$. The ascending vertical line is $y = x$ and consists of the points $(0, 0)$ and $(1, 1)$. The descending vertical line is $y = x + 1$ and consists of the points $(0, 1)$ and $(1, 0)$.

This plane may look different from the plane given in Fig. 4.9 but they are exactly the same plane with a slightly different representation.

We shall now show how a projective plane can be constructed from a finite field.

Instead of looking at points as elements in F^2, we shall look at points as elements in F^3. However, we only want a specified subset of these points.

We shall say that two elements of F^3, (a, b, c) and (a', b', c') are equivalent, written as

$$(a, b, c) \equiv (a', b', c') \tag{4.2}$$

if there exists a non-zero $\lambda \in F$ with $a = \lambda a', b = \lambda b', c = \lambda c'$. For example, in $\mathbb{F}_5^3$ the elements that are equivalent to $(1, 2, 3)$ are $\{(1, 2, 3), (2, 4, 1), (3, 1, 4), (4, 3, 2)\}$.

The points set $\mathcal{P} = (F^3 - \{(0, 0, 0)\})/ \equiv$. This simply means that the points are the equivalence classes of $F^3 - \{(0, 0, 0)\}$ formed from the equivalence relation $\equiv$. We usually just take a specific element of each class to represent the point.

As an example we shall show the points in $(\mathbb{F}_3^3 - \{(0, 0, 0)\})/ \equiv$. The points are

$$\{(0, 0, 1), (0, 1, 0), (0, 1, 1), (0, 1, 2), (1, 0, 0), (1, 0, 1), (1, 0, 2),$$
$$(1, 1, 0), (1, 1, 1), (1, 1, 2), (1, 2, 0), (1, 2, 1), (1, 2, 2)\}.$$

The lines are defined in the same way, that is, $\mathcal{L} = (F^3 - \{(0, 0, 0)\})/ \equiv$. In some texts the lines are written as column vectors so that there is something to distinguish lines and points. We shall distinguish them by writing points as the vector (a, b, c) and lines as the vector $[a, b, c]$.

Incidence for the plane is given by the following. The point (a, b, c) is on the line $[d, e, f]$ if and only if $ad + be + cf = 0$. It is clear why we had to eliminate $(0, 0, 0)$ since it would be incident with every line.

For example, for the plane of order 3, the point $(1, 2, 2)$ is on the line $[1, 1, 0]$ since $1(1) + 2(1) + 2(0) = 1 + 2 = 0$. In fact the point $(1, 2, 2)$ is incident with the four lines $[0, 1, 2]$, $[1, 0, 1]$, $[1, 1, 0]$, and $[1, 2, 2]$.

Theorem 4.8 *If* $\Pi = (\mathcal{P}, \mathcal{L}, \mathcal{I})$ *with* $\mathcal{P} = \mathcal{L} = (F^3 - \{(0, 0, 0)\})/ \equiv$ *and* (a, b, c) *is incident with* $[d, e, f]$ *if and only if* $ad + be + cf = 0$, *then* Π *is a projective plane of order* $|F|$.

Proof Take two distinct points (a, b, c) and (a', b', c'). We know from elementary linear algebra that over any field the system of equations

$$ax + by + cz = 0$$
$$a'x + b'y + c'z = 0$$

has a family of solutions with one degree of freedom provided that (a', b', c') is not a multiple of (a, b, c), which by the way the points are defined, we know this one is not the multiple of another, since, in that case they would be the same point. This means that we have at least one line that is incident with both points, all that remains is to show that any two vectors $[x, y, z]$ satisfying these two equations must be multiples of each other. We know that $[0, 0, 0]$ is a particular solution to this system of equations so we know that the family of solutions is a vector space of dimension 1 and hence every vector in that space is of the form $\lambda[d, e, f]$ for some vector $[d, e, f]$. This means that $[d, e, f]$ is the unique line through these two points.

To show that any two lines meet in a unique point, simply reverse the words line and point and the round parentheses with the square parentheses. $\qquad\square$

As an example, we shall construct the projective plane of order 2. We have

$$\mathcal{P} = \{(0, 0, 1), (0, 1, 0), (0, 1, 1), (1, 0, 0), (1, 0, 1), (1, 1, 0), (1, 1, 1)\}, \qquad (4.3)$$

$$\mathcal{L} = \{[0, 0, 1], [0, 1, 0], [0, 1, 1], [1, 0, 0], [1, 0, 1], [1, 1, 0], [1, 1, 1]\}. \qquad (4.4)$$

The incidence is as follows:

$$[0,0,1] \leftrightarrow \{(1,0,0),(0,1,0),(1,1,0)\};$$
$$[0,1,0] \leftrightarrow \{(1,0,0),(0,0,1),(1,0,1)\};$$
$$[0,1,1] \leftrightarrow \{(1,0,0),(0,1,1),(1,1,1)\};$$
$$[1,0,0] \leftrightarrow \{(0,1,0),(0,0,1),(0,1,1)\};$$
$$[1,0,1] \leftrightarrow \{(0,1,0),(1,0,1),(1,1,1)\};$$
$$[1,1,0] \leftrightarrow \{(0,0,1),(1,1,0),(1,1,1)\};$$
$$[1,1,1] \leftrightarrow \{(1,1,0),(1,0,1),(0,1,1)\}.$$

Recall that Lemma 3.1 states that finite fields exist for all orders p^e where p is a prime and $e > 1$. The technique we have just developed gives that if we have a field, we can construct an affine and projective plane of the same order as the field. This gives us the following important theorem.

Theorem 4.9 *Affine and projective planes exist for all orders of the form p^e where p is a prime and $e > 1$.*

Exercises

1. Verify all of the remaining counting information in Theorem 4.4 for the construction given in Theorem 4.7. That is, count the number of lines, points on a line, lines through a point, and parallel classes, and make sure they match what is given in the theorem.
2. Find the points and lines of the affine plane of order 3 formed from the field of order 3, $\mathbb{F}_3 = \{0,1,2\}$. Then make a graphic representation of the affine plane of order 3.
3. Prove that $\equiv$ defined in Eq. 4.2 is an equivalence relation on $\mathbb{F}^3$.
4. Find all elements equivalent to $(3,2,6)$ in $\mathbb{F}_7^3$ using the relation given in Eq. 4.2.
5. Prove that $|\mathcal{P}| = |(\mathbb{F}_n^3 - \{(0,0,0)\})/\equiv)| = n^2 + n + 1$.
6. Label the points and lines on Fig. 4.6 so that the incidence is correct.
7. Construct the projective plane of order 3 from the field of order 3. Namely, list the 13 lines and the 4 points on each.
8. Construct an affine and projective plane of order 4 from the field of order 4.

4.4   Connection Between Affine Planes and MOLS

The two main problems we have studied up to this point are the existence of a complete set of MOLS and the existence of finite affine and projective planes. They may have seemed to be quite different and, in fact, their origins are quite different. However, it turns out that they are really the same problem. We shall now show this connection between a complete set of Mutually Orthogonal Latin Squares of order n and an affine plane of order n.

Theorem 4.10 *A complete set of MOLS of order n exist if and only if an affine plane of order n exists.*

Proof For both directions of the proof associate the n^2 points of the plane with the n^2 coordinates of the Latin squares. We can describe the points as (a, b) where $a, b \in \{0, 1, \ldots, n-1\} = \mathbb{Z}_n$.

Assume we have a complete set of MOLS of order n, that is, $n-1$ MOLS of order n on the alphabet $\{0, 1, \ldots, n-1\}$. The first two parallel classes have the lines that are the horizontal and vertical lines of the grid. That is, the c-th line in the first parallel class contains the n points $(c, 0), (c, 1), \ldots, (c, n-1)$ and the c-th line in the second parallel class contains the n points $(0, c), (1, c), \ldots, (n-1, c)$. Let $L^k = (L_{ij})$ be the k-th Latin square, then the $(k+2)$-nd parallel class has lines corresponding to the symbols of the square. That is, the c-th line is incident with the point (i, j) if and only if $L_{ij} = c$. By construction, these n lines are parallel. Moreover, since any two lines corresponding to the symbols c and c' from different parallel classes are from orthogonal Latin squares, we know that the formed Graeco-Latin square has the pair (c, c') exactly once, so the lines meet exactly once. Through any point on the grid there are $n+1$ lines corresponding to the first two parallel classes and the $n-1$ Latin squares. This means that taking any point p there are $n+1$ lines through p each with $n-1$ points other than p and so there are $(n+1)(n-1) = n^2 - 1$ points on the plane that are connected to p with a line. Hence, there is a line through any two points. Therefore, we have shown that we have constructed an affine plane of order n.

If we assume we have an affine plane of order n then we use the first two parallel classes to describe the grid. That is, the point (i, j) is the point of intersection of the i-th line of the first parallel class and the j-th line of the second parallel class. Then we reverse the construction, $L^k_{i,j} = c$ if and only if the c-th line of the $(k+2)$-nd parallel class is incident with the line (i, j). It is evident that these are Latin squares since each line intersects each line in the first two parallel classes exactly once so each symbol occurs in each row and column exactly once. Since any two lines from different parallel classes intersect exactly once it shows that any two squares are orthogonal since the pair (c, c') appears exactly once.   □

As an example we show two MOLS of order 3 and the affine plane of order 3 (Fig. 4.11).

$$\begin{pmatrix} 1 & 2 & 3 \\ 3 & 1 & 2 \\ 2 & 3 & 1 \end{pmatrix} \begin{pmatrix} 1 & 2 & 3 \\ 2 & 3 & 1 \\ 3 & 1 & 2 \end{pmatrix}$$

Fig. 4.11 Affine plane of
order 3

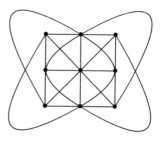

Notice that this theorem, and the knowledge that there is no solution to the 36 officer problem, gives that there is no affine plane of order 6. However, if one knew that there were no plane of order 6, then this does not solve the 36 officer problem.

Exercises

1. Use the Latin square of order 2 to draw the affine plane of order 2.
2. Draw the affine plane of order 4 using the three MOLS of order 4. It may be helpful to draw the third, fourth, and fifth parallel classes in different colors.

4.5 Fundamental Question

We make the following definition general to include both affine and projective planes. The definition applies to any incidence structure with points, lines, and an incidence relation between them.

Definition 4.3 Two incidence structures $D = (\mathcal{P}, \mathcal{L}, \mathcal{I})$ and $D' = (\mathcal{P}', \mathcal{L}', \mathcal{I}')$ are isomorphic if and only if there exists a bijection $\Phi : D \to D'$, where $\Phi : \mathcal{P} \to \mathcal{P}'$ and $\Phi : \mathcal{L} \to \mathcal{L}'$, with $(p, L) \in \mathcal{I}$ if and only if $(\Phi(p), \Phi(L)) \in \mathcal{I}'$. In other words, the isomorphism sends points to points and lines to lines bijectively with the property that p is incident with L if and only if $\Phi(p)$ is incident with $\Phi(L)$.

We say that two planes are isomorphic if and only if there exists an isomorphism between them. It is natural to think that two isomorphic planes are really the same planes with different representations.

Theorem 4.11 *Let Π and Σ be isomorphic planes (affine or projective), and let I_Π and I_Σ be their respective incidence matrices. Then I_Π can be transformed into I_Σ by permuting the rows and columns.*

Proof Let Φ be the isomorphism between Π and Σ. Then if the rows correspond to the points p_i and the columns to L_i then permute the rows and columns by Φ, that is, the rows and columns are now

$$\Phi(p_1), \Phi(p_2), \ldots, \Phi(p_{n^2+n+1})$$

and

$$\Phi(L_1), \Phi(L_2), \ldots, \Phi(L_{n^2+n+1})$$

which gives the incidence matrix of Σ.   □

We have seen that if $\Pi = (\mathcal{P}, \mathcal{L}, \mathcal{I})$ is a projective plane then its dual $\Pi^D = (\mathcal{L}, \mathcal{P}, \mathcal{I})$ is a projective plane. There are examples of projective planes that are not isomorphic to their dual, but if the plane is isomorphic to its dual then the incidence matrix can have its rows and columns permuted to be the transpose of the incidence matrix.

The fundamental question of finite planes is the following.

Fundamental Question: For which orders n do there exist finite affine and projective planes. For orders where planes exist, how many non-isomorphic planes are there?

At present, no one has found a plane of non-prime power order. We do know that there exist planes for all prime power orders as we can construct them from finite fields. The first orders where there are planes which are not isomorphic to the plane constructed from the finite field is at order 9. For order 9 there are four projective planes. Since we know there are not two MOLS of order 6 we know that there is no plane of order 6. An infinite number of orders are eliminated in the Bruck-Ryser Theorem which we shall see later in Theorem 6.27.

The first order for which its existence could not be determined by the above information was order 10. For many years, the existence of a plane of order 10 was a large open question. The proof that there was no plane of order 10 was given in [58]. The proof relied heavily on a large (year long!) computer computation. Basically, it was shown that if such a plane existed then there would have to exist a doubly even self-dual binary code of length 112 with minimum weight 12 and no vectors of weight 16. The computer computation basically showed that no such code could exist by showing that certain configurations do not exist.

The two major conjectures for finite planes are the following.

Conjecture 4.1 *There exists projective planes of order n if and only if n is a prime power.*

Conjecture 4.2 *All projective planes of prime order are isomorphic.*

For this first conjecture, we know that planes exist for all prime powers. Namely, from any finite field we can construct a plane. Moreover, the non-Desarguesian planes that we know, all have the same orders as Desarguesian planes. Some conjecture that this case is always true, that is if a plane exists then a Desarguesian plane of the same order exists. This conjecture is equivalent to the conjecture, discussed in a previous section, that the only complete sets of MOLS occur for prime power orders.

The second conjecture arises from the fact that no non-Desarguesian plane of prime order has yet to be discovered. It would correspond to the conjecture that the only complete set of MOLS of prime order is isomorphic to the set constructed from $\mathbb{F}_p$.

Exercises

1. Show that any two affine planes of order 2 are isomorphic.
2. Show that any two projective planes of order 2 are isomorphic.
3. Show that any affine plane of order 3 must be Desarguesian by showing that any pair of MOLS of order 3 must be equivalent to the ones arising from the construction based on the field of order 3, $\mathbb{F}_3$.

Graphs

5

5.1   Königsberg Bridge Problem

In an earlier chapter, we described how Euler began the study of Latin squares by examining the 36 officer problem. Euler also began the study of graph theory by examining the Königsberg bridge problem. In a short article [33], he started a branch of combinatorics that has become one of the largest and most studied branches of discrete mathematics. Moreover, it has found numerous applications in a variety of fields. It has also been suggested that this paper began the study of topology as well. We shall begin with an investigation of this problem and develop some of the more elementary aspects of graph theory.

Unlike the 36 officer problem, which Euler did not solve, he did solve the Königsberg bridge problem quite easily and gave general solutions for all similar problems. The problem was whether one could walk over each of the bridges of Königsberg, Prussia exactly once. The city is now called Kalingrad and is in Russia. The bridges at that time were arrayed as in the following diagram. Since then bridges have been added so it no longer looks like this (Fig. 5.1).

There are seven bridges and four pieces of land (the two banks and the two islands). He described the situation in Fig. 5.2 and asked whether a tour could be found over each bridge exactly once.

The reasoning to solve this problem is quite simple. At each point (piece of land) there are bridges that connect it to other points. Unless the point is the beginning or end of the tour then there must be an even number of bridges at that point. This is quite simple since if you arrive at a point, that is, neither the beginning nor the end, then you must leave on a different bridge than you came. Hence, the number of bridges at each of these middle points must be even. If the beginning is also the end then the number of bridges at that point must be even as well. If the beginning is not

S. T. Dougherty, *Combinatorics and Finite Geometry*,
Springer Undergraduate Mathematics Series,

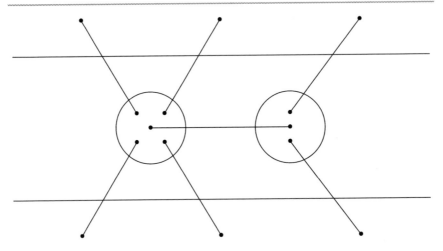

Fig. 5.1 The bridges at Königsberg

Fig. 5.2 Graph of the
bridges at Königsberg

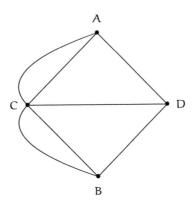

the end then there must be an odd number of bridges at those two points, since there is exactly one unmatched bridge at those points.

Examining this diagram we see that at A there are three bridges, at B there are three bridges, at C there are five bridges, and at D there are three bridges. Clearly there can be no such tour across the bridges of Königsberg where each bridge is crossed exactly once. We shall now generalize these ideas.

The kind of graph denoted here is called a multigraph, while a graph usually refers to a graph with no repeated edges. It is also possible that the term graph can refer to a family to which both of these objects, as well as others, belong. We can now give the definition of a multigraph.

Definition 5.1 A multigraph is a structure $G = (V, E)$ where V is a set of vertices, E is a set of edges, and each edge is represented by $(\{a, b\}, i)$ where a and b are vertices connected by the edge and i is used to distinguish multiple edges. We say that two vertices are adjacent if they are connected by an edge.

The degree of a vertex is the number of edges incident with it. Notice that if a vertex is connected to itself, called a loop, this edge contributes 2 to the degree of the vertex. We denote the degree of a vertex v by $\deg(v)$.

Lemma 5.1 *Let* $G = (V, E)$ *be a multigraph, then* $2|E| = \sum_{v \in V} \deg(v)$.

Proof This follows from that fact that each edge contributes 1 to the degree of two vertices. ☐

The real power of graph theory is not in the rigorous definitions. For example, we could have defined multigraph in a variety of ways. But rather it is that the diagram form of the graph is highly intuitive. Thinking of a graph in terms of its representation is often the best way to come to solutions for a host of combinatorial problems.
We make the following definition.

Definition 5.2 A tour in a graph $G = (V, E)$ is an ordered k-tuple

$$(e_1, e_2, \ldots, e_k),$$

$e_i = \{v_i, w_i\} \in E$, where $w_i = v_{i+1}$ and $e_i \neq e_j$ if $i \neq j$. An Euler tour is a tour such that each edge of the graph appears in the tour exactly once. An Euler cycle is an Euler tour where the starting vertex of the tour is also the ending vertex, i.e., $v_1 = w_k$. In this definition, we say that the tour is from the vertex v_1 to w_k.

The Königsberg bridge problem is then to find an Euler tour on the graph given in Fig. 5.2. As we have seen this is impossible and we shall generalize this result with two theorems.
We are not really interested in the existence of Euler tours if the graph is not connected, that is, if the graph is made up of distinct parts. We make the rigorous definition as follows.

Definition 5.3 A connected graph is a graph in which for any two vertices v and w there is a tour from v to w. Otherwise, we say that the graph is disconnected.

It follows immediately from the definition that a disconnected graph is made up of connected components.

Theorem 5.1 *Let* $G = (V, E)$ *be a connected graph. There exists an Euler cycle if and only if every vertex has even degree.*

Proof The fact that every vertex must have even degree is easy to see. Namely, if a tour goes to a vertex w , i.e., $e_i = \{v, w\}$, then it must leave that vertex, i.e., $e_{i+1} = \{w, x\}$. The starting vertex is also the ending vertex so the same applies there.

We shall show that this is a sufficient condition. Let G be a graph such that the degree of every vertex is even. Pick any vertex v_1 and choose an edge e_1 that is of the form $\{v_1, v_2\}$. Then choose an edge e_2 of the form $\{v_2, v_3\}$. The process terminates because the graph is finite and as long as the last edge is not of the form $\{v_k, v_1\}$ then there will be additional choices since the degree of every vertex is even (intuitively, that is, if the tour arrives at a vertex it can also leave). If this cycle is an Euler cycle then we are done. If not then delete the edges of the tour and what remains is a graph where every vertex has even degree. Choose one of the vertices that we have visited that still has a non-zero degree (we can do this because the graph is connected). Apply this technique again starting at this vertex. Continue this technique until an Euler cycle is constructed. □

What the proof of this theorem shows is that it is quite easy to construct an Euler cycle simply by choosing edges nearly at random.

Theorem 5.2 *Let G $=$ (V, E) be a connected graph. There exists an Euler tour that is not an Euler cycle if and only if there are exactly two vertices with odd degree.*

Proof Add an edge between the two vertices of odd degree and then apply Theorem 5.1. Then remove the added edge from the cycle to produce the Euler tour. □

We note that in applying this theorem any Euler tour must begin at one of the points with odd degree and end at the other with odd degree.

A variation on the idea of Euler tours and cycles is the notion of the Hamiltonian tours and cycles, named after William Rowan Hamilton. See [45,46] for early work on the subject. We now give their definitions.

Definition 5.4 A Hamiltonian tour is a tour that hits every vertex exactly once. A Hamiltonian cycle hits every vertex exactly once except for the final vertex which was also the beginning vertex.

Consider the following graph.

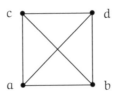

The tour $\{a, b\}, \{b, c\}, \{c, d\}$ is a Hamiltonian tour and $\{a, b\}, \{b, c\}, \{c, d\}$, $\{d, a\}$ is a Hamiltonian cycle. Notice that these do not involve every edge. The tour $\{a, b\}, \{b, d\}, \{d, c\}$ is a also a Hamiltonian tour and $\{a, b\}, \{b, d\}, \{d, c\}, \{c, a\}$ is also a Hamiltonian cycle. Notice that there are numerous Hamiltonian tours in this graph.

It is easy to see that any Hamiltonian cycle can be changed into a Hamiltonian tour by removing one of its edges. However, a Hamiltonian tour can be changed into a Hamiltonian cycle only when its endpoints are adjacent.

A graph is Hamiltonian-connected if for every pair of vertices there is a Hamiltonian tour between the two vertices.

There is not a complete characterization of Hamiltonian tours like there is for Euler tours. However, Dirac [27] proved the following in 1952. We leave the proof as an exercise.

Theorem 5.3 *A graph with no multiple edges on* $n \geq 3$ *vertices has a Hamiltonian tour if the degree of every vertex is at least* $\frac{n}{2}$.

Proof See Exercise 8.   □

The idea of a tour on a graph has also been used in a variety of popular notions. For example, consider the set of all people on the earth. Let these people be the vertices of the graph. Let there be an edge between two people if they know each other. This leads to the popular notion of six degrees of separation. The idea is that between any two people (vertices) there is a tour of distance less than or equal to 6.

For mathematicians, they are concerned with the collaboration graph. The vertices are the people who have published a mathematics article. Two authors are connected if they have ever coauthored a paper together. The most important tours for mathematicians are tours that go from themselves to the great twentieth-century mathematician Paul Erdös. The distance of this tour is known as an Erdös number. The lower the number the closer you are to Erdös. The American Mathematical Society's webpage has a function that will calculate the distance of the tour between any two authors. Of course, it is not true that all authors have such a tour, since there is a significant number of authors who have never collaborated. However, an extremely large portion of the graph is connected and the distance between two authors is often surprisingly low.

Exercises

1. Use a modified algorithm from the proof of Theorem 5.1 to find an Euler tour in the following graph. This exact problem is often a game played by children, namely, to draw this figure without going over any of the lines twice.

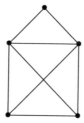

2. Consider the following three graphs:

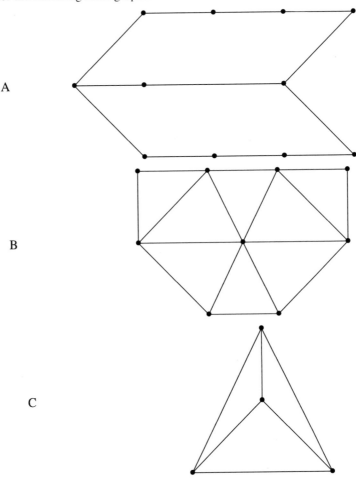

A

B

C

    a. Give the degree of each vertex in each graph.
    b. Determine which (if any) of the three graphs has an Euler cycle.
    c. Determine which (if any) of the three graphs has an Euler tour.
    d. Determine which (if any) of the three graphs has a Hamiltonian cycle.

3. Construct a graph on six vertices that has neither an Euler tour nor an Euler cycle.
4. Let A and B be disjoint sets of vertices in a graph where every vertex of A is connected exactly once to every vertex of B and no vertex of A is connected to another vertex of A and no vertex of B is connected to another vertex of B. Prove conditions on the cardinalities of A and B for Euler cycles and Euler tours to exist.
5. Construct a connected graph on five vertices that has a Hamiltonian tour and a graph on four vertices that does not have a Hamiltonian tour.

6. Prove that the number of Hamiltonian cycles on the complete graph, which is the graph where any two distinct edges are connected, with n vertices is $\frac{(n-1)!}{2}$.
7. Determine for which complete graphs, which is the graph where any two distinct vertices are connected, Euler tours and Euler cycles exist.
8. Prove Theorem 5.3.
9. Prove that for all $n > 1$, there exists a graph on n vertices with no Euler cycle.
10. Find all Hamiltonian tours starting from a given vertex in the following cube graph:

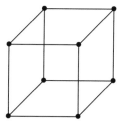

5.2   Simple Graphs

We shall now restrict ourselves to the most commonly studied family of graphs, namely, simple graphs. A simple graph is a graph with no repeated edges and no loops. Throughout the remainder of the text if we say graph it means simple graph. Hence, we can define a graph as follows.

Definition 5.5 A simple graph is $G = (V, E)$, where V is the set of vertices, and E is the set of edges E, where $\{a, b\} \in E$, $a \neq b$, indicates that a and b are connected by an edge.

We begin with a counting theorem.

Theorem 5.4 *The number of graphs on a set of* n *vertices is* $2^{\frac{n(n-1)}{2}}$.

Proof There are $C(n, 2) = \frac{n(n-1)}{2}$ ways of picking two vertices and each pair is either connected or not connected. $\qquad\square$

Example 5.1 If $n = 2$ then $2^{\frac{n(n-1)}{2}} = 1$. Graphs A and B are the two possible graphs on two vertices.

A   B

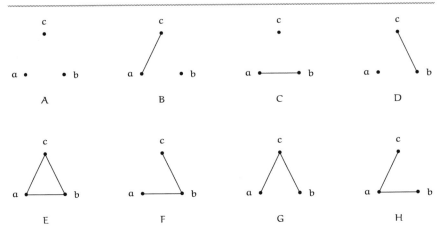

Fig. 5.3 Graphs on three vertices

Fig. 5.4 Discrete graph on
five vertices

Example 5.2 If $n = 3$ then $2^{\frac{n(n-1)}{2}} = 8$. Hence, there are eight possible simple graphs on three vertices and they are given in Fig. 5.3.

There are several important families of simple graphs. We shall describe some of those families.

- The discrete graph on n vertices, D_n, is the graph consisting of n vertices and no edges. The discrete graph on five vertices is given in Fig. 5.4.
- The complete graph on n vertices, K_n, is the graph for which any two distinct vertices are connected by an edge. It is immediate that the degree of any vertex in the complete graph is $n - 1$. The complete graph on five vertices is given in Fig. 5.5.
- The linear graph on n vertices, L_n, is the connected graph on n vertices consisting of a single Euler tour with no repeated vertices. We give an example of L_7 in Fig. 5.6.
- A cycle graph on n vertices, C_n, is the connected graph on n vertices consisting of a single Euler cycle with no repeated vertices. Each vertex in this graph has degree 2. We give the example of C_6 in Fig. 5.7.
- The wheel graph, W_{n+1}, is formed from the cycle graph C_n by adding an additional vertex that is connected to every other vertex. The graph W_7 is given in Fig. 5.8.

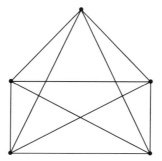

Fig. 5.5 Complete graph on five vertices

Fig. 5.6 The linear graph on seven vertices

Fig. 5.7 The cycle graph on six vertices

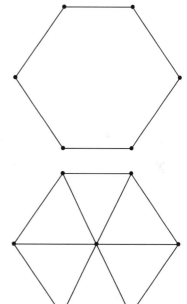

Fig. 5.8 The wheel graph with seven vertices

Let A and B be subsets of the set of vertices V with $A \cup B = V$ and $A \cap B = \emptyset$. The sets A and B are said to be a partition of V. The graph $G = (V, E)$ is said to be a bipartite graph if $v \in A$ implies $w \in B$, whenever $\{v, w\} \in E$. That is, if two vertices are connected then they must be in different sets of the partition.

If each vertex in A is connected to a vertex in B, then it is said to be the complete bipartite graph $K_{m,n}$ with $m = |A|, n = |B|$. We give the complete bipartite graph $K_{4,3}$ in Fig. 5.9. Here $A = \{a, b, c, d\}$ and $B = \{e, f, g\}$.

We can generalize the notion of bipartite as follows. Let $A_1, A_2, \ldots, A_k$ be a partition of the set of vertices V. The graph $G = (V, E)$ is said to be a k-partite

Fig. 5.9 The graph $K_{4,3}$

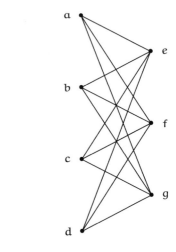

Fig. 5.10 The graph $K_{2,3,2}$

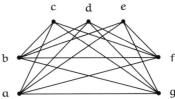

graph if $v \in A_i$ implies $w \notin A_i$, whenever $\{v, w\} \in E$. That is, if two vertices are connected then they must be in different sets of the partition.

If each vertex in A_i is connected to each vertex in A_j when $i \neq j$ then this graph is said to be the complete k-partite graph. We give the complete tripartite graph $K_{2,3,2}$ in Fig. 5.10. Here $A_1 = \{a, b\}$, $A_2 = \{c, d, e\}$, and $A_3 = \{f, g\}$.

A graph is said to be regular if the degree of each vertex is the same. For example, the complete graph, K_n, is a regular graph since each vertex has degree $n - 1$. The cycle graph, C_n, is regular since each vertex has degree 2. The wheel graph, W_n, is not, in general, since one vertex has degree $n - 1$ and the rest have degree 3. However, W_4 is regular since $n - 1 = 3$. The linear graph L_n, $n > 2$, is not regular since each vertex has degree 2 except the first and the last which both have degree 1.

Example 5.3 A Latin square graph is constructed as follows. Given a Latin square L of order n, the vertices are the n^2 coordinates of L. Then two vertices (a, b) and (c, d) are connected with an edge if $a = c$, $b = d$, or $L_{a,b} = L_{c,d}$. That is, they are connected if they are in the same row, column, or share the same symbol. Each vertex is connected to $n - 1$ vertices from its column, $n - 1$ vertices from its row, and $n - 1$ vertices from its symbol. Therefore, a Latin square graph is a regular graph of degree $3(n - 1)$.

Let $G = (V, E)$ be a graph then $G' = (V', E')$ is a subgraph of G if $V' \subseteq V$ and $E' \subseteq E$.

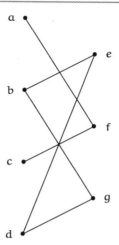

Fig. 5.11 Subgraph of $K_{4,3}$

Example 5.4 The graph in Fig. 5.11 is a subgraph of the complete bipartite graph $K_{4,3}$.

Let $G = (V, E)$ be a graph. Let R be an equivalence relation on the set V. Then let $G_R = (V_R, E_R)$ be the graph formed where V_R are the equivalence classes on V under the relation R and $\{[v], [w]\} \in E_R$ if and only if there exists $a \in [v], b \in [w]$ with $\{a, b\} \in E$. The graph G_R is called the quotient graph of G under the relation R.

Example 5.5 Let G be the graph given below on the set of vertices $V = \{a, b, c, d, e, f\}$. Let R be the equivalence relation on V induced by the partition $\{\{a, b\}, \{c, d\}, \{e, f\}\}$. Then G_R is the quotient graph given in Fig. 5.12.

Two graphs $G = (V, E)$ and $G' = (V', E')$ are isomorphic graphs if there exists a bijection $f : V \to V'$ such that $\{v, w\}$ is an edge in E if and only if $\{f(v), f(w)\}$ is an edge in E'. In this case, the map f is called an isomorphism.

Example 5.6 The three graphs in Fig. 5.13 are isomorphic.

Theorem 5.5 *Let $G = (V, E)$ and $G' = (V', E')$ be isomorphic graphs with isomorphism ϕ. If $v \in V$, then $\deg(v) = \deg(\phi(v))$.*

Proof Let d be the degree of v. If the complete list of vertices to which v is connected is $v_1, v_2, \ldots, v_d$, then the complete list of vertices to which $\phi(v)$ is connected is $\phi(v_1), \phi(v_2), \ldots, \phi(v_d)$. The result follows. $\qquad\square$

The next corollary follows immediately from Theorem 5.5.

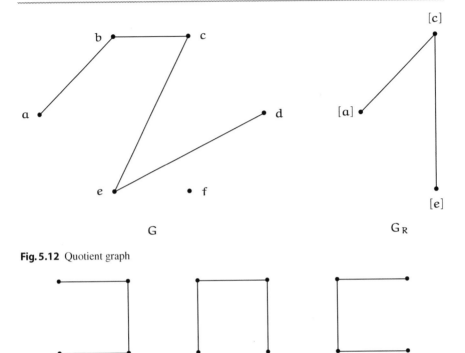

Fig. 5.12 Quotient graph

Fig. 5.13 Isomorphic graphs

Corollary 5.1 *Let* $G = (V, E)$ *and* $G' = (V', E')$ *be isomorphic graphs then the list of degrees of the vertices of* V *is exactly the list of degrees of the vertices of* V'.

Proof The list

$$\deg(v_1), \deg(v_2), \ldots, \deg(v_n)$$

is identical to the list

$$\deg(\phi(v_1)), \deg(\phi(v_2)), \ldots, \deg(\phi(v_n)).$$

The result follows.   □

We illustrate this with the following example.

Example 5.7 It follows from Corollary 5.1 that no two of the following three graphs are isomorphic, since the degree lists are $(2, 2, 2, 2)$, $(1, 2, 2, 1)$, and $(1, 3, 2, 2)$, respectively. Notice that even though the first and the third have the same number of vertices and edges they are not isomorphic.

Theorem 5.6 *If* $G = (V, E)$ *and* $G' = (V', E')$ *are isomorphic graphs then* G *is a bipartite graph if and only if* G' *is a bipartite graph.*

Proof Assume G is a bipartite graph and let f be the isomorphism from G to G'. Let A and B be the bipartite partition of the vertices of G. Let $A' = f(A)$ and $B' = f(B)$. Then A' and B' partition the vertices of G'. Assume $\{v', w'\} \in E'$, then $v' = f(v)$ and $w' = f(w)$ for some $v, w \in V$. Then $\{v, w\} \in E$. Without loss of generality, this implies that $v \in A$ and $w \in B$ since G is bipartite. Then $f(v) \in A'$ and $f(w) \in B'$. Hence, G' is a bipartite graph.

The other direction follows from Exercise 10 using f^{-1} as the isomorphism. □

Exercises

1. Prove that if G and G' are isomorphic then G has an Euler tour if and only if G' has an Euler tour.
2. Determine the number of edges in the complete graph on n vertices.
3. Prove that the complete graph K_n has an Euler cycle if and only if n is odd. Find such an Euler cycle on K_5. Prove that the only complete graph with an Euler tour is K_2.
4. Prove that if G and G' are isomorphic then G is connected if and only if G' is connected.
5. Determine the number of edges in the wheel graph W_n and the number of edges in the cycle graph C_n.
6. Determine the number of distinct bipartite graphs on a set with n vertices.
7. If A and B partition the set of vertices with $|A| = a$ and $|B| = b$ then determines the number of edges in the corresponding complete bipartite graph.
8. Prove that isomorphism is an equivalence relation on the set of all graphs.
9. We found eight simple graphs on three vertices in Example 5.2. Partition these eight graphs into equivalence classes induced by isomorphism.
10. Prove that if f is an isomorphism from G to G' then f^{-1} is an isomorphism from G' to G.
11. Let G and G' be two bipartite graphs on the same points sets with the same bipartite partition. Define G_R and G'_R to be the quotient graphs formed by the relation that two vertices are related if and only if they are in the same set of the partition. Prove that G_R and G'_R are isomorphic.
12. Extend Theorem 5.6 to k-partite graphs.

13. Find two graphs with five vertices and an equal number of edges that are not isomorphic.
14. Prove that any two graphs with the same number of vertices and with no edges must be isomorphic. Determine the number of isomorphisms between them.
15. Prove that any two graphs with the same number of vertices and with a complete set of edges must be isomorphic. Determine the number of isomorphisms between them.

5.3   Colorings of Graphs

We now move to a highly intuitive aspect of graph theory, namely, the coloring of graphs. It arises from some natural questions, namely, how many colors are needed to color a political map so that no two adjacent states have identical colors. It leads to one of the most interesting and controversial theorems of the twentieth century, namely, the four color map theorem. We begin with some definitions.

Definition 5.6 A coloring of a graph $G = (V, E)$ is a map from V to a set C such that if $\{v, w\} \in E$ then $f(v) \neq f(w)$.

Intuitively it means that no two connected vertices have the same color.

Definition 5.7 The chromatic number of a graph is the minimum number of colors needed to color a graph. This is denoted by $\chi(G)$.

The only graph with $\chi(G) = 1$ would be a graph with no edges. It is also obvious that assigning each vertex a different color will give a coloring. The complete graph on n vertices requires n colors. These two results give the standard bounds, that is, $1 \leq \chi(G) \leq n$ for graphs with n vertices. We shall consider a few examples.

- Discrete graph:
  The discrete graph on n vertices, D_n, has chromatic number 1 since each vertex can be colored the same.
- Complete graph:
  The complete graph on n vertices, K_n, has chromatic number n since each vertex is connected to every other vertex. Therefore, no two vertices can have the same color.
- Linear graph:
  The linear graph L_n, with $n \geq 2$ has chromatic number 2 since the vertices are alternatively colored.

- Bipartite graph:
  A bipartite graph has chromatic number 2, that is, one color for each set in the partition. Since no two vertices in the same set of the partition are connected then they can all have the same color.
- Cycle graph:
  The cycle graph, C_n, has chromatic number 2 if n is even and 3 if n is odd. If n is even the colors are simply alternated around the cycle. If n is odd this technique will not work since the last color would be the same as the first and these two vertices are adjacent. Hence, you need an additional color. We give the examples of C_4 and C_5 which illustrate the point.

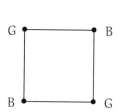
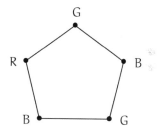

- Wheel graph
  The wheel graph, W_n, has chromatic number 4 if n is even and 3 if n is odd. By removing the center vertex from the wheel graph we obtain the cycle graph. If n is even then we have a cycle graph with oddly many vertices which requires three colors. The additional vertex is adjacent with ever other vertex and so requires a color distinct from the rest. If n is odd then we have a cycle graph with evenly many vertices which requires two colors. Then, the same argument gives that we need three colors to color the graph. We illustrate it with the following example of W_5 and W_6.

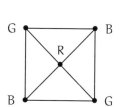
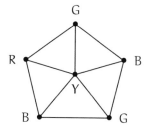

We shall now show how to determine the chromatic number of a disconnected graph by determining the chromatic number of its components.

Theorem 5.7 *Let G be a graph with components* $G_1, G_2, \ldots, G_s$. *Then* $\chi(G) = \max\{\chi(G_i)\}$.

Proof Since no vertex in G_i is connected with any vertex in G_j when $i \neq j$, then the graph can be colored with as many colors that are needed to color the component with the largest chromatic number.   □

Example 5.8 Consider the following disconnected graph.

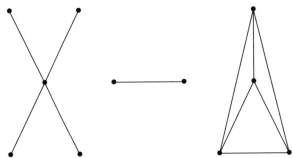

The first component has chromatic number 2, the second component has chromatic number 2, and the third component has chromatic number 4. Hence, the chromatic number of the graph is 4.

A graph is said to be a planar graph if it can be drawn in $\mathbb{R}^2$ such that no two edges (when viewed as curves in $\mathbb{R}^2$) intersect at a point that is not a vertex. Of course, this does not mean that every rendering of a graph is planar but rather that there is rendering that is planar.

For example, the following are isomorphic graphs where one is given without intersecting edges and one is not.

 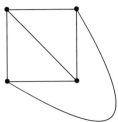

One of the most famous problems of the twentieth century was the four color map problem. The problem was brought to public attention in 1854, most probably by Francis or Frederick Guthrie [41]. For an early description by Cayley, see [23]. Essentially, it asks whether a standard political map can be colored so that no neighbors have the same color with only four colors. It is assumed that all countries are connected, that is, that there is a tour on the map between any two countries. The problem can easily be transformed into a graph theory question. That is, take each country as a vertex and connect two countries with an edge if they share more than a single point. The problem then becomes the following conjecture.

Conjecture 5.1 Any planar graph be colored with four colors.

Of course, not all planar graphs require four colors but the question simply asks if four colors will always suffice.

The solution of the problem did not occur until 1976 when it was solved by Appel and Haken assisted by Koch. A description of the solution can be found in [1–3].

One drawback of this solution is that an extremely large number of cases were eliminated by computer—far more than could be examined by hand. This aspect of the proof caused a bit of controversy over whether this solution should indeed be considered a proof in the traditional sense. Since this early computer proof, numerous problems have been solved with an extensive computer search. One of the most famous cases would be the proof of the non-existence of the projective plane of order 10, see [58]. The difficulty is that unlike standard proofs, the veracity of the proof is difficult to verify. Lam has suggested that mathematicians take a cue from physical scientists and only accept the proof with independent verification. The great difficulty here is that few people would be willing to undertake such an arduous task simply to prove someone else's result correct.

We can now examine the number of ways a graph can be colored with a given set of colors rather than how many colors are necessary to color it. Let G be a graph and let $P_G(n)$ be the number of ways to make a coloring of G with n or fewer colors. Certainly, this is a function, but it may not be as obvious that this is a polynomial for all graphs G. We refer to this as the chromatic polynomial of the graph and denote it by $P_G(x)$.

- Discrete graph:
  The discrete graph on n vertices, D_n, has chromatic polynomial $P_G(x) = x^n$, since each vertex can be colored any of the x colors.
- Complete graph:
  The complete graph on n vertices, K_n, has chromatic polynomial $P_G(x) = x(x - 1)(x - 2) \cdots (x - n + 1)$ since each vertex that is colored lowers the number for each other vertex by 1. For example, $P_{K_5}(x) = x(x - 1)(x - 2)(x - 3)(x - 4)$.
- Linear graph:
  The linear graph L_n with $n \geq 2$ has chromatic polynomial $P_G(x) = x(x - 1)^{n-1}$ since you have x choices for the first and then $x - 1$ for each of the following.

Example 5.9 Consider the complete bipartite graph $K_{2,3}$ given in Fig. 5.14.

Assume there are x colors to color the graph. If the two on the left side are colored the same then there are x choices for this, and each of those on the right has $x - 1$ choices. If the two on the left side are colored differently then there are x choices for the first and $x - 1$ for the second leaving $x - 2$ for each of those on the right. This gives that the chromatic polynomial is $x(x - 1)^3 + x(x - 1)(x - 2)^3 = x(x - 1)(x^3 - 5x^2 + 10x - 7)$.

Fig. 5.14 The complete
bipartite graph $K_{2,3}$

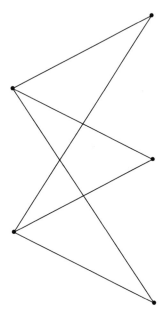

Theorem 5.8 *Let G be a graph with components* $G_1, G_2, \ldots, G_s$. *Then*

$$P_G(x) = \prod_{i=1}^{s} P_{G_i}(x). \tag{5.1}$$

Proof Any coloring of G_i can be matched with any coloring of G_j with $i \neq j$ since
no vertex of G_i is connected to any vertex in G_j. The result follows.   □

Example 5.10 Consider the following disconnected graph:

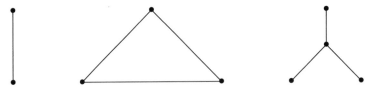

The first component has chromatic polynomial $x(x-1)$, the second has chromatic
polynomial $x(x-1)(x-2)$, and the third has chromatic polynomial $x(x-1)^3$.
Therefore, the chromatic polynomial of the graph is $x^3(x-1)^5(x-2)$.

Theorem 5.9 *Let* $G = (V, E)$ *be a simple graph, let* $e = \{a, b\} \in E$, *and let* G^e
be the graph $(V, E - \{e\})$. *Then let R be the equivalence relation on V formed by
equating* a *and* b. *Then*

$$P_G(x) = P_{G^e}(x) - P_{G_R}(x). \tag{5.2}$$

Proof Given a coloring of G^e, either a and b have the same color or they have different colors. If they have different colors then it is a coloring of G. If they are of the same color then it is a coloring of G_R. This gives that $P_G(x) + P_{G_R}(x) = P_{G^e}(x)$ which gives the result. $\qquad\square$

Example 5.11 Consider the following graph, which is the complete graph on three vertices.

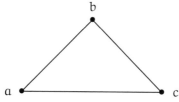

This graph has chromatic polynomial $P_G(x) = x(x-1)(x-2)$.

The graph G^e is

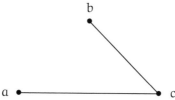

and has chromatic polynomial $P_{G^e}(x) = x(x-1)^2$.

The graph P_{G_R} is

a •——————————• c

and has chromatic polynomial $x(x-1)$. Then

$$P_{G^e}(x) - P_{G_R}(x) = x(x-1)^2 - x(x-1)$$
$$= x(x-1)(x-1-1)$$
$$= x(x-1)(x-2) = P_G(x).$$

Exercises

1. Find the chromatic number of the following graph:

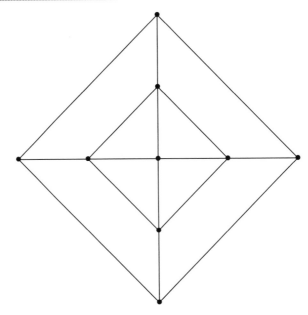

2. Take a political map of one of the inhabited continents and construct its planar graph G. Determine the chromatic number $\chi(G)$ of this map. Which continent(s) have the highest chromatic number and which have the lowest chromatic number?

3. Construct a planar graph that requires four colors, that is, a graph with chromatic number 4.

4. Determine the chromatic polynomial of the complete tripartite graph $K_{2,3,2}$.

5. Determine the chromatic polynomial of the complete bipartite graph $K_{2,2}$.

6. Draw the Latin square graph for the Latin square

$$\begin{pmatrix} 1\ 2\ 3 \\ 2\ 3\ 1 \\ 3\ 1\ 2 \end{pmatrix}.$$

   Determine its chromatic number.

7. Let $\Delta(G)$ be the maximum degree of any vertex in the graph. Prove that $\chi(G) \leq \Delta(G) + 1$. This can be done by ordering the vertices by degree and assigning an available color to each vertex in this order. This is known as greedy coloring.

8. Prove that the complete bipartite graph $K_{3,3}$ is not planar. This is a puzzle often played by children. That is, they draw three houses and three utilities and try to connect each utility to each house without crossing the connecting lines.

5.4   Directed Graphs and Relations

In addition to multigraphs and simple graphs, there are also directed graphs. In a directed graph, the edges have a direction, so there is a difference between an edge starting at vertex a and ending at vertex b and an edge starting at vertex b and ending at vertex a. A directed graph $G = (V, E)$ is a set of vertices V and a set of edges E where $(a, b) \in E$ if and only if there is an edge from a to b. Notice that the edges here are not sets but ordered pairs since (a, b) and (b, a) are not the same edge. We begin with an example of a directed graph on six vertices in Fig. 5.15.

For a directed graph, we need more general definitions of degree. For a vertex v, we define $degree_{in}(v) = |\{(w, v) \mid (w, v) \in E\}|$ and $degree_{out}(v) = |\{(w, v) \mid (v, w) \in E\}|$.

The adjacency matrix M for a graph $G = (V, E)$ is indexed by V and $M_{a,b} = 1$ if $(a, b) \in E$ and 0 otherwise (Table 5.1).

Notice that the number of ones in the row corresponding to a vertex v is $degree_{out}(v)$ and the number of ones in the column corresponding to a vertex v is $degree_{in}(v)$. Note also that the matrix is not symmetric.

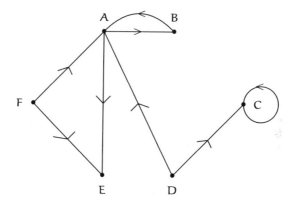

Fig. 5.15 Directed graph on six vertices

Table 5.1 Adjacency matrix for the directed graph

	A	B	C	D	E	F
A	0	1	0	0	1	0
B	1	0	0	0	0	0
C	0	0	1	0	0	0
D	1	0	1	0	0	0
E	0	0	0	0	0	0
F	1	0	0	0	1	0

Each edge adds to the degree_{in} and degree_{out} of exactly one vertex. Thus, for a directed graph $G = (V, E)$, we have

$$2|E| = \sum_{v \in V} (\text{degree}_{in}(v) + \text{degree}_{out}(v)).$$

Recall that a relation on a set A is a subset of $A \times A$.

Theorem 5.10 *There is a bijective correspondence between the set of relations on a set A of cardinality n and the set of directed graphs on a set of n-vertices.*

Proof The correspondence is easy to describe. That is, the n vertices correspond to the n elements of A. Then (a, b) is in the relation if and only if (a, b) is an edge in the directed graph. □

Example 5.12 Consider the relation on the set $\{a, b, c, d\}$ given by

$$R = \{(a, a), (a, c), (c, a), (d, d)\}.$$

Then its corresponding graph is the following:

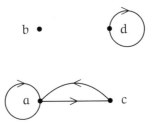

Likewise we have the following theorem.

Theorem 5.11 *There is a bijective correspondence between the set of binary square matrices of order n and the set of directed graphs on a set of n-vertices.*

Proof The correspondence is easy to describe. That is, the n vertices correspond to the n rows and columns of a matrix M. Then $M_{a,b} = 1$ if and only if (a, b) is an edge in the directed graph. □

Example 5.13 Given the graph in Example 5.12, its matrix M is indexed by a, b, c, d and is

$$M = \begin{pmatrix} 1 & 0 & 1 & 0 \\ 0 & 0 & 0 & 0 \\ 1 & 0 & 0 & 0 \\ 0 & 0 & 0 & 1 \end{pmatrix}.$$

We see from these two theorems that there is a bijective correspondence between directed graphs, binary matrices, and relations on a set.

We shall describe some properties of relations and how they correspond to the directed graph. A relation R on a set A is reflexive if $(a, a) \in R$ for all $a \in A$. In terms of the graph this means that at every vertex there is a loop. For the matrix, this means that there is a 1 in every entry in the diagonal.

A relation is symmetric if $(a, b) \in R$ implies $(b, a) \in R$. For the graph, this means that if there is an edge going from a to b then there is an edge from b to a. For the matrix, this means that the matrix is symmetric, that is, $M = M^T$.

A relation is transitive if (a, b) and (b, c) are in R then $(a, c) \in R$. For the graph, this means that if there is an edge from a to b and an edge from b to c then there is an edge from a to c. For the matrix, this property is not as easily seen.

A tour in a directed graph $G = (V, E)$ is an ordered k-tuple $(e_1, e_2, \ldots, e_k)$, $e_i = (v_i, w_i) \in E$, where $w_i = v_{i+1}$ and $e_i \neq e_j$ if $i \neq j$. In this definition, we say that the tour is from the vertex v_1 to w_k.

We can mimic the definition of a Hamiltonian tour and cycle. The only difference is that a tour must respect the direction of an edge. This means that a tour can contain the edge from a to b only if there is an edge starting at a and ending at b. Namely, in the tour you must follow the direction of the edge.

A complete directed graph is the graph such that for any vertex a and b, (a, b) is an edge in the graph. Note that this graph has more than $C(n, 2)$ edges, which is the number of edges for the simple complete graph. It has $2C(n, 2)$ edges, since each direction must have an edge as well. This is assuming that there are no loops, meaning no edge connects a vertex to itself. If we allow loops then there are $2C(n, 2) + n$ edges.

Exercises

1. Draw the directed graph for the relation

   $$\{(0, 1), (1, 2), (0, 0), (2, 4), (4, 3), (3, 4), (2, 2), (4, 1), (4, 2)\}$$

   on the set $\{0, 1, 2, 3, 4\}$. Give the in and out degrees of each vertex.

2. There are eight subsets of the set $\{reflexive, symmetric, transitive\}$, given an example of a directed graph that corresponds exactly to each of these subsets. Namely, the graph has the property if and only if it is in the subset.

3. Prove that in a directed graph corresponding to a transitive relation, that if there is a tour from vertex v to vertex w, then (v, w) is an edge in the graph.

4. Mimic the proofs in Theorems 5.1 and 5.2 to determine when there exists Euler tours and Euler cycles in a directed graph. Hint: Consider $degree_{in}$ and $degree_{out}$ for each vertex.

5. Prove that the number of Hamiltonian cycles on the complete directed graph with n vertices is $(n - 1)!$.

Higher Dimensional Finite Geometry

6

We shall now turn our attention to higher dimensional geometries using the theory of linear algebra as our setting. This approach can be used for finite or infinite geometry although it is often not presented this way in most undergraduate geometry classes. It is, however, a very powerful technique which reduces many difficult geometric proofs to very simple algebraic proofs.

As an example of viewing geometry in algebraic terms, consider the real Euclidean plane. Here, we can consider the points as elements $(x, y) \in \mathbb{R}^2$. The lines on this plane can be considered as one-dimensional subspaces (lines through the origin) and their cosets (translates). A parallel class consists of all translates of a subspace. For example, the subspace may consist of all (x, y) satisfying $y = mx$. Then we consider all points in the space $\{(0, b) + (x, y)\}$ where (x, y) satisfy $y = mx$. This new set is all the points that satisfy the equation $y = mx + b$. Therefore, the lines are one-dimensional vector spaces and their cosets and the usual algebraic manipulations can be done to them to prove geometric results.

In three dimensions, the planes are two-dimensional subspaces and their translates. This is usually described by taking the set of all (x, y, z) perpendicular to some vector $\langle a, b, c \rangle$, which describes a subspace of dimension 2 in this space, and then specifying a point that it must pass through which determines which coset of this space we want. If we are in n-dimensional space $\mathbb{R}^n$, then the points are the elements of the space, the lines are one-dimensional subspaces, their cosets and the planes are two-dimensional subspaces and their cosets, and any k-dimensional subspace and its cosets are k-dimensional geometries. We say that a $n - 1$-dimensional geometry contained in n-dimensional space is a hyperplane.

A similar construction can be done to construct projective geometry, which we shall describe in detail later. We shall begin our discussion by recalling some of the necessary definitions and theorems from linear algebra that we shall need.

© The Editor(s) (if applicable) and The Author(s), under exclusive license
to Springer Nature Switzerland AG 2020
S. T. Dougherty, *Combinatorics and Finite Geometry*,
Springer Undergraduate Mathematics Series,
https://doi.org/10.1007/978-3-030-56395-0_6

6.1   Linear Algebra

We begin by recalling the definition of a vector space over a field F. In this text, we generally use a finite field but the same definition applies to any field.

Definition 6.1 A vector space V over a field F is a non-empty set V with addition and scalar multiplication such that for all $\mathbf{v}, \mathbf{w}, \mathbf{u} \in V$ and all $\alpha, \beta \in F$, we have

1. (Closure) $\mathbf{v} + \mathbf{w} \in V$;
2. (Commutativity) $\mathbf{v} + \mathbf{w} = \mathbf{w} + \mathbf{v}$;
3. (Associativity) $\mathbf{v} + (\mathbf{w} + \mathbf{u}) = (\mathbf{v} + \mathbf{w}) + \mathbf{u}$;
4. (Additive Identity) there exists $\mathbf{0} \in V$ with $\mathbf{0} + \mathbf{v} = \mathbf{v} + \mathbf{0} = \mathbf{v}$;
5. (Inverses) there exists $-\mathbf{v} \in V$ with $\mathbf{v} + (-\mathbf{v}) = (-\mathbf{v}) + \mathbf{v} = \mathbf{0}$;
6. (Closure) $\alpha\mathbf{v} \in V$;
7. (Associativity) $(\alpha\beta)\mathbf{v} = \alpha(\beta\mathbf{v})$;
8. (Distributive Property) $(\alpha + \beta)\mathbf{v} = \alpha\mathbf{v} + \beta\mathbf{v}$;
9. (Scalar Identity) $1\mathbf{v} = \mathbf{v}$;
10. (Distributive Property) $\alpha(\mathbf{v} + \mathbf{w}) = \alpha\mathbf{v} + \alpha\mathbf{w}$.

Notice that we never used that F is a field in the definition. It is, however, vital in the theory that we are going to pursue. If the field F were replaced with a ring then these same axioms are used to define a module. This small change in the definition makes enormous changes in the results. For example, consider the ring $\mathbb{Z}_4$. The space $\{0, 2\}$ is a non-trivial module over this ring contained in the ring. This does not happen for fields. There are no non-trivial subspaces of a field.

Definition 6.2 We say that W is a subspace of a vector space V if W is a vector space and $W \subseteq V$.

Example 6.1 A line through the origin in a subspace of $\mathbb{R}^2$ and a plane through the origin in a subspace of $\mathbb{R}^3$.

Example 6.2 The space consisting of $\{(0, 0), (1, 1)\}$ is a subspace of $\mathbb{F}_2^n$.

Theorem 6.1 *Let V be a vector space and let $W \subseteq V$. Then, if $\alpha\mathbf{v} + \beta\mathbf{w} \in W$ for all $\mathbf{v}, \mathbf{w} \in W$ and all $\alpha, \beta \in F$, then W is a subspace of V.*

Proof Assume $\alpha\mathbf{v} + \beta\mathbf{w} \in W$ for all $\mathbf{v}, \mathbf{w} \in W$ and all $\alpha, \beta \in F$. Axioms 2, 3, 7, 8, 9, and 10 all follow from the fact that $W \subseteq V$. Axiom 1 follows by taking $\alpha = \beta = 1$. Axiom 4 follows by taking $\alpha = \beta = 0$. Axiom 5 follows by taking $\alpha = -1$ and $\beta = 0$. Axiom 6 follows by taking $\beta = 0$. Therefore, W is a vector space. □

Example 6.3 Let $\mathbf{v} \in V$, V a vector space. Then the set $W = \{\alpha\mathbf{v} \mid \alpha \in F\}$ is a subspace of V. We have $\alpha\mathbf{v} + \beta\mathbf{v} = (\alpha + \beta)\mathbf{v} \in W$. Then by Theorem 6.1, W is a subspace.

Definition 6.3 A set of vectors $\mathbf{v}_1, \dots, \mathbf{v}_n$ is a spanning set of V if every vector in V can be expressed as $\sum_{i=1}^{n} \alpha_i \mathbf{v}_i$.

We shall denote the set spanned by vectors $\mathbf{v}_1, \dots, \mathbf{v}_n$ by $\langle \mathbf{v}_1, \dots, \mathbf{v}_n \rangle$. Note that a spanning set is not necessarily minimal. For example, $\mathbb{R}^2 = \langle (1,0), (0,1), (1,1) \rangle$ and hence $\{(1,0), (0,1), (1,1)\}$ is a spanning set for $\mathbb{R}^2$; however, $\{(1,0), (0,1)\}$ is also a spanning set for $\mathbb{R}^2$.

Example 6.4 Let $\mathbf{e}_i$ be the vector of length n, that is, 1 in the i-th coordinate and 0 elsewhere. The set $\mathbf{e}_1, \mathbf{e}_2, \dots, \mathbf{e}_n$ is a spanning set for F^n where F is any field.

Definition 6.4 A set of vectors $\{\mathbf{v}_1, \dots, \mathbf{v}_n\}$ is linearly independent if $\sum \alpha_i \mathbf{v}_i = 0$ implies $\alpha_i = 0$ for all i.

Example 6.5 The set $\{(1,2), (2,1)\}$ is not linearly independent over $\mathbb{F}_3$ since $1(1,2) + 1(2,1) = (0,0)$. The set $\{(1,0), (0,1)$ is linearly independent over any field since $\alpha(1,0) + \beta(0,1) = (0,0)$ implies $(\alpha, \beta) = (0,0)$ and then $\alpha = 0$ and $\beta = 0$.

Definition 6.5 If a set $\{\mathbf{v}_1, \dots, \mathbf{v}_n\}$ is a linearly independent spanning set for V then $\{\mathbf{v}_1, \dots, \mathbf{v}_n\}$ is a basis for V. We then say that V has dimension n.

Example 6.6 The set $\mathbf{e}_1, \mathbf{e}_2, \dots, \mathbf{e}_n$ defined in Example 6.4 is a basis for F^n for any field F.

We note that the basis in the previous example is not the only possible basis. For example, $\{(1,0), (1,1)\}$ is also a basis for $\mathbb{F}_2^2$.

Theorem 6.2 *A dimension k vector space over the field $\mathbb{F}_q$ has q^k elements.*

Proof Let $\mathbf{v}_1, \mathbf{v}_2, \dots, \mathbf{v}_k$ be a basis for a k-dimensional vector space over $\mathbb{F}_q$. Then each element of the form $\sum_{i=1}^{k} \alpha_i \mathbf{v}_i$ is distinct since the vectors are linearly independent. For each α_i there are q choices which gives q^k vectors. □

Example 6.7 Consider the two linearly independent vectors over $\mathbb{F}_3$: $(1,1,0)$, $(0,1,2)$. The space generated by these two vectors consists of all vectors of the form $\alpha(1,1,0) + \beta(0,1,2)$ where $\alpha, \beta \in \mathbb{F}_3$. Hence, there are nine vectors in $\langle (1,1,0), (0,1,2) \rangle$. Specifically, they are $\{(0,0,0), (1,1,0), (2,2,0), (0,1,2), (1,2,2), (2,0,2), (0,2,1), (1,0,1), (2,1,1)\}$.

We can now give a theorem that counts the number of subspaces of a given vector space over a finite field.

Theorem 6.3 *Let* V *be an* n-*dimensional vector space over* $\mathbb{F}_q$. *The number of subspaces of* V *of dimension* k *is*

$$\frac{(q^n - 1)(q^n - q) \cdots (q^n - q^{k-1})}{(q^k - 1)(q^k - q) \cdots (q^k - q^{k-1})}. \tag{6.1}$$

Proof In V there are q^n elements and so there are $q^n - 1$ ways of picking a non-zero element. That non-zero element generates a one-dimensional space with q elements. This means there are $q^n - q$ ways of picking an element not in that space. These two elements generate a space with q^2 elements giving $q^n - q^2$ ways of picking an element not in that space. Continuing, by induction, we see that the numerator counts the number of ways of picking k linearly independent vectors from the space. The bottom counts, in the same manner, the number of different bases that this k-dimensional subspace has. □

Example 6.8 Assume $q = 2, n = 4$ and $k = 2$. Then the number of two-dimensional subspaces of a four-dimensional space over $\mathbb{F}_2$ is

$$\frac{(2^4 - 1)(2^4 - 2)}{(2^2 - 1)(2^2 - 2)} = \frac{15(14)}{(3)(2)} = 35.$$

The number of k-dimensional subspaces of an n-dimensional space over $\mathbb{F}_q$ is denoted by $\begin{bmatrix} n \\ k \end{bmatrix}_q$. The notation is close to the notation for binomial coefficients and there are corresponding identities, such as the following theorem.

Theorem 6.4 *Let* n, k *be integers with* $0 \leq k \leq n$, *then*

$$\begin{bmatrix} n \\ k \end{bmatrix}_q = \begin{bmatrix} n - 1 \\ k - 1 \end{bmatrix}_q + q^k \begin{bmatrix} n - 1 \\ k \end{bmatrix}_q. \tag{6.2}$$

Proof Let V be a k-dimensional vector space over $\mathbb{F}_q$. Let $\mathbf{x}$ be a vector in V. We can write V as $V = \langle \mathbf{x} \rangle \oplus W$ where W is a $k - 1$-dimensional subspace of V. Every k-dimensional subspace either contains $\langle \mathbf{x} \rangle$ or does not contain it. If it does contain $\langle \mathbf{x} \rangle$ then it is of the form $\langle \mathbf{x} \rangle \oplus W'$ where W' is a $k - 1$-dimensional subspace of W. Hence, there are precisely $\begin{bmatrix} n - 1 \\ k - 1 \end{bmatrix}_q$ such subspaces.

Let $\mathbf{w}_1, \mathbf{w}_2, \ldots, \mathbf{w}_k$ denote a basis for a k-dimensional subspace of W. There are $\begin{bmatrix} n - 1 \\ k \end{bmatrix}_q$ such subspaces. Let $v_i = w_i + \alpha_i \mathbf{x}$. Then $\mathbf{v}_1, \mathbf{v}_2, \ldots, \mathbf{v}_k$ form a basis for a k-dimensional subspace of V that is not contained in W. Since there are q choices for each α_i there are q^k such vector spaces. This accounts for all k-dimensional subspaces of V which gives the identity.   □

Example 6.9 Let $q = 5$. Then

$$\begin{bmatrix} 4 \\ 2 \end{bmatrix}_5 = \frac{(5^4 - 1)(5^4 - 5)}{(5^2 - 1)(5^2 - 5)} = 806,$$

$$\begin{bmatrix} 3 \\ 2 \end{bmatrix}_5 = \frac{(5^3 - 1)(5^3 - 5)}{(5^2 - 1)(5^2 - 5)} = 31,$$

$$\begin{bmatrix} 3 \\ 1 \end{bmatrix}_5 = \frac{(5^3 - 1)}{(5 - 1)} = 31.$$

The recursion gives

$$\begin{bmatrix} 4 \\ 2 \end{bmatrix}_5 = \begin{bmatrix} 3 \\ 1 \end{bmatrix}_5 + 5^2 \begin{bmatrix} 3 \\ 2 \end{bmatrix}_5 = 31 + (25)(31) = 806. \tag{6.3}$$

This recurrence leads to a Pascal type triangle. Label the rows by n and the columns by k, then we get the following tables for $\begin{bmatrix} n \\ k \end{bmatrix}_q$ when $q = 2$ and $q = 3$.

The values of $\begin{bmatrix} n \\ k \end{bmatrix}_q$ **when** $q = 2$

$$\begin{array}{ccccccc}
1 \\
1 & 1 \\
1 & 3 & 1 \\
1 & 7 & 7 & 1 \\
1 & 15 & 35 & 15 & 1 \\
1 & 31 & 155 & 155 & 31 & 1 \\
1 & 63 & 651 & 1395 & 651 & 63 & 1
\end{array} \tag{6.4}$$

The values of $\begin{bmatrix} n \\ k \end{bmatrix}_q$ **when** $q = 3$

$$\begin{array}{ccccccc}
1 \\
1 & 1 \\
1 & 4 & 1 \\
1 & 13 & 13 & 1 \\
1 & 40 & 130 & 40 & 1 \\
1 & 121 & 1210 & 1210 & 121 & 1 \\
1 & 364 & 11011 & 33880 & 11011 & 364 & 1
\end{array} \tag{6.5}$$

We can also do a strictly computational proof of Theorem 6.4. We have

$$
\begin{aligned}
\begin{bmatrix} n \\ k \end{bmatrix}_q - \begin{bmatrix} n-1 \\ k-1 \end{bmatrix}_q &= \frac{(q^n-1)(q^n-q)\cdots(q^n-q^{k-1})}{(q^k-1)(q^k-q)\cdots(q^k-q^{k-1})} - \begin{bmatrix} n-1 \\ k-1 \end{bmatrix}_q \\
&= \frac{(q^n-1)q(q^{n-1}-1)q(q^{n-1}-q)\cdots q(q^{n-1}-q^{k-2})}{(q^k-1)q(q^{k-1}-1)q(q^{k-1}-q)\cdots q(q^{k-1}-q^{k-2})} - \begin{bmatrix} n-1 \\ k-1 \end{bmatrix}_q \\
&= \frac{q^n-1}{q^k-1} \begin{bmatrix} n-1 \\ k-1 \end{bmatrix}_q - \begin{bmatrix} n-1 \\ k-1 \end{bmatrix}_q \\
&= \left(\frac{q^n-1}{q^k-1} - 1 \right) \begin{bmatrix} n-1 \\ k-1 \end{bmatrix}_q \\
&= \left(\frac{q^n-q^k}{q^k-1} \right) \begin{bmatrix} n-1 \\ k-1 \end{bmatrix}_q \\
&= q^k \left(\frac{q^{n-k}-1}{q^k-1} \right) \begin{bmatrix} n-1 \\ k-1 \end{bmatrix}_q \\
&= q^k \begin{bmatrix} n-1 \\ k \end{bmatrix}_q .
\end{aligned}
$$

This gives that

$$
\begin{bmatrix} n \\ k \end{bmatrix}_q - \begin{bmatrix} n-1 \\ k-1 \end{bmatrix}_q = q^k \begin{bmatrix} n-1 \\ k \end{bmatrix}_q .
$$

It follows that

$$
\begin{bmatrix} n \\ k \end{bmatrix}_q = \begin{bmatrix} n-1 \\ k-1 \end{bmatrix}_q + q^k \begin{bmatrix} n-1 \\ k \end{bmatrix}_q .
$$

Examining the Pascal type triangle, we note that all of the elements were congruent to 1 (mod q). We prove this fact as an easy consequence of the recursion.

Corollary 6.1 *The number of k-dimensional subspaces of $\mathbb{F}_q^n$, $k \le n$, is congruent to* 1 (mod q).

Proof The result follows by induction, noting that by the recursion, we have

$$
\begin{bmatrix} n \\ k \end{bmatrix}_q = \begin{bmatrix} n-1 \\ k-1 \end{bmatrix}_q + q^k \begin{bmatrix} n-1 \\ k \end{bmatrix}_q . \tag{6.6}
$$

It follows that

$$
\begin{bmatrix} n \\ k \end{bmatrix}_q \equiv \begin{bmatrix} n-1 \\ k-1 \end{bmatrix}_q \quad (\text{mod } q). \tag{6.7}
$$

Then we note that $\begin{bmatrix} n \\ 1 \end{bmatrix}_q \equiv 1$ (mod q), for $n \ge 1$ and we have the result. □

Example 6.10 Consider the values in Table 6.5. The values

$$1, 4, 13, 40, 130, 121, 121, 364, 11011, 33880$$

are all congruent to 1 (mod 3).

One of the most powerful tools in the study of linear algebra is the linear transformation. A linear transformation $T : V \rightarrow W$, where V and W are vector spaces over a field F, is a map that satisfies, for $\mathbf{v}, \mathbf{w} \in V$, $\alpha \in F$,

$$T(\mathbf{v} + \mathbf{w}) = T(\mathbf{v}) + T(\mathbf{w});$$
$$T(\alpha\mathbf{v}) = \alpha T(\mathbf{v}).$$

The kernel of a transformation T is given by

$$\mathrm{Ker}(T) = \{\mathbf{v} \mid \mathbf{v} \in V \text{ and } T(\mathbf{v}) = \mathbf{0}\}. \tag{6.8}$$

The image is

$$\mathrm{Im}(T) = \{\mathbf{w} \mid \text{there exists } \mathbf{v} \text{ in } V \text{ with } T(\mathbf{v}) = \mathbf{w}\}. \tag{6.9}$$

We shall now give two important lemmas.

Lemma 6.1 *If* T *is a linear transformation* $T : V \rightarrow W$ *then* $\mathrm{Ker}(T)$ *is a subspace of* V.

Proof Let $\mathbf{v}, \mathbf{w} \in \mathrm{Ker}(T)$, $\alpha, \beta \in F$. Then $T(\alpha\mathbf{v} + \beta\mathbf{w}) = \alpha T(\mathbf{v}) + \beta T(\mathbf{w}) = \alpha\mathbf{0} + \beta\mathbf{0} = \mathbf{0}$ which gives that $\alpha\mathbf{v} + \beta\mathbf{w} \in \mathrm{Ker}(T)$ which gives that $\mathrm{Ker}(T)$ is a subspace of V. $\square$

Lemma 6.2 *If* T *is a linear transformation* $T : V \rightarrow W$ *then* $\mathrm{Im}(T)$ *is a subspace of* W.

Proof Let $\mathbf{v}, \mathbf{w} \in \mathrm{Im}(T)$, $\alpha, \beta \in F$. Then $\mathbf{v} = T(\mathbf{v}')$ and $\mathbf{w} = T(\mathbf{w}')$ for some $\mathbf{v}', \mathbf{w}' \in V$. Then $T(\alpha\mathbf{v}' + \beta\mathbf{w}') = \alpha T(\mathbf{v}') + \beta T(\mathbf{w}') = \alpha\mathbf{v} + \beta\mathbf{w} \in \mathrm{Im}(T)$. Therefore, $\mathrm{Im}(T)$ is a subspace of W. $\square$

Example 6.11 Let $T : \mathbb{F}_2^3 \rightarrow \mathbb{F}_2^3$, where

$$T((a, b, c)) = \begin{pmatrix} 1 & 1 & 0 \\ 0 & 1 & 1 \\ 1 & 0 & 1 \end{pmatrix} \begin{pmatrix} a \\ b \\ c \end{pmatrix}.$$

Then

$$\mathrm{Ker}(T) = \{(0, 0, 0), (1, 1, 1)\},$$

which is a subspace of $\mathbb{F}_2^3$ and

$$\text{Im}(T) = \{(0,0,0), (0,1,1), (1,1,0), (1,0,1)\},$$

which is also a subspace of $\mathbb{F}_2^3$.

This brings us to one of the most important theorems of linear algebra.

Theorem 6.5 *Let* $T : V \to W$ *be a linear transformation. Then*

$$\dim(\text{Ker}(T)) + \dim(\text{Im}(T)) = \dim(V). \tag{6.10}$$

Proof Let $\mathbf{w}_1, \mathbf{w}_2, \ldots, \mathbf{w}_k$ be a basis for the image, with $\mathbf{v}_i \in V$ and $T(\mathbf{v}_i) = \mathbf{w}_i$. Let $\mathbf{u}_1, \mathbf{u}_2, \ldots, \mathbf{u}_s$ be a basis for the kernel of T. We shall show that

$$\{\mathbf{v}_1, \mathbf{v}_2, \ldots, \mathbf{v}_k, \mathbf{u}_1, \mathbf{u}_2, \ldots, \mathbf{u}_s\}$$

is a basis for V.

If $\mathbf{v} \in V$ then $T(\mathbf{v})$ is in the image of T. This implies that $T(\mathbf{v}) = \sum \beta_i \mathbf{w}_i$. Then we have

$$T(\mathbf{v}) = T\left(\sum \beta_i \mathbf{v}_i\right)$$

$$T(\mathbf{v}) - T\left(\sum \beta_i \mathbf{v}_i\right) = 0$$

$$T\left(\mathbf{v} - \sum \beta_i \mathbf{v}_i\right) = 0,$$

which implies that $\mathbf{v} - \sum \beta_i \mathbf{v}_i \in \text{Ker}(T)$. This gives that $\mathbf{v} - \sum \beta_i \mathbf{v}_i = \sum \gamma_i \mathbf{u}_i$ and $\mathbf{v}$ is in the span of $\{\mathbf{v}_1, \mathbf{v}_2, \ldots, \mathbf{v}_k, \mathbf{u}_1, \mathbf{u}_2, \ldots, \mathbf{u}_s\}$. Hence it spans V.

Next we shall show that the set is linearly independent. Assume

$$\sum \alpha_i \mathbf{v}_i + \sum \beta_i \mathbf{u}_i = 0.$$

Then we have

$$T\left(\sum \alpha_i \mathbf{v}_i + \sum \beta_i \mathbf{u}_i\right) = 0$$

and $\sum \alpha_i \mathbf{v}_i + \sum \beta_i \mathbf{u}_i \in \text{Ker}(T)$. This implies that $\alpha_i = 0$ for all i, since if it were not then it would not be in the kernel. Then $\mathbf{0} = \sum \alpha_i \mathbf{v}_i + \sum \beta_i \mathbf{u}_i = \sum \beta_i \mathbf{u}_i \in \text{ker}(T)$ but we know that $\{\mathbf{u}_1, \mathbf{u}_2, \ldots, \mathbf{u}_s\}$ are linearly independent so $\beta_i = 0$ for all i. Hence the set $\{\mathbf{w}_1, \mathbf{w}_2, \ldots, \mathbf{w}_k, \mathbf{u}_1, \mathbf{u}_2, \ldots, \mathbf{u}_s\}$ is a basis for V, which gives that $\dim(V) = k + s = \dim(\text{Im}(T)) + \dim(\text{Ker}(T))$. $\qquad\square$

Example 6.12 In Example 6.11, the ambient space had dimension 3, the kernel had dimension 1, and the image had dimension 2. This illustrates the theorem as $3 = 1 + 2$.

We shall define an inner product similar to the one used on the real numbers. Let F be a field. On the space $\mathbb{F}_q^n$ define the inner product

$$[\mathbf{v}, \mathbf{w}] = \sum v_i w_i. \tag{6.11}$$

This inner product is called the Euclidean inner product.

If two vectors $\mathbf{v}, \mathbf{w}$ have $[\mathbf{v}, \mathbf{w}] = 0$ then we say that $\mathbf{v}$ and $\mathbf{w}$ are orthogonal.

Let V be a subspace of $\mathbb{F}_q^n$. Define $V^\perp = \{\mathbf{w} \mid \mathbf{w} \in \mathbb{F}_q^n$ and $[\mathbf{v}, \mathbf{w}] = 0$ for all $\mathbf{v} \in V\}$.

Lemma 6.3 *Let* $\mathbf{u}, \mathbf{v}, \mathbf{w} \in \mathbb{F}_q^n$, $\alpha \in \mathbb{F}_q$, *where* $\mathbb{F}_q$ *is the field of order* q. *Then* $[\mathbf{v} + \mathbf{u}, \mathbf{w}] = [\mathbf{v}, \mathbf{w}] + [\mathbf{u}, \mathbf{w}]$ *and* $[\alpha \mathbf{v}, \mathbf{w}] = \alpha[\mathbf{v}, \mathbf{w}] = [\mathbf{v}, \alpha \mathbf{w}]$.

Proof Let $\mathbf{u}, \mathbf{v}, \mathbf{w} \in \mathbb{F}_q^n$. Then

$$[\mathbf{v} + \mathbf{u}, \mathbf{w}] = \sum (v_i + u_i) w_i$$
$$= \sum v_i w_i + u_i v_i = \sum v_i w_i + \sum u_i v_i$$
$$= [\mathbf{v}, \mathbf{w}] + [\mathbf{u}, \mathbf{w}].$$

Next,

$$[\alpha \mathbf{v}, \mathbf{w}] = \sum \alpha v_i w_i = \alpha \sum v_i w_i$$
$$= \alpha[\mathbf{v}, \mathbf{w}].$$

Similarly,

$$[\mathbf{v}, \alpha \mathbf{w}] = \sum v_i \alpha w_i = \alpha \sum v_i w_i$$
$$= \alpha[\mathbf{v}, \mathbf{w}]. \qquad \square$$

Lemma 6.4 *Let* $V \subseteq \mathbb{F}_q^n$ *be a vector space, then* $V^\perp$ *is a subspace of* $\mathbb{F}_q^n$.

Proof Let $\mathbf{v}, \mathbf{w} \in V^\perp$. Then, let $\mathbf{u} \in V$,

$$[\alpha \mathbf{v} + \beta \mathbf{w}, \mathbf{u}] = \alpha[\mathbf{v}, \mathbf{u}] + \beta[\mathbf{w}, \mathbf{u}] = \alpha 0 + \beta 0 = \mathbf{0}. \tag{6.12}$$

Therefore, $\alpha \mathbf{v} + \beta \mathbf{w} \in V^\perp$ and therefore $V^\perp$ is a subspace of $\mathbb{F}_q^n$. $\qquad \square$

We use these lemmas to prove the following.

Theorem 6.6 *Let V be a subspace of $\mathbb{F}_q^n$ with basis $\{\mathbf{v}_1, \mathbf{v}_2, \ldots, \mathbf{v}_k\}$. Then $\mathbf{w} \in V^\perp$ if and only if $[\mathbf{w}, \mathbf{v}_i] = 0$ for all i.*

Proof If $\mathbf{v} = \sum \alpha_i \mathbf{v}_i \in V$ and $\mathbf{w} \in \mathbb{F}_q^n$, we have

$$[\mathbf{v}, \mathbf{w}] = [\sum \alpha_i \mathbf{v}_i, \mathbf{w}]$$
$$= \sum \alpha_i [\mathbf{v}_i, \mathbf{w}].$$

Therefore, if $[\mathbf{v}_i, \mathbf{w}] = 0$ for all i then $\mathbf{w} \in V^\perp$.

If $\mathbf{w} \in V^\perp$ then $[\mathbf{w}, \mathbf{v}_i] = 0$ for all i since $\mathbf{v}_i \in V$. $\square$

Let $\{\mathbf{v}_1, \mathbf{v}_2, \ldots, \mathbf{v}_k\}$ be a basis for V, where V is a subspace of $\mathbb{F}_q^n$. Let M be the k by n matrix whose i-th row is $\mathbf{v}_i$. We know from Exercise 16 that $T(\mathbf{v}) = M\mathbf{v}$ is a linear transformation. Notice that $T(\mathbf{w}) = 0$ if and only if $[\mathbf{w}, \mathbf{v}_i] = 0$ for all i. This gives that $\mathrm{Ker}(T) = V^\perp$. It is clear that $\mathbf{w} \in \mathrm{Im}(T)$ if and only if $\mathbf{w} \in V$ since $T(\alpha_1, \alpha_2, \ldots, \alpha_n) = \sum \alpha_i \mathbf{v}_i$.

Using Theorem 6.5, we have proven the following.

Theorem 6.7 *Let V be a subspace of $\mathbb{F}_q^n$, then $\dim(V) + \dim(V^\perp) = n$.*

We can use this result to prove the following theorem, which was evident when noticing the symmetry in the Pascal type triangle.

Theorem 6.8 *Let n and k be positive integers. Then*

$$\begin{bmatrix} n \\ k \end{bmatrix}_q = \begin{bmatrix} n \\ n-k \end{bmatrix}_q.$$

Proof If V is a vector space in $\mathbb{F}_q^n$ of dimension k then $V^\perp$ is a vector space in $\mathbb{F}_q^n$ of dimension $n - k$. Therefore, there is a bijection between the set of k-dimensional vector spaces in $\mathbb{F}_q^n$ and the set of $n - k$-dimensional vector spaces in $\mathbb{F}_q^n$ given by $f(V) = V^\perp$. It follows that the number of k-dimensional subspaces in $\mathbb{F}_q^n$ is equal to the number of $n - k$-dimensional subspaces in $\mathbb{F}_q^n$. Therefore, $\begin{bmatrix} n \\ k \end{bmatrix}_q = \begin{bmatrix} n \\ n-k \end{bmatrix}_q$.

$\square$

Again, we can give a purely computational proof. We have that

$$\begin{bmatrix} n \\ k \end{bmatrix}_q = \frac{(q^n - 1)(q^n - q) \cdots (q^n - q^{k-1})}{(q^k - 1)(q^k - q) \cdots (q^k - q^{k-1})}.$$

Factoring out all possible q from the numerator and denominator gives

$$\begin{bmatrix} n \\ k \end{bmatrix}_q = \frac{(q^n - 1)(q^{n-1} - 1) \cdots (q^{n-k+1} - 1)}{(q^k - 1)(q^{k-1} - 1) \cdots (q - 1)}. \tag{6.13}$$

Similarly, we have

$$\begin{bmatrix} n \\ n-k \end{bmatrix}_q = \frac{(q^n - 1)(q^n - q) \cdots (q^n - q^{n-k-1})}{(q^{n-k} - 1)(q^{n-k} - q) \cdots (q^{n-k} - q^{n-k-1})}.$$

Factoring out all possible q from the numerator and denominator gives

$$\begin{bmatrix} n \\ n-k \end{bmatrix}_q = \frac{(q^n - 1)(q^{n-1} - 1) \cdots (q^{k+1} - 1)}{(q^{n-k} - 1)(q^{n-k-1} - 1) \cdots (q - 1)}. \tag{6.14}$$

To show that the fraction in Eq. 6.13 is equal to the fraction in Eq. 6.14, simply cross multiply and get $\prod_{i=1}^{n}(q^i - 1)$. This gives the result.

Example 6.13 Let $n = 5$ and $k = 2$. Then

$$\begin{bmatrix} 5 \\ 2 \end{bmatrix}_3 = \frac{(3^5 - 1)(3^5 - 3)}{(3^2 - 1)(3^2 - 3)} = \frac{(3^5 - 1)(3^4 - 1)}{(3^2 - 1)(3 - 1)}$$

and

$$\begin{bmatrix} 5 \\ 3 \end{bmatrix}_3 = \frac{(3^5 - 1)(3^5 - 3)(3^5 - 3^2)}{(3^3 - 1)(3^3 - 3)(3^3 - 3^2)} = \frac{(3^5 - 1)(3^4 - 1)(3^3 - 1)}{(3^3 - 1)(3^2 - 1)(3 - 1)}.$$

It is clear that the two fractions are equal.

Let V be a subspace of $\mathbb{F}_q^n$ then a coset of the subspace is $\mathbf{w} + V = \{\mathbf{w} + \mathbf{v} \mid \mathbf{v} \in V\}$. We note that if $\mathbf{w} \in V$ then $\mathbf{w} + V = V$. We now show that the coset of a vector space is either the vector space itself or disjoint from the vector space.

Theorem 6.9 *If* $\mathbf{w} \notin V$ *then* $(\mathbf{w} + V) \cap V = \emptyset$.

Proof Assume $\mathbf{v} \in \mathbf{w} + V \cap V$. Then $\mathbf{v} = \mathbf{v}' + \mathbf{w}$ for some $\mathbf{v}' \in V$, which gives $\mathbf{w} = \mathbf{v} - \mathbf{v}'$ and then $\mathbf{w} \in V$ which contradicts the assumption that $\mathbf{w} \notin V$. $\square$

We shall now show that the cardinalities of all cosets of a vector space have the same cardinality as the vector space.

Theorem 6.10 *Let* V *be a subspace of* $\mathbb{F}_q^n$. *Then, for* $\mathbf{w} \in \mathbb{F}_q^n$, $|V| = |\mathbf{w} + V|$.

Proof Consider the map $f : V \rightarrow \mathbf{w} + V$, where $f(\mathbf{v}) = \mathbf{w} + \mathbf{v}$. We shall show that this map is a bijection which shows that $|V| = |\mathbf{w} + V|$. Let $\mathbf{v}_1, \mathbf{v}_2$ be elements of V. If $f(\mathbf{v}_1) = f(\mathbf{v}_2)$ then $\mathbf{w} + \mathbf{v}_1 = \mathbf{w} + \mathbf{v}_2$ and therefore $\mathbf{v}_1 = \mathbf{v}_2$. Therefore, f is an injection. Given an element of $\mathbf{w} + V$, it is of the form $\mathbf{w} + \mathbf{v}$ by definition and so $f(\mathbf{v}) = \mathbf{w} + \mathbf{v}$ and the map f is a surjection. Therefore, f is a bijection and we have the result. $\qquad \square$

Finally, we have that the ambient space consists of the vector space and its cosets, all of which have the same cardinality.

Corollary 6.2 *The cosets of a vector space form a partition of the ambient space.*

Proof The result follows from the fact that any two cosets are disjoint or identical and that every element $\mathbf{v}$ is in $\mathbf{v} + V$, where V is the vector space. $\qquad \square$

Example 6.14 A line through the origin is necessarily a vector subspace of a Euclidean plane. Its cosets are the lines in a parallel class. These lines partition the plane.

Example 6.15 Consider the subspace of $\mathbb{F}_3^2$ of dimension 1, $V = \{(0,0), (1,2), (2,1)\}$. There are two cosets of this vector space $(1,1) + V = \{(1,1), (2,0), (0,2)\}$ and $(2,2) + V = \{(2,2), (0,1), (1,0)\}$. These three sets partition the space $\mathbb{F}_3^2$.

Exercises

1. Determine if the following set of vectors is linearly independent over $\mathbb{F}_3$:

$$(1,2,1), (2,0,1), (2,2,0).$$

2. Prove that $\mathbb{F}_q^n$ is a vector space for any field F.
3. Find a basis for $\mathbb{F}_q^n$ for any field F.
4. Prove that any vector space has a basis.
5. Prove that $\{\mathbf{v}_1, \mathbf{v}_2, \ldots, \mathbf{v}_k\}$ is a linearly dependent set if and only if $\mathbf{v}_i$ can be written as a linear combination of $\mathbf{v}_j$, where $j \neq i$.
6. Prove that no basis of a vector space V over $\mathbb{F}_q$ is unique unless $q = 2$ and $\dim(V) = 1$.
7. Prove that any linearly independent set of vectors in a vector space V can be extended to a basis for V.
8. Prove that any spanning set for a vector space V contains a basis for V.
9. Prove that the set $V = \{(x,y) \mid y = mx\}$ is a subspace of $\mathbb{F}_q^3$ for any $m \in F$. Determine the dimension of V.

10. Prove that the set $V = \{(x, y, z) \mid \alpha x + \beta y + \gamma z = 0\}$ is a subspace of $\mathbb{F}_q^3$ for any $\alpha, \beta, \gamma \in F$. Determine the dimension of V. Determine what geometric object this vector is.
11. Prove that the intersection of two vector spaces is a vector space.
12. Prove that any two bases for a vector space have the same number of vectors.
13. Prove that $\mathbb{F}_q^n$ has dimension n.
14. Prove the identity in Theorem 6.4, by computing the right side and adding them with a common denominator.
15. Prove that if T is a linear transformation $T : V \to W$ then $T(\mathbf{0}) = \mathbf{0}$.
16. Let $V = \mathbb{F}_q^n$ and let M be an r by n matrix with entries from a field F. Show that $T : \mathbb{F}_q^n \to \mathbb{F}_q^r$ defined by $T(v) = Mv$ is a linear transformation.

6.2   Affine Geometry

We shall now describe classical Euclidean geometry in terms of the linear algebra presented in the previous section. Euclidean refers to the fact that in these geometries there exists parallel lines on a plane. The term affine is usually used instead of Euclidean and we shall adopt this term.

While we can begin with any n-dimensional vector space over the field F, we shall begin with the space $\mathbb{F}_q^n$ for simplicity. We shall describe the n-dimensional affine geometry over the field $\mathbb{F}_q$ which is denoted by $AG_n(\mathbb{F}_q)$.

Definition 6.6 If a set is either a k-dimensional subspace or a coset of a k-dimensional vector space then we say that it is a k-flat.

Affine Geometry $AG_n(\mathbb{F}_q)$

- The points of $AG_n(\mathbb{F}_q)$ are the elements of $\mathbb{F}_q^n$.
- The lines of the space are the one-dimensional subspaces and their cosets, namely, the 1-flats.
- The planes of the space are the two-dimensional subspaces and their cosets, namely, the 2-flats.
- The k-dimensional geometries are the k-dimensional subspaces and their cosets, namely, the k-flats.
- Hyperplanes are $n - 1$-dimensional subspaces and their cosets, namely, $n - 1$-flats.
- A geometric object A is incident with a geometric object B if and only if $A \subset B$ or $B \subset A$. Namely, incidence is given by set-theoretic containment.

We note that by Corollary 6.2, the k-flats of a single subspace partition the ambient space for any k.

As an example we shall take a look at the space $AG_3(\mathbb{F}_2)$. There are $2^3 = 8$ points in the space. They are

$$(0,0,0), (0,0,1), (0,1,0), (0,1,1), (1,0,0), (1,0,1), (1,1,0), (1,1,1).$$

By Theorem 6.7, we can find the two-dimensional subspaces by taking the orthogonals of the one-dimensional subspaces corresponding to the seven non-zero points. By Theorem 6.2 we know that each two-dimensional subspace has four points in it. Theorem 6.10 gives that the cardinality of a coset is the same as the cardinality of the subspace. Therefore, we know there can be only one coset for each two-dimensional subspace, specifically the complement of the subspace in the space $\mathbb{F}_2^3$. We could have counted the number of two-dimensional subspaces using Theorem 6.3 which, with $n = 3, k = 2$ and $q = 2$, would give that there are

$$\frac{(2^3 - 1)(2^3 - 2)}{(2^2 - 1)(2^2 - 2)} = \frac{7(6)}{3(2)} = 7$$

subspaces of dimension 2. This gives that there are 14 planes in $AG_3(\mathbb{F}_2)$, 7 corresponding to the subspaces and 7 to their cosets. We list them by describing the subspace as an orthogonal and then listing its complement.

$$P_1 = \langle(0,0,1)\rangle^\perp = \{(1,1,0), (1,0,0), (0,1,0), (0,0,0)\}$$

$$P_2 = P_1^c = \{(1,1,1), (1,0,1), (0,1,1), (0,0,1)\}$$

$$P_3 = \langle(0,1,0)\rangle^\perp = \{(1,0,0), (0,0,1), (1,0,1), (0,0,0)\}$$

$$P_4 = P_3^c = \{(1,1,0), (0,1,1), (1,1,1), (0,1,0)\}$$

$$P_5 = \langle(0,1,1)\rangle^\perp = \{(1,0,0), (0,1,1), (1,1,1), (0,0,0)\}$$

$$P_6 = P_5^c = \{(1,1,0), (0,0,1), (1,0,1), (0,1,0)\}$$

$$P_7 = \langle(1,0,0)\rangle^\perp = \{(0,1,1), (0,1,0), (0,0,1), (0,0,0)\}$$

$$P_8 = P_7^c = \{(1,1,1), (1,1,0), (1,0,1), (1,0,0)\}$$

$$P_9\langle(1,0,1)\rangle^\perp = \{(0,1,0), (1,0,1), (1,1,1), (0,0,0)\}$$

$$P_{10} = P_9^c = \{(1,1,0), (0,0,1), (0,1,1), (0,1,1)\}$$

$$P_{11} = \langle(1,1,0)\rangle^\perp = \{(0,0,1), (1,1,0), (1,1,1), (0,0,0)\}$$

$$P_{12} = P_{11}^c = \{(1,0,1), (0,1,0), (0,1,1), (1,0,0)\}$$

Table 6.1 Incidence matrix for the points and planes in $AG_3(\mathbb{F}_2)$

	$(0,0,0)$	$(0,0,1)$	$(0,1,0)$	$(0,1,1)$	$(1,0,0)$	$(1,0,1)$	$(1,1,0)$	$(1,1,1)$
P_1	1	0	1	0	1	0	1	0
P_2	0	1	0	1	0	1	0	1
P_3	1	1	0	0	1	1	0	0
P_4	0	0	1	1	0	0	1	1
P_5	1	0	0	1	1	0	0	1
P_6	0	1	1	0	0	1	1	0
P_7	1	1	1	1	0	0	0	0
P_8	0	0	0	0	1	1	1	1
P_9	1	0	1	0	0	1	0	1
P_{10}	0	1	0	1	1	0	1	0
P_{11}	1	1	0	0	0	0	1	1
P_{12}	0	0	1	1	1	1	0	0
P_{13}	1	0	0	1	0	1	1	0
P_{14}	0	1	1	0	1	0	0	1

$$P_{13} = \langle (1,1,1) \rangle^{\perp} = \{(1,1,0),(0,1,1),(1,0,1),(0,0,0)\}$$

$$P_{14} = P_{13}^c = \{(1,1,1),(0,1,0),(1,0,0),(0,0,1)\}.$$

Looking at the entries in Table 6.1 we see that there are 8 points, 14 planes and each point is on exactly 7 planes and each plane has exactly 4 points on it.

The lines in the space are the one-dimensional subspaces and their cosets. A one-dimensional subspace has two elements and so each subspace has four cosets (including itself), since four mutually exclusive sets of size 2 would make eight points in total. The number of one-dimensional subspaces is seven since there is one corresponding to each non-zero point. Alternatively, we could have used Theorem 6.3 which, with $n = 3, k = 1$ and $q = 2$, would give that there are

$$\frac{(2^3 - 1)}{(2^1 - 1)} = \frac{7}{1} = 7$$

subspaces of dimension 1. Hence, there are 28 lines in $AG_3(\mathbb{F}_2)$.

A plane is a two-dimensional subspace or a coset of a two-dimensional subspace. Given two planes their intersection can either be two dimensional, if they coincide, zero dimensional if they meet in a point or if they are parallel (or skew in higher dimensions), or one dimensional if it is a line. For example, consider the following two planes in $AG_5(\mathbb{F}_2)$, $\pi_1 = \langle (1,0,0,0,0),(0,1,0,0,0) \rangle$ and $\pi_2 = \langle (0,0,1,0,0),(0,0,0,1,0) \rangle$. It is clear that these two planes have only the origin in common. Consider the coset of π_2 that is $\pi_2 + (0,0,0,0,1)$. This plane does not intersect π_1 at all nor is it a coset of π_1.

Theorem 6.11 *Let* p_1, p_2 *be any two points in* $AG_n(\mathbb{F}_q)$. *Then there is a unique line through* p_1 *and* p_2.

Proof Let $\mathbf{v}$ be the vector $(p_1 - p_2)$ and let $\mathbf{w}$ be the vector $(p_1 - \mathbf{0})$. Let $V = \langle \mathbf{v} \rangle$. Then V is a one-dimensional subspace. Let $L = \mathbf{w} + V$. Then $p_1 = \mathbf{w} + 0\mathbf{v} \in \mathbf{w} + V$ and $p_2 = \mathbf{w} - \mathbf{v} \in \mathbf{w} + V$. Therefore, $p_1, p_2 \in L$, and L is a line.

If M is any other line containing p_1 and p_2 then M must be of the form $\mathbf{u} + (p_1 - p_2)$. The cosets of V are disjoint, therefore since M and L are not disjoint, we have that $L = M$.   □

Example 6.16 Consider the two points $(0, 1, 2)$ and $(2, 1, 1)$ in $AG_3(\mathbb{F}_3)$. We have that the difference of the two points gives the vector $\mathbf{v} = (2, 0, 2)$ and so the vector space $V = \langle (2, 0, 2) \rangle = \{(0, 0, 0), (2, 0, 2), (1, 0, 1)\}$. The line is $(0, 1, 2) + V = \{(0, 1, 2), (2, 1, 1), (1, 1, 0)\}$. Notice the line has three points and contains the two points $(0, 1, 2)$ and $(2, 1, 1)$.

We can give the standard geometric definition of parallel and skew lines.

Definition 6.7 Two lines that are on the same plane that are disjoint or equal are said to be parallel. Two lines that are not on the same plane that are disjoint are said to be skew lines.

Theorem 6.12 *Any three non-collinear points in* $AG_n(\mathbb{F}_q)$ *are incident with a plane.*

Proof Let p_1, p_2, and p_3 be three non-collinear points. Let $\mathbf{v}$ be the vector $(p_1 - p_2)$, $\mathbf{w}$ be the vector $(p_1 - p_3)$, and $\mathbf{u}$ be the vector $(p_1 - \mathbf{0})$. Since the points are not collinear we have that $V = \langle \mathbf{v}, \mathbf{w} \rangle$ is a two-dimensional subspace of $\mathbb{F}_q^n$. Then let $P = \mathbf{u} + V$. We have $p_1 = \mathbf{u} + 0\mathbf{v}$, $p_2 = \mathbf{u} - \mathbf{v}$ and $p_3 = \mathbf{u} - \mathbf{w}$. Therefore, all three points are on P and P is the coset of a two-dimensional subspace. Hence, the three points are on a plane.   □

Theorem 6.13 *Any two lines in* $AG_n(\mathbb{F}_q)$ *either intersect at a point are parallel or skew.*

Proof Assume that the two lines are neither skew nor parallel. If the two lines share two points p_1 and p_2, then there is a unique line through p_1 and p_2 giving that the lines are equal. Hence, they can only intersect in a point.   □

We shall now count the number of k-dimensional subspaces of affine space.

Theorem 6.14 *The number of* k-*dimensional spaces in* $AG_n(\mathbb{F}_q)$ *is* $\begin{bmatrix} n \\ k \end{bmatrix}_q q^{n-k}$.

Proof A k-dimensional space is either a k-dimensional subspace of $\mathbb{F}_q^n$ or a coset of such a space. There are $\begin{bmatrix} n \\ k \end{bmatrix}_q$ such subspaces and each has q^{n-k} distinct cosets. This gives the result. □

Example 6.17 In $AG_n(\mathbb{F}_q)$, we have the following counts of spaces:

- Let $k = 0$. The number of points is q^n.
- Let $k = 1$. The number of lines is $\frac{q^n-1}{q-1}q^{n-1}$.
- Let $k = 2$. The number of planes is $\frac{(q^n-1)(q^n-q)}{(q^2-1)(q^2-q)}q^{n-2}$.
- Let $k = n - 1$. The number of hyperplanes is $\begin{bmatrix} n \\ k \end{bmatrix}_q q^{n-(n-1)} = \frac{q^n-1}{q-1}q$.

We note that, in general, the number of k-dimensional spaces is not equal to the number of $n - k$-dimensional spaces.

The Game of Set

The game of Set™ is a well-known mathematics game. The game evolved from a study of genetics rather than from mathematics. However, it is really an idea that we have just discussed. It is played with a collection of cards with four attributes: shape (ovals, squiggles, diamonds), color (red, purple, green), number (one, two, three), and shading (solid, striped, outlined). A set consists of three cards in which each of the cards' features is the same on each card or is different on each card. To begin 12 cards are placed down and the first player that finds a set calls "set" and removes the three cards in that set. These cards are then replaced and the play continues until the entire deck is used. There are 81 cards in the deck.

Notice that given any two cards in a Set deck there is a unique card that completes the set (prove this in Exercise 2). It follows that there are 1080 sets (see Exercise 4). Consider each of the three choices for each attributes as 0, 1, or 2 assigned arbitrarily. Then each card represents an element in $\mathbb{F}_3^4$, which gives 81 cards. Therefore, we can associate each card with a point in $AG_4(\mathbb{F}_3)$.

In $\mathbb{F}_3^4$ there are $\begin{bmatrix} 4 \\ 1 \end{bmatrix}_3 = 40$ one-dimensional subspaces. Each of these has 27 cosets in $\mathbb{F}_3^4$. Therefore, there are 1080 lines in $AG_4(\mathbb{F}_3)$. This is, of course, the exact number of sets in the game. Moreover, we know that any two cards complete to a unique set and any two points complete to a unique line. Therefore, the game of Set consists in trying to identify lines in $AG_4(\mathbb{F}_3)$. For a complete description of the mathematics of this game see "The joy of Set. The many mathematical dimensions of a seemingly simple card game" [65].

Exercises

1. Give the incidence matrix for points and lines in $AG_2(\mathbb{F}_3)$.
2. Prove that given any two cards in a Set deck there is a unique card that completes the set.
3. Determine the number of points, lines, and planes in $AG_3(\mathbb{F}_3)$.
4. Prove that there are 1080 sets in the game of Set.
5. Determine the number of points and hyperplanes in $AG_4(\mathbb{F}_5)$.
6. Prove that the construction of $AG_2(\mathbb{F}_q)$ given in this section is equivalent to the construction of the affine plane given in the section on planes constructed from fields.
7. Prove that every point in $AG_n(\mathbb{F}_q)$ is on a k-flat.
8. Give an example of skew lines in $AG_4(\mathbb{F}_2)$.
9. Give an example of parallel planes in $AG_4(\mathbb{F}_2)$.

6.3   Projective Geometry

We shall now describe projective geometry, which has a canonical description in terms of linear algebra as well.

To describe n-dimensional projective space, $PG_n(\mathbb{F}_q)$, we use the ambient space $\mathbb{F}_q^{n+1}$, that is, $n + 1$-dimensional space over the field F. The k-dimensional projective space will be described in terms of vectors spaces of this ambient space. In general, the projective dimension of a geometric object is 1 less than the dimension as a vector space of the associated space.

Projective Geometry $PG_n(\mathbb{F}_q)$

- The points in $PG_n(\mathbb{F}_q)$ are the one-dimensional subspaces of $\mathbb{F}_q^{n+1}$.
- The lines in $PG_n(\mathbb{F}_q)$ are the two-dimensional subspaces of $\mathbb{F}_q^{n+1}$.
- The planes in $PG_n(\mathbb{F}_q)$ are the three-dimensional subspaces of $\mathbb{F}_q^{n+1}$.
- The hyperplanes in $PG_n(\mathbb{F}_q)$ are the n-dimensional subspaces of $\mathbb{F}_q^{n+1}$.
- A projective geometric object A is incident with a projective object B if A is a subspace of B or B is a subspace of A.

In general, an object with projective dimension s corresponds to an $s + 1$-dimensional subspace. Notice that there is no mention of flats or cosets here. In projective space, all of the objects are subspaces of $\mathbb{F}_q^{n+1}$, that is, cosets are not projective geometric objects.

Theorem 6.3 gives that the number of k-dimensional subspaces of a vector space of dimension n is

$$\frac{(q^n - 1)(q^n - q)\cdots(q^n - q^{k-1})}{(q^k - 1)(q^k - q)\cdots(q^k - q^{k-1})}. \tag{6.15}$$

This gives the following theorem.

Theorem 6.15 *In* $PG_n(\mathbb{F}_q)$, *the number of objects with projective dimension* k *is*

$$\begin{bmatrix} n+1 \\ k+1 \end{bmatrix}_q = \frac{(q^{n+1}-1)(q^{n+1}-q)\cdots(q^{n+1}-q^k)}{(q^{k+1}-1)(q^{k+1}-q)\cdots(q^{k+1}-q^k)}. \tag{6.16}$$

Proof We note that an object with projective dimension k is a subspace of dimension $k+1$ in an $n+1$-dimensional space. Then Eq. 6.15 gives the result. □

As an example, the number of points ($k=0$) in $PG_2(\mathbb{F}_q)$ is $\frac{q^3-1}{q-1} = q^2+q+1$ and the number of lines ($k=1$) is $\frac{(q^3-1)(q^3-q)}{(q^2-1)(q^2-q)} = q^2+q+1$. Consider the points in $PG_2(\mathbb{F}_3)$. They are the one-dimensional vector spaces in $\mathbb{F}_3^3$. The number of points is 3^2+3+1 and they are $\langle(0,0,1)\rangle$, $\langle(0,1,0)\rangle$, $\langle(0,1,1)\rangle$, $\langle(0,1,2)\rangle$, $\langle(1,0,0)\rangle$, $\langle(1,0,1)\rangle$, $\langle(1,0,2)\rangle$, $\langle(1,1,0)\rangle$, $\langle(1,1,1)\rangle$, $\langle(1,1,2)\rangle$, $\langle(1,2,0)\rangle$, $\langle(1,2,1)\rangle$, and $\langle(1,2,2)\rangle$.

Example 6.18 In $PG_4(\mathbb{F}_3)$, we have the following counts of spaces:

- There are $\begin{bmatrix} 5 \\ 1 \end{bmatrix}_3 = 121$ points.
- There are $\begin{bmatrix} 5 \\ 2 \end{bmatrix}_3 = 1210$ lines.
- There are $\begin{bmatrix} 5 \\ 3 \end{bmatrix}_3 = 1210$ planes.
- There are $\begin{bmatrix} 5 \\ 4 \end{bmatrix}_3 = 121$ hyperplanes.

Incidence is given by set-theoretic containment, so that two spaces A and B are incident if and only if $A \subset B$ or $B \subset A$. We can now count the number of such incidences.

If we fix the bigger space and count how many of the smaller spaces are on it we can use the following slight adjustment of the previous theorem.

Theorem 6.16 *In* $PG_n(\mathbb{F}_q)$ *the number of objects with projective dimension* k *in a projective space of dimension* s, *with* $s > k$, *is*

$$\frac{(q^{s+1}-1)(q^{s+1}-q)\ldots(q^{s+1}-q^k)}{(q^{k+1}-1)(q^{k+1}-q)\ldots(q^{k+1}-q^k)}. \tag{6.17}$$

Proof We are simply counting the number of $k+1$-dimensional subspaces of an $s+1$-dimensional vector space. □

If we fix the smaller space and count how many bigger spaces contain it then we use the following theorem.

Theorem 6.17 *If A has projective dimension* k *in* $PG_n(F_q)$, *then the number of* $s > k$ *projective dimensional objects it is incident with is*

$$\frac{(q^{n+1} - q^{k+1})(q^{n+1} - q^{k+2}) \cdots (q^{n+1} - q^s)}{(q^{s+1} - q^{k+1})(q^{s+1} - q^{k+2}) \cdots (q^{s+1} - q^s)}. \tag{6.18}$$

Proof We are looking for the number of $s + 1$-dimensional subspaces of $\mathbb{F}_q^{n+1}$ containing a specific subspace of dimension $k + 1$. So the number of ways of picking an element not in the $k + 1$-dimensional space is $q^{n+1} - q^{k+1}$. Then we pick an element not in the space generated by the $k + 1$-dimensional space and the new vector. We continue by induction until we have chosen $(s + 1) - (k + 1) = s - k$ vectors that are linearly independent over the $k + 1$-dimensional subspace. The denominator counts the number of different basis for the larger space over the smaller one. $\qquad\square$

As a simple example, we can count the number of lines that are incident with a point, i.e., $s = 1$, $k = 0$, $n = 2$. Equation 6.18 gives that the number is $\frac{q^3 - q}{q^2 - q} = q + 1$ as expected.

Theorem 6.18 *Through any two points in* $PG_n(\mathbb{F}_q)$ *there exists a unique line.*

Proof Let p, q be two points with $p = \langle \mathbf{v} \rangle$, $q = \langle \mathbf{w} \rangle$ with $p \neq q$, i.e., $\mathbf{v} \notin \langle \mathbf{w} \rangle$, $\mathbf{w} \notin \langle \mathbf{v} \rangle$. Then the two-dimensional subspace $\langle \mathbf{v}, \mathbf{w} \rangle$ contains p and q. Any other line through p and q must be two dimensional and contain $\mathbf{v}$ and $\mathbf{w}$ and hence must be this line. $\qquad\square$

It is possible for two lines to not meet in $PG_n(\mathbb{F}_q)$. For example, in $PG_4(\mathbb{F}_2)$, the lines $\langle (1, 0, 0, 0, 0), (0, 1, 0, 0, 0) \rangle$ and $\langle (0, 0, 0, 1, 0), (0, 0, 0, 0, 1) \rangle$ do not share a common one-dimensional subspace.

It is true, however, that in any plane the lines must meet. That is, given any 2 two-dimensional spaces, they cannot have only a trivial intersection since together they would generate a four-dimensional space contained in a three-dimensional space. They cannot have a two-dimensional intersection since they would be the same line, so their intersection must be a one-dimensional subspace, that is, a point. This idea generalizes to the following.

Theorem 6.19 *Two objects of projective dimensions* s *and* s' *must intersect if they are contained in a space of projective dimension* $s + s'$.

Proof Let M and M' be s and s' projective dimensional objects. Then M is an $s + 1$-dimensional subspace and M' is an $s' + 1$-dimensional subspace. If they have only a trivial intersection then together they generate an $s + s' + 2$-dimensional space, but they are contained in an $s + s' + 1$-dimensional subspace which is a contradiction. Hence, they must have a non-trivial intersection. $\qquad\square$

As an example, consider the space $PG_3(\mathbb{F}_2)$. In this space, there are 15 points, 35 lines, and 15 planes. There are 5 lines through each point and 3 points on each line. Each plane consists of 7 points and through each point there are 7 planes. Through each line there are 3 planes and each plane contains 8 lines.

These seemingly very abstract ideas about affine and projective planes are actually very useful in a practical engineering sense. In [4] (and elsewhere), it is described how to use these spaces in the construction of error-correcting codes.

Theorem 6.20 *The number of objects with projective dimension* r *is equal to the number of objects with projective dimension* $n - r - 1$ *in* $PG_n(\mathbb{F}_q)$.

Proof Each projective r-dimensional object corresponds to an $r + 1$-dimensional vector space. The orthogonal of this vector space has dimension $n + 1 - (r + 1) = n - r$. These $n - r$-dimensional subspaces correspond to projective $n - r - 1$-dimensional objects. Then, since the map that sends a vector space to its orthogonal is a bijection, we have that the number of objects with projective dimension r is equal to the number of objects with projective dimension $n - r - 1$ in $PG_n(\mathbb{F}_q)$. $\qquad\square$

This theorem gives that the number of points is equal to the number of lines on the projective plane $PG_2(\mathbb{F}_q)$.

Example 6.19 In $PG_3(\mathbb{F}_2)$, there are 15 points, 35 lines, and 15 planes. Each line has 3 points on it and each plane has 7 points and 7 lines. Each line is incident with 3 planes.

Theorem 6.21 *Any two lines in* $PG_2(\mathbb{F}_q)$ *must intersect.*

Proof Let L and M be two lines. These correspond to two-dimensional vector spaces V and W. If the lines do not intersect then the dimension of the intersection of V and W must be 0. However, in this case, $\langle V, W \rangle$ has dimension 4 but sits inside $\mathbb{F}_q^3$ which is a contradiction. Therefore, L and M must intersect. $\qquad\square$

Theorem 6.22 *Any three non-collinear distinct points are incident with a unique plane in* $PG_n(\mathbb{F}_q)$.

Proof Let $\langle \mathbf{u} \rangle, \langle \mathbf{v} \rangle, \langle \mathbf{w} \rangle$ be three non-collinear points. Since they are distinct $\langle \mathbf{u}, \mathbf{v} \rangle$ is a two-dimensional subspace. Then since they are collinear $\langle \mathbf{u}, \mathbf{v}, \mathbf{w} \rangle$ cannot be two dimensional so it must be three dimensional. Therefore, $\langle \mathbf{u}, \mathbf{v}, \mathbf{w} \rangle$ is a plane containing all three points. $\qquad\square$

Exercises

1. Give the incidence matrix of points and lines in $PG_2(\mathbb{F}_3)$.
2. Give the incidence matrix of points and planes in $PG_3(\mathbb{F}_2)$.
3. Count the number of lines in $PG_3(\mathbb{F}_5)$.
4. Determine the number of points, lines, planes, and the number of each incidences for $PG_3(\mathbb{F}_3)$.
5. Prove that every point in $PG_n(\mathbb{F}_q)$ is contained in a k-dimensional object.
6. Determine the number of planes containing a given line in $PG_3(\mathbb{F}_q)$.
7. Determine how large n must be there for two planes to exist that do not intersect in $PG_n(\mathbb{F}_q)$.
8. Prove that the construction of $PG_2(\mathbb{F}_q)$ given in this section is equivalent to the construction of the projective plane given in the section on planes constructed from fields.

6.4   Desargues' Theorem

One of the most interesting and important configurational theorems for projective planes is Desargues' theorem. Desargues' Theorem does not apply in all projective planes. Those planes in which it does hold are known as Desarguesian planes. These are precisely the classical planes that we constructed from finite fields, that is, $PG_2(\mathbb{F}_q)$.

Theorem 6.23 (Desargues' Theorem) *Let* p, q, r *and* p', q', r' *be two sets of distinct non-collinear points in the projective plane* $PG_2(\mathbb{F}_q)$. *That is,* p, q, r *and* p', q', r' *are triangles. Denote the line opposite a point by its capital letter. Let* L_1 *denote the line through* p *and* p', L_2 *denote the line through* q *and* q', *and* L_3 *denote the line through* r *and* r'. *The lines* L_1, L_2 *and* L_3 *meet in a point if and only if the intersection points of* P *with* P', Q *with* Q' *and* R *with* R' *are incident with a line.*

Proof We assume that all points and lines in the theorem are distinct to avoid the trivial cases. The duality principle implies that we need only prove one direction of the if and only if theorem.

We assume the lines L_1, L_2, and L_3 meet in a point. That is, we assume $L_1 \cap L_2 \subseteq L_3$. Recall that each line is a two-dimensional subspace of $\mathbb{F}_q^3$. Let $\mathbf{v} \in R \cap R'$, that is, $\mathbf{v}$ is a vector in the one-dimensional subspace that is contained in both the space R and R', where R is the line through p and q and R' is the line through p' and q'. The vector $\mathbf{v}$ can be expressed as $\mathbf{v}_p + \mathbf{v}_q$, where $\mathbf{v}_p$ is a vector in the one-dimensional space p and $\mathbf{v}_q$ is a vector in the one-dimensional space q. This is because $\mathbf{v}$ is on the line between p and q. In the same way, the vector $\mathbf{v}$ can be expressed as $\mathbf{v}_{p'} + \mathbf{v}_{q'}$ where $\mathbf{v}_{p'}$ is a vector in the one-dimensional space p' and $\mathbf{v}_{q'}$ is a vector in the one-dimensional space q'. Then we have

$$\mathbf{v}_p - \mathbf{v}_{p'} = \mathbf{v}_{q'} - \mathbf{v}_q \in \langle p, p' \rangle \cap \langle q, q' \rangle = L_1 \cap L_2 \subseteq L_3 = \langle r, r' \rangle.$$

This follows since $\mathbf{v}_p - \mathbf{v}_{p'} \in L_1$ and $\mathbf{v}_{q'} - \mathbf{v}_q \in L_2$ and $\mathbf{v}_p - \mathbf{v}_{p'} = \mathbf{v}_{q'} - \mathbf{v}_q$ so the vector is in the intersection. We know the intersection is contained in L_3 which is generated by r and r'. Thus, for some $\mathbf{v}_r \in r, \mathbf{v}_{r'} \in r'$ we have

$$\mathbf{v}_p - \mathbf{v}_{p'} = \mathbf{v}'_q - \mathbf{v}_q = \mathbf{v}_r + \mathbf{v}_{r'}.$$

We shall show that $\mathbf{v} = \mathbf{v}_p + \mathbf{v}_q = (\mathbf{v}_p - \mathbf{v}_r) + (\mathbf{v}_q + \mathbf{v}_r) \in \langle (Q \cap Q'), (P \cap P') \rangle$. First, $\mathbf{v}_p - \mathbf{v}_r = \mathbf{v}_{p'} + \mathbf{v}_{r'}$ and so $\mathbf{v}_p - \mathbf{v}_r \in Q \cap Q'$. Then, $\mathbf{v}_q + \mathbf{v}_r = \mathbf{v}_{q'} - \mathbf{v}_{r'}$ therefore $\mathbf{v}_q + \mathbf{v}_r \in P \cap P'$. We have $(\mathbf{v}_p - \mathbf{v}_r) + (\mathbf{v}_q + \mathbf{v}_r) = \mathbf{v}_p + \mathbf{v}_q \in R \cap R'$. Therefore, the intersection of R and R' lies on the line between the intersection of Q with Q' and the intersection of P with P' and we have the result. □

We note that although the theorem is geometric in its statement, its proof is highly algebraic.

We can now show that Desargues' theorem is true in any space which has at least three dimensions. We give the proof as it was given in [43].

Theorem 6.24 (Multidimensional Desargues' Theorem) *Let* p, q, r *and* p', q', r' *be two sets of distinct non-collinear points, that is,* p, q, r *and* p', q', r' *are triangles, such that the triangles lie on two different planes. Denote the line opposite a point by its capital letter. Let* L_1 *denote the line through* p *and* p', L_2 *denote the line through* q *and* q' *and* L_3 *denote the line through* r *and* r'. *Assume the lines* L_1, L_2 *and* L_3 *meet in a point* t, *then the intersection points of* P *with* P', Q *with* Q' *and* R *with* R' *are incident with a line.*

Proof The line between p and q and the line between p' and q' both lie in the plane through t, p, and q and intersect in a point s_1. Similarly, the line between p and r and the line between p' and r' intersect in a point s_2 and the line between r and q and the line between r' and q' intersect in a point s_3. The points s_1, s_2, and s_3 must be incident both with the plane containing p, q, and r and the plane containing p', q', and r' and therefore on M which is the line of intersection of these two planes. Therefore, the lines through s_3 and s_2 and through s_1 and s_2 and s_1 and s_3 coincide with this intersection and so the points s_1, s_2, and s_3 are collinear. □

There is another theorem that is equivalent to Desargues' theorem which is also sometimes used to characterize those planes that come from a finite field. This theorem is known as Pappus' theorem. We state the theorem.

Theorem 6.25 *Let* L *and* M *be two distinct lines in the projective plane* $PG_2(\mathbb{F}_q)$, *where* p_1, p_2, *and* p_3 *are distinct points on* L *and* q_1, q_2, *and* q_3 *are distinct points on* M, *where none of these six points are the point of intersection of* L *and* M. *Let*

r_1 *be the point of intersection of the line through* p_1 *and* q_2 *and the line through* q_1 *and* p_2; *let* r_2 *be the point of intersection of the line through* p_1 *and* q_3 *and the line through* q_1 *and* p_3; *and let* r_3 *be the point of intersection of the line through* p_2 *and* q_3 *and the line through* q_2 *and* p_3. *Then the points* $r_1, r_2,$ *and* r_3 *are collinear.*

The configuration in the theorem is known as the Pappus configuration.

Any plane in which Desargues' theorem holds is known as Desarguesian and any plane in which Pappus' theorem holds is known as Pappian.

The following can be found in Pickert's 1955 text [67] or more recently in Stevenson's 1972 text [83].

Theorem 6.26 *Every Pappian plane is Desarguesian.*

Exercises

1. Verify that the unique projective plane of order 2 satisfies both Desargues' theorem and Pappus' theorem.
2. Verify that the unique projective plane of order 3 satisfies both Desargues' theorem and Pappus' theorem.

6.5   The Bruck–Ryser Theorem

There are many open problems in terms of the existence of finite projective planes. Constructively, we know that we can create a projective plane of order n if n is a power of a prime, but there is really only one strong theorem eliminating orders as possible orders for projective planes, namely, the Bruck–Ryser theorem. This theorem first appeared in [17]. We shall describe this proof in this section.

Rather than giving the original proof by Bruck and Ryser, we shall follow the proof given in Hughes and Piper which is simpler and requires less knowledge of number theory.

We state two lemmas that we will require. Their proofs can be found in any number theory book.

Lemma 6.5 *If* n *is a positive integer then* n *can be written as* $n = a^2 + b^2 + c^2 + d^2$, *where* $a, b, c,$ *and* d *are integers.*

Lemma 6.6 *If* n *can be written as* $n = q^2 + r^2$ *where* q, r *are rational numbers then* n *can be written as* $n = a^2 + b^2$ *where* a, b *are integers.*

We can now state and prove the well-known Bruck–Ryser theorem first proven in [17].

Theorem 6.27 (Bruck–Ryser Theorem) *Let* $n \equiv 1$ *or* $2 \pmod 4$. *If there exists a projective plane of order* n *then* $n = a^2 + b^2$ *for some integers* a, b.

Proof Let $\Pi = (\mathcal{P}, \mathcal{L}, \mathcal{I})$ be a projective plane of order n with $\mathcal{P} = \{p_1, \ldots, p_{n^2+n+1}\}$ and $\mathcal{L} = \{L_1, \ldots, L_{n^2+n+1}\}$.

Let x_i be an indeterminant and write

$$\Lambda_i = \sum_{j,(p_j,L_i)\in\mathcal{I}} x_j.$$

That is, we sum over all j for which p_j is incident with the line L_i. Hence, each Λ_i is the sum of $n + 1$ distinct x_j.

Then

$$\sum_{k=1}^{n^2+n+1} \Lambda_k^2 = (n+1) \sum_{k=1}^{n^2+n+1} x_k^2 + 2 \sum_{k\neq j} x_k x_j. \tag{6.19}$$

This is easy to see since each x_j appears in $n + 1$ different Λ_i. Hence, when the Λ_i are squared there are $n + 1$ occurrences of the square. Then x_i and x_j occur in the same Λ_i exactly once so when squaring you get $x_i x_j$ twice in the summation.

Then pulling out one $\sum_{k=1}^{n^2+n+1} x_k^2$ we get the following:

$$\sum_{k=1}^{n^2+n+1} \Lambda_k^2 = (n) \sum_{k=1}^{n^2+n+1} x_k^2 + \left(\sum_{k=1}^{n^2+n+1} x_k\right)^2. \tag{6.20}$$

Adding $nx_{n^2+n+2}^2$ to each side we get

$$\sum_{k=1}^{n^2+n+1} \Lambda_k^2 + nx_{n^2+n+2}^2 = (n) \sum_{k=1}^{n^2+n+2} x_k^2 + \left(\sum_{k=1}^{n^2+n+1} x_k\right)^2. \tag{6.21}$$

For the remainder of the section, let $n \equiv 1$ or $2 \pmod 4$ then $n^2 + n + 1 \equiv 3 \pmod 4$.

Lemma 6.5 gives that there are integers a, b, c, d with $n = a^2 + b^2 + c^2 + d^2$. Let M be the following matrix:

$$\begin{pmatrix} a & b & c & d \\ -b & a & d & -c \\ -c & -d & a & b \\ -d & c & -b & a \end{pmatrix}. \tag{6.22}$$

We note that $\det(M) = n^2$. The matrix M gives a linear transformation defined by $\Phi(v) = vM$.

We say that an integer t is represented by (w, x, y, z) if $t = w^2 + x^2 + y^2 + z^2$.

The following is an easy computation to verify

$$(a^2 + b^2 + c^2 + d^2)(w^2 + x^2 + y^2 + z^2)$$
$$= (aw - bx - cy - dz)^2 + (bw + ax - dy + cz)^2$$
$$+ (cw + dx + ay - bz)^2 + (dw - cx + by + az)^2.$$

Using this equation, we see that if t is represented by (w, x, y, z) then $(w, x, y, z)M$ represents tn.

Let (y_1, y_2, y_3, y_4) be the image of (x_1, x_2, x_3, x_4) under the linear transformation given by M, that is,

$$(y_1, y_2, y_3, y_4) = (x_1, x_2, x_3, x_4)M.$$

We have

$$n(x_1^2 + x_2^2 + x_3^2 + x_4^2) = (y_1^2 + y_2^2 + y_3^2 + y_4^2). \tag{6.23}$$

The determinant of M is non-zero and so there exists an inverse matrix M^{-1} which also gives a linear transformation. There is no guarantee that entries in M^{-1} are integers, but it is true that they must all be rational numbers. What we have is that

$$(y_1, y_2, y_3, y_4)M^{-1} = (x_1, x_2, x_3, x_4).$$

This gives that each x_i is written as a linear combination of the y_i where the coefficients in the linear combination are rational numbers.

Applying Eqs. 6.20 and 6.23 to the indeterminants

$$y_1, y_2, y_3, y_4, x_5, x_6, \ldots, x_{n^2+n+2},$$

we get

$$\sum_{k=1}^{n^2+n+1} \Lambda_k^2 + nx_{n^2+n+2}^2 = y_1^2 + y_2^2 + y_3^2 + y_4^2 + n\sum_{k=5}^{n^2+n+2} x_i^2 + \left(\sum_{k=1}^{n^2+n+1} x_i\right)^2. \tag{6.24}$$

Notice that $n^2 + n + 2 \equiv 0 \pmod 4$ so we can repeat this process and arrive at

$$\sum_{k=1}^{n^2+n+1} \Lambda_k^2 + nx_{n^2+n+2}^2 = \sum_{k=1}^{n^2+n+2} y_i^2 + \left(\sum_{k=1}^{n^2+n+1} y_i\right)^2, \tag{6.25}$$

where the indeterminates in Λ are y_i.

If the coefficient of y_1 in Λ_1 is -1 then put $\Lambda_1 = -y_1$ and otherwise put $L_1 = y_1$. This can be solved to get y_1 as a linear expression in the remaining y_i. Substituting into Eq. 6.25 we get

$$\sum_{k=2}^{n^2+n+1} \Lambda_k^2 + nx_{n^2+n+2}^2 = \sum_{k=2}^{n^2+n+2} y_i^2 + \left(\sum_{k=1}^{n^2+n+1} y_i\right)^2. \tag{6.26}$$

Continuing this we have

$$nx_{n^2+n+2}^2 = y_{n^2+n+2}^2 + \left(\sum_{k=1}^{n^2+n+1} y_i \right)^2. \qquad (6.27)$$

This gives that there are rational numbers α and β with

$$n\alpha^2 = 1 + \beta^2$$
$$n = \frac{1}{\alpha^2} + \left(\frac{\beta}{\alpha}\right)^2.$$

Lemma 6.6 now implies that n can be written as the sum of two integer squares, which gives the theorem. □

This theorem eliminates many possible orders. For example, 14 is 2 (mod 4) and 14 is not the sum of two squares, and hence there is no plane of order 14. Notice that a plane of order 10 is not eliminated by the theorem but there is no plane of order 10. Hence, the theorem is not a biconditional.

Exercises

1. Determine which orders less than 50 the Bruck–Ryser theorem eliminates.
2. Prove that the Bruck–Ryser theorem eliminates all planes of order 2k where $k \equiv 3$ (mod 4).
3. Verify that the determinant of the matrix M in (6.22) is n^2.
4. Prove that the Bruck–Ryser theorem cannot eliminate a plane if the order is p^e, where p is a prime and $e > 0$.

6.6 Arcs and Ovals

In any finite plane, we are always interested in describing geometric configurations which may occur in the plane. We shall now describe an interesting and important collection of configurations which occur in planes. These are known as arcs and ovals.

Definition 6.8 A k-arc is a set of k points in a plane such that no three are collinear. An arc of maximal size is said to be an oval.

As an example, consider the projective plane of order 2. In this plane with seven points, any collection of four points that are a compliment of a line has the property

Fig. 6.1 The projective
plane of order 3

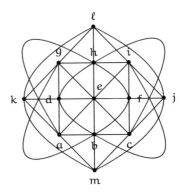

that no three of them are on a line. Hence, in this plane, there are exactly seven
ovals. We shall see shortly that this case is quite different from the case for planes
of odd order. Specifically, the maximum size of an arc in a plane of odd order is
different than the maximum size of a plane of even order. We exhibit this result in
the following theorem.

Theorem 6.28 *The maximum size of an arc in a plane of order* n *is* $n + 2$ *if* n *is
even and* $n + 1$ *if* n *is odd.*

Proof Let A be the set of points of an arc and let p be a point in the arc. For all $q \in A$
there is a line through p and q. These lines are distinct because no line intersects
A in more than two places. Since there are $n + 1$ lines through q it shows that
$|A| \leq n + 2$.

If $|A| = n + 2$ then by the above argument no line can intersect A exactly once.
Namely, it must either be disjoint or intersect twice. Let q be a point not in A.
If there are α lines through q that intersect A twice, then consider the set $A_q =
\{(p, L) \mid p \in A$ and the line L through p and q intersects A twice $\}$. Since there
are α lines through q that intersect the arc twice, then $|A_q| = 2\alpha$. Counting in a
different way there are $n + 2$ points in A and each has a corresponding element in
A_q. Therefore, $2\alpha = n + 2$ which gives $n = 2(\alpha - 1)$ and n must be even. $\square$

We call an $n + 1$ arc in a plane of odd order an oval and an $n + 2$ arc in a plane
of even order a hyperoval.

While it is true that no point can be added to one of these arcs and still be an arc,
there are arcs of smaller size that no point can be added to them.

Hyperovals are very important in the study of finite planes for a variety of reasons.
As an example of this importance, we note that one of the most important steps in
proving the non-existence of a projective plane of order 10 was showing that such a
plane would have no hyperoval.

Consider the diagram of the projective plane of order 3 in Fig. 6.1.

Fig. 6.2 4-arc in the projective plane of order 3

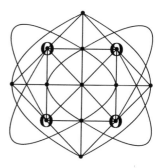

The 13 lines are

$$\{a, b, c, j\}, \{d, e, f, j\}, \{g, h, i, j\}$$
$$\{a, d, g, \ell\}, \{b, e, h, \ell\}, \{c, f, i, \ell\}$$
$$\{c, e, g, m\}, \{a, f, h, m\}, \{b, d, i, m\}$$
$$\{a, e, i, k\}, \{c, d, h, k\}, \{b, f, g, k\}$$
$$\{j, k, \ell, m\}.$$

The points a, c, g, i form a 4-arc illustrated in Fig. 6.2.

Definition 6.9 A conic in a projective plane is the set of points satisfying a non-degenerate (not the product of two linear equations) homogeneous polynomial of degree 2.

Theorem 6.29 *A conic in* $PG_2(\mathbb{F}_q)$ *has* $q + 1$ *points.*

Proof Consider the three coordinates x_1, x_2, x_3 in $\mathbb{F}_q^3$. Given two independent variables in a non-degenerate homogeneous polynomial of degree 2, there are q^2 solutions in $\mathbb{F}_q^3$. Eliminating the all-zero vector and then dividing by $q - 1$ for the non-zero multiples we get $\frac{q^2-1}{q-1} = q + 1$ projective points satisfying the equation. □

Example 6.20 The points in $PG_2(\mathbb{F}_3)$ are

$$\{(0,0,1), (0,1,0), (0,1,1), (0,1,2), (1,0,0), (1,0,1), (1,0,2),$$

$$(1,1,0), (1,1,1), (1,1,2), (1,2,0), (1,2,1), (1,2,2)\}.$$

- The conic $x_1x_2 + x_2x_3 + x_1x_3 = 0$ contains the points $\{(0,0,1), (0,1,0), (1,0,0), (1,1,1)\}$.
- The conic $x_1^2 + x_2^2 + x_3^2 = 0$ contains the points $\{(1,1,1), (1,1,2), (1,2,1), (1,2,2)\}$.

- The conic $x_1^2 + x_2 x_3 = 0$ contains the points $\{(0,0,1), (0,1,0), (1,1,2), (1,2,1)\}$.
- The equation $(x_1 + x_2 + x_3)(x_1 - x_2 - x_3) = x_1^2 - x_2^2 - x_3^2 + x_2 x_3$ contains the points

$$(0,1,2), (1,0,2), (1,1,1), (1,2,0), (1,0,1), (1,1,0), (1,2,2).$$

That is, it contains the points $(0,1,2), (1,0,2), (1,1,1), (1,2,0)$ of the first line and the points $(0,1,2), (1,0,1), (1,1,0), (1,2,2)$ of the second line. Hence, it is degenerate.

Theorem 6.30 *Any non-degenerate conic in* $PG(\mathbb{F}_q)$, q *odd, is an oval.*

Proof Any conic that contains at least three points of a line contains the entire line and so is degenerate. Therefore, any line meets the conic in either $0, 1$, or 2 points. Therefore, it is an oval.   □

We see that all non-degenerate conics are ovals. But it is not at all obvious that ovals should be conics. In [52], it was conjectured by Järnefelt and Kustaanheimo that all ovals in a desarguesian finite projective plane of odd order were in fact conics. Some mathematicians were not convinced by this conjecture. In fact, the review for this paper (MR0054979) written by Marshall Hall states the following:

> It is conjectured that in a plane with $p^2 + p + 1$ points a set of $p + 1$ points, no three on a line, will form a quadric. The reviewer finds this conjecture implausible.

This conjecture was later proved by Segre in [77]. In the review of this paper, (MR0071034) also written by Marshall Hall, he states the following.

> If, when n is odd, we call $n + 1$ points, no three on a line, an oval, then it was conjectured by Järnefelt and Kustaanheimo that in a Desarguesian plane of odd order n, an oval is necessarily a conic. This conjecture is shown to be true in this paper. The method of proof is ingenious.

He ends with the following:

> The fact that this conjecture seemed implausible to the reviewer seems to have been at least a partial incentive to the author to undertake this work. It would be very gratifying if further expressions of doubt were as fruitful.

We shall reproduce the proof given by Segre with the notation changed to match that of the text.

We begin with some lemmas.

Lemma 6.7 *Let* $\mathbb{F}_q$ *be a finite field of order* q, *then the product of all non-zero elements in* $\mathbb{F}_q$ *is* -1.

Proof In a field, there are at most two solutions to a polynomial of degree 2. The equation $x^2 - 1 = 0$ has solutions 1 and -1. Therefore, no other element is its own inverse except for 1 and -1. If the characteristic of the field is 2 then $1 = -1$ is a repeated root as $x^2 + 1 = (x + 1)^2$ in this case.

Each non-zero element of the field must also have a multiplicative inverse. Therefore, in $\prod_{\alpha \in \mathbb{F}_q^*} \alpha$ each element is multiplied by its multiplicative inverse except for -1. This gives that $\prod_{\alpha \in \mathbb{F}_q^*} \alpha = -1$. □

This lemma is a generalization of Wilson's theorem which states that $\prod_{i=1}^{p-1} i \equiv -1 \pmod{p}$ when p is a prime.

We let $\Pi = PG(2, \mathbb{F}_q)$ with q odd. Let $\mathcal{O}$ be an oval in Π, that is, $\mathcal{O}$ consists of $q + 1$ points, no three of which are collinear. Take an arbitrary point p in Π. The oval $\mathcal{O}$ has a tangent at this point (where a tangent is a line that meets the oval exactly once). We know from Theorem 3 in [69] that no three tangents of $\mathcal{O}$ meet at a point.

Lemma 6.8 *Let* $\Pi = PG(2, \mathbb{F}_q)$, *q odd and let* $\mathcal{O}$ *be an oval in* Π. *Every inscribed triangle of* $\mathcal{O}$ *and its circumscribed triangle are in perspective.*

Proof As usual we denote the points of Π as $\{(x_1, x_2, x_3) \mid x_i \in \mathbb{F}_q\} - \{(0, 0, 0)\}/\equiv$ where $(x_1, x_2, x_3) \equiv (x_1', x_2', x_3')$ if and only if $x_i = \lambda x_i'$ for some non-zero $\lambda \in \mathbb{F}_q$.

Without loss of generality we identify the inscribed triangle with the triangle consisting of points $p_1 = (1, 0, 0)$, $p_2 = (0, 1, 0)$, and $p_3 = (0, 0, 1)$. Denote the tangents of $\mathcal{O}$ at p_i by L_i where L_1 is given by $x_2 = k_1 x_3$, L_2 is given by $x_3 = k_2 x_1$, and L_3 is given by $x_1 = k_3 x_2$, where k_i is a non-zero element of the field $\mathbb{F}_q$.

Let $c = (c_1, c_2, c_3)$ be any of the remaining $q - 2$ points of the oval distinct from p_1, p_2 and p_3. Then we have that $c_1 c_2 c_3 \neq 0$ since otherwise a line would intersect the oval in more than two points. Additionally, the lines $\overline{p_1 c}$, $\overline{p_2 c}$, and $\overline{p_3 c}$ have equations of the form $x_2 = \lambda_1 x_3$, $x_3 = \lambda_2 x_1$, and $x_1 = \lambda_3 x_2$ where the λ_i are distinct from the k_i and are non-zero. The coefficients are then determined by

$$\lambda_1 = c_2 c_3^{-1}, \quad \lambda_2 = c_3 c_1^{-1}, \quad \lambda_3 = c_1 c_2^{-1}.$$

Then we have

$$\lambda_1 \lambda_2 \lambda_3 = c_2 c_3^{-1} c_3 c_1^{-1} c_1 c_2^{-1} = 1. \qquad (6.28)$$

Conversely, if λ_1 denotes any of the $q - 2$ non-zero elements of the field distinct from k_1, the line $x_2 = \lambda_1 x_3$ meets the oval at p_1 and some other point denoted by r which is distinct from the p_i. Then, the coefficients λ_2 and λ_3 in the equations $x_3 = \lambda_2 x_1$ and $x_1 = \lambda_3 x_2$ of the lines $\overline{p_2 r}$ and $\overline{p_3 r}$ are functions of λ_1 connected by the fact that $\lambda_1 \lambda_2 \lambda_3 = 1$, which take each of the non-zero of values of the field distinct from k_2 and k_3, respectively. Multiplying the $q - 2$ equations obtained in this manner, we have that

$$\left(\prod_{\alpha \in \mathbb{F}_q^*} \alpha \right)^3 = k_1 k_2 k_3.$$

Lemma 6.7 gives that $(\prod_{\alpha \in \mathbb{F}_q^*} \alpha) = -1$. Therefore, we have that

$$k_1 k_2 k_3 = -1. \tag{6.29}$$

This gives that the point of intersection of L_2 and L_3 is $(k_3, 1, k_2 k_3)$; the point of intersection of L_3 and L_1 is $(k_3 k_1, k_1, 1)$; and the point of intersection of L_1 and L_2 is $(1, k_1 k_2, k_2)$. These three points are joined to p_1, p_2, and p_3, respectively, by the lines $x_3 = k_2 k_3 x_2$, $x_1 = k_3 k_1 x_3$, and $x_2 = k_1 k_2 x_1$. By Eq. 6.29, these lines concur at the point $(1, k_1 k_2, -k_2)$ which is the center of perspective of the triangles.   □

We can now prove Segre's famous theorem.

Theorem 6.31 (Segre's Theorem) *Every oval of* $PG(2, \mathbb{F}_q)$, *with q odd, is a conic.*

Proof We retain all of the notations given in Lemma 6.8. With reference to Lemma 6.8, we can, without loss of generality, assume that the lines concur at the point $(1, 1, 1)$, that is, we assume $k_1 = k_2 = k_3 = -1$.

We can denote the line L_b tangent to the oval at the point b by $b_1 x_1 + b_2 x_2 + b_3 x_3 = 0$. This tangent line contains the point b but does not contain the points p_1, p_2, or p_3. Then let

$$\beta_1 = b_1 - b_2 - b_3,$$
$$\beta_2 = -b_1 + b_2 - b_3,$$
$$\beta_3 = -b_1 - b_2 + b_3.$$

It follows that

$$b_1 c_1 + b_2 c_2 + b_3 c_3 = 0 \tag{6.30}$$

and

$$b_1 b_2 b_3 \beta_1 \beta_2 \beta_3 \neq 0. \tag{6.31}$$

Lemma 6.8 gives that the triangles $\triangle bp_2 p_3$ and $\triangle bp_2 p_3$ are in perspective. This gives that

$$\begin{vmatrix} c_3 - c_2 & c_1 + c_3 & -c_1 - c_2 \\ b_1 - b_3 & b_2 & 0 \\ b_1 - b_2 & 0 & b_3 \end{vmatrix} = 0.$$

Then, we have $b_2(c_1 + c_2) = b_3(c_1 + c_3)$. The inscribed triangles $\triangle bp_3 p_1$, $\triangle bp_1 p_2$ and their circumscribed triangles give

$$b_3(c_2 + c_3) = b_1(c_2 + c_1)$$
$$b_1(c_3 + c_1) = b_2(c_3 + c_2).$$

These last three equations, Eqs. 6.30 and 6.31, imply that

$$c_2c_3 + c_3c_1 + c_1c_2 = 0.$$

Therefore, the $q - 2$ points lie on the conic $4x_2x_3x_3x_1 + x_1x_2 = 0$, which also contains the points p_1, p_2, and p_3. Therefore, there are $q + 1$ points on it and therefore the oval must coincide with it. □

Theorem 6.32 *Assume there is a hyperoval in a projective plane of even order. Any line in this projective plane of even order is either secant to the hyperoval or disjoint from the hyperoval.*

Proof Let $p_1, p_2, \ldots, p_{n+2}$ be the points on a hyperoval. Assume there is a line L that meets the hyperoval only once. Without loss of generality assume that L is incident with p_1. Through p_1 and p_j, with $j \neq 1$, there is a unique line, call it M_j, that does not intersect any other point on the hyperoval. Then we have that $L, M_2, M_3, \ldots, M_{n+2}$ are lines through p_1 and there are $n + 2$ lines through p_1 which is a contradiction. Therefore, no line is tangent to the hyperoval. □

Let π be a projective plane of even order. Fix a line to be the line at infinity. We refer to those hyperovals that are secant to the line at infinity as hyperbolic hyperovals and to those that are disjoint from the line at infinity as elliptic hyperovals.

Definition 6.10 A Singer cycle is a map $\psi : \Pi \to \Pi$ such that ψ maps points to points, lines to lines and preserves incidence, that is, an automorphism, such that for any two points p and p' in Π there exists an i such that $\psi^i(p) = p'$ and ψ^{n^2+n+1} is the identity.

A Singer cycle is known to exist in any Desarguesian plane. In other words, a Singer cycle is a permutation of the points on the plane preserving incidence. This was first shown in [80].

We can count the number of hyperbolic and elliptic hyperovals in a Desarguesian plane under a Singer cycle.

Theorem 6.33 *In a Singer cycle of a hyperoval in a Desarguesian projective plane of even order, there are $\frac{(n+1)(n+2)}{2}$ hyperovals that are hyperbolic and $\frac{n(n-1)}{2}$ hyperovals that are elliptic for any choice of L_∞.*

Proof Let $\mathfrak{O}$ be a hyperoval in Π, where Π is a Desarguesian plane of even order. Let ψ be the Singer cycle. Then consider the set $\{\psi^i(\mathfrak{O})\}$, which is the set of images of a hyperoval in a Singer cycle with $\psi^0(\mathfrak{O}) = \mathfrak{O}$. Each point on the hyperoval $\mathfrak{O}$ will have $(n + 1)$ values of i such that ψ^i maps it to a point on L_∞. Each of these values of i actually sends two points of $\mathfrak{O}$ to L_∞ since any line is either disjoint or secant to a hyperoval.

Fig. 6.3 Projective plane of
order 2

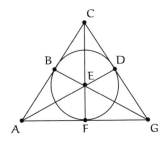

Now we can count the hyperovals. There are $\frac{(n+1)(n+2)}{2}$ hyperovals that are hyperbolic in a Singer cycle, leaving $n^2 + n + 1 - \frac{(n+1)(n+2)}{2} = \frac{n(n-1)}{2}$ hyperovals that are elliptic in a Singer cycle. □

Consider the projective plane of order 2 in Fig. 6.3.
There are exactly seven hyperovals in the projective plane, namely,

$$\{D, E, F, G\}$$
$$\{A, B, E, F\}$$
$$\{B, C, D, E\}$$
$$\{B, C, F, G\}$$
$$\{A, C, D, F\}$$
$$\{A, B, D, G\}$$
$$\{A, C, E, G\}$$

Given any of these hyperovals the others are in the image of the Singer cycle of that hyperoval. Take as L_∞ the line $\{BDF\}$ then the theorem gives that there are $\frac{(n+1)(n+2)}{2} = 6$ hyperovals that are hyperbolic and $\frac{n(n-1)}{2} = 1$ hyperovals that are elliptic. The unique elliptic hyperoval in this case is $\{A, C, E, G\}$.

The number of hyperovals increases greatly as the order of the plane increases. For example, there are 168 hyperovals in the projective plane of order 4. Using any line as L_∞ of this projective plane, there are 120 hyperbolic hyperovals and 48 elliptic hyperovals.

Exercises

1. Verify that no line meets the point set $\{a, c, g, i\}$ three times in the projective plane of order 3.
2. Prove the existence of a Singer cycle in a Desarguesian plane. Hint: View the points as elements of $\mathbb{F}_q^3 / \equiv$ as in the construction of Desarguesian planes. Take a field of order $|\mathbb{F}_q|^3$ that is a Galois extension of $\mathbb{F}_q$ and as such can be viewed as a vector space over $\mathbb{F}_q$. Then consider the action formed by multiplication of the elements of this field extension by a single element.

3. Find another arc in the projective plane of order 3.
4. Use Segre's theorem to find the conic for the oval given in Exercise 2.

6.7   **Baer Subplanes**

We shall now look at projective planes which are contained in other projective planes. This topic is also an important topic in the study of infinite projective planes.

If $\Pi = (\mathcal{P}, \mathcal{L}, \mathcal{I})$ is a projective plane of order n then $\Sigma = (\mathcal{P}', \mathcal{L}', \mathcal{I}')$ is a subplane of Π if $\mathcal{P}' \subseteq \mathcal{P}$, $\mathcal{L}' \subseteq \mathcal{L}$, $\mathcal{I}' \subseteq \mathcal{I}$, and Σ is a projective plane.

The following theorem is due to Bruck (see [17]).

Theorem 6.34 *If Σ is a subplane of order m of a projective plane Π of order n, then $n \geq m^2 + m$ or $n = m^2$.*

Proof Let ℓ be a line in $\mathcal{L}'$. Notice that ℓ is also a line in $\mathcal{L}$ but it has more points in Π that are incident with it. In Σ, ℓ has $m + 1$ points and hence there are $n - m$ points on ℓ in Π that are not on ℓ in Σ. Any two lines in Σ meet in Σ. Hence, for every line in Σ there are $n - m$ points that are not in $\mathcal{P}'$. Hence, there are $(m^2 + m + 1)(n - m)$ points in $\mathcal{P} - \mathcal{P}'$ that have a line of Σ passing through them. Notice that there are $n^2 + n + 1$ points in $\mathcal{P}$, $m^2 + m + 1$ in $\mathcal{P}'$ and so we have

$$n^2 + n + 1 \geq (m^2 + m + 1) + (m^2 + m + 1)(n - m)$$
$$n^2 + n + 1 \geq m^2 + m + 1 + m^2 n - m^3 + mn - m^2 + n - m$$
$$n^2 \geq m^2 n - m^3 + mn$$
$$0 \geq -n^2 + m^2 n - m^3 + mn$$
$$0 \geq (m^2 - n)(n - m).$$

We know that $n > m$ and so this gives that $n \geq m^2$.

We can have equality, that is, $n = m^2$ and then each of the inequalities above becomes an equation.

Assume that we do not have equality, that is, $n > m^2$. Then there exists a point $q \in \mathcal{P}$ such that no line of Σ is incident with it. Hence, any line through q can have at most one point from $\mathcal{P}'$ on it, since if there were two then the line through these two points would be a line from Σ passing through q. Now if r is a point of Σ then there is a line through q and r in Π, meaning that the number of lines through q is at least as large as the size of $\mathcal{P}'$ which is $m^2 + m + 1$. This gives that

$$n + 1 \geq m^2 + m + 1$$
$$n \geq m^2 + m,$$

which gives the result.   □

In the case when $n = m^2$, the subplane is known as a Baer subplane. The following proof can be found in [18].

Theorem 6.35 *If Σ is a Baer subplane of Π then every point of the Π lies on a line of the Baer subplane Σ and every line of the plane Π contains a point of the Baer subplane Σ.*

Proof Given the duality of points and lines in a projective plane, we only need to prove one of the statements. We shall prove the first.

Let Π be a plane of order m^2 and Σ be a Baer subplane of order m. Any line in the plane Π contains at most $m + 1$ points that are in Σ, and therefore it contains at least one point which is not in Σ. Let L be a line of Π and let p be a point on L that is not in the subplane Σ. Since Σ is a subplane, there can be at most one line of Σ incident with the point p (otherwise the two lines of Σ would meet in a point not in Σ giving a contradiction).

Given any point in Σ, it must have a line of Π through it and p since Π is a plane. Therefore, the $m^2 + m + 1$ points of Σ are contained in the $m^2 + 1$ lines of Π through p. If the line L contained no point of Σ, then the lines of Π incident with p could account for at most $(m + 1) + (m^2 - 1) = m^2 + m$ points of the subplane Σ. Namely, there is at most one point of Σ meeting it in $m + 1$ points, 1 meeting it at 0 points, and $m^2 - 1$ meeting it at 1 point. Since there are $m^2 + m + 1$ points of Σ this is a contradiction. This gives that the line L contains a point of Σ and gives our result.   □

Definition 6.11 A blocking set $\mathcal{B}$ in a projective plane is a subset of the points of the plane such that every line of Π contains at least one point in $\mathcal{B}$ and one point not in $\mathcal{B}$.

The proof of Theorem 6.35 gives the following corollary.

Corollary 6.3 *If Σ is a Baer subplane of Π then Σ is a blocking set of Π.*

Exercises

1. Let Π be a projective plane with subplane Σ. Prove that if there is an isomorphism from Π to the plane Π' then Π contains a subplane isomorphic to Σ.
2. Find the Baer subplane in the projective plane of order 4.
3. Prove that in the Desarguesian plane $PG_2(\mathbb{F}_{q^2})$ there is a Baer subplane.
4. Determine for which orders there exist subplanes of $PG_2(\mathbb{F}_q)$.

6.8   Translation Planes and Non-Desarguesian Planes

We begin with a definition that is required to define translation planes.

Definition 6.12 Let Π be a finite projective plane. Let P be a point on the plane and let L be a line. We define a central collineation with center P and axis L to be an isomorphism that fixes the points of L and the lines through the point P.

We say that the collineation is an elation if the point p is incident with the line L. For a description of groups see Chap. 10.

Theorem 6.36 *The set of all central collineations with center P and axis L forms a group with function composition.*

Proof It is immediate that the identity map is a central collineation. We know that function composition is associative.

To show closure, let σ and τ be two central collineations with center P and axis L. Then, if Q is a point on L then $\sigma(\tau(Q)) = \sigma(Q) = Q$ and if M is a line through P then $\sigma(\tau(M)) = \sigma(M) = M$.

Let σ be a central collineation with center P and axis L. Then, if Q is a point on L then $\sigma(Q) = Q$ and then $Q = \sigma^{-1}(\sigma(Q)) = \sigma^{-1}(Q)$. If M is a line through P then $\sigma(M) = M$ and then $M = \sigma^{-1}(\sigma(M)) = \sigma^{-1}(M)$. Therefore, the inverse map of a central collineation with center P and axis L is a central collineation with center P and axis L. This gives that the set is a group. $\square$

Recall that a group G acts transitively on a set A if for every $a, b \in A$, there is a group element $\sigma \in G$ with $\sigma(a) = b$.

Definition 6.13 If the group of elations acts transitively on the points of the affine plane $\pi = \Pi^L$ then the line L is called a translation line and a projective plane with such a line is said to be a translation plane.

Given a translation plane Π with translation line L, the affine plane $\pi = \Pi^L$ is said to be an affine translation plane.

We shall now describe some planes which do not satisfy Desargues' theorem. We have seen from Theorem 6.24 that any plane that can be embedded in three dimensions is a Desarguesian plane. There are, however, planes that are not Desarguesian. It is known that every plane of order less than or equal to 8 is a Desarguesian plane, but there are three non-Desarguesian planes of order 9.

Definition 6.14 A finite ternary ring R is a set R together with two distinguished elements 0 and 1 and a ternary operation $T(a, b, c) = ab + c$ satisfying the following:

1. For all $a \in R$, we have $T(1, a, 0) = T(a, 1, 0) = a$.
2. For all $a, b \in R$, we have $T(a, 0, b) = T(0, a, b) = b$.
3. For all $a, b, c \in R$, $T(a, b, y) = c$ has a unique solution for the variable y.
4. For all $a, b, c, d \in R$, with $a \neq c$, the equations $T(x, a, b) = T(x, c, d)$ have a unique solution for $x \in R$.

If we relax the definition to include infinite ternary rings we require another axiom.

Theorem 6.37 *If R is a ternary ring, then R gives rise to an affine plane with the usual construction.*

Proof The points of the plane are (x, y) where x and y are elements of R. For each, $x \in R$ there is a line consisting of all points (x, y) where y is any element of R. All other lines are of the form $y = T(x, a, b)$, which corresponds to the line $y = xa + b$. It is easy to see that this forms an affine plane of order $|R|$.   □

This theorem leads to the following important theorem.

Theorem 6.38 *If R is a ternary ring, then R gives rise to a projective plane.*

Proof Theorem 6.37 gives that we can form an affine plane π of order $|R|$ from R, then simply make the projective completion to form a projective plane Π of order $|R|$.   □

If the ternary ring is a finite field then the plane is Desarguesian. Moreover, it is known that if the ternary ring is not a finite field then the plane is not Desarguesian.

We can now give a definition of a near field which will be a ternary ring which we can construct more easily to form non-Desarguesian projective planes.

Definition 6.15 A near-field $(Q, +, *)$ is a set Q with two operations $+$ and $*$ such that the following hold:

1. (additive closure) If $a, b \in Q$ then $a + b \in Q$.
2. (additive associativity) For all $a, b, c \in Q$ we have $(a + b) + c = a + (b + c)$.
3. (additive identity) There exists an element $0 \in Q$ such that $0 + x = x + 0 = x$ for all $x \in Q$.
4. (additive inverses) For all $a \in Q$ there exists $b \in Q$ with $a + b = b + a = 0$.
5. (additive commutativity) For all $a, b \in Q$ we have $a + b = b + a$.
6. (multiplicative closure) If $a, b \in Q$ then $a * b \in Q$.
7. (multiplicative associativity) For all $a, b, c \in F$ we have $(a * b) * c = a * (b * c)$.
8. (multiplicative identity) There exists an element $1 \in Q$ such that $1 * x = x * 1 = x$ for all $x \in Q$.
9. (multiplicative inverse) For all $a \in Q - \{0\}$, there exists b with $a * b = 1$.

10. (unique solution) For all $a, b, c \in Q$, $a \neq b$ there is exactly one solution to $xa - xb = c$.
11. (right distributive) For all $a, b, c \in Q$ we have $(b + c) * a = b * a + c * a$.

Notice that it is not necessarily a division ring since it only has right distribution and it is not necessarily a field for the same reason and also because the multiplication is not necessarily commutative. Of course, division rings and fields are examples of near fields.

A near field is a ternary ring with function $T(a, b, c) = ab + c$. Therefore, if we can construct a near field then we can construct a non-Desarguesian projective plane.

We shall now construct a near-field J_9 with elements $\{0, 1, -1, i, -i, j, -j, k, -k\}$. Defining equations for the near field are given by

$$ijk = -1,$$
$$i^2 = j^2 = k^2 = -1,$$
$$j = 1 + i,$$
$$k = 1 - i.$$

The addition and multiplication tables for this near field are given in (6.32) and (6.33). We note that the multiplication is non-commutative.

$$
\begin{array}{c|ccccccccc}
+ & 0 & 1 & -1 & i & -i & j & -j & k & -k \\
\hline
0 & 0 & 1 & -1 & i & -i & j & -j & k & -k \\
1 & 1 & -1 & 0 & j & k & -k & -i & -j & i \\
-1 & -1 & 0 & 1 & -k & -j & i & k & -i & j \\
i & i & j & -k & -i & 0 & k & -1 & 1 & -j \\
-i & -i & k & -j & 0 & i & 1 & -k & j & -1 \\
j & j & -k & i & k & 1 & -j & 0 & -1 & -i \\
-j & -j & -i & k & -1 & -k & 0 & j & i & 1 \\
k & k & -j & -i & 1 & j & -1 & i & -k & 0 \\
-k & -k & i & j & -j & -1 & -i & 1 & 0 & k \\
\end{array}
\qquad (6.32)
$$

$$
\begin{array}{c|ccccccccc}
* & 0 & 1 & -1 & i & -i & j & -j & k & -k \\
\hline
0 & 0 & 0 & 0 & 0 & 0 & 0 & 0 & 0 & 0 \\
1 & 0 & 1 & -1 & i & -i & j & -j & k & -k \\
-1 & 0 & -1 & 1 & -i & i & -j & j & -k & k \\
i & 0 & i & -i & -1 & 1 & k & -k & -j & j \\
-i & 0 & -i & i & 1 & -1 & -k & k & j & -j \\
j & 0 & j & -j & -k & k & -1 & 1 & i & -i \\
-j & 0 & -j & j & k & -k & 1 & -1 & -i & i \\
k & 0 & k & -k & j & -j & -i & i & -1 & 1 \\
-k & 0 & -k & k & -j & j & i & -i & 1 & -1 \\
\end{array}
\qquad (6.33)
$$

Corollary 6.4 *The near-field* J_9 *gives a non-Desarguesian projective plane of order 9.*

Proof Apply Theorem 6.38 to the near-field J_9.   □

This plane is known as the Hall plane. It was first described in [43]. We shall give a geometric construction that also gives rise to this plane as well as many others.

Let Π be the Desarguesian projective plane $PG_2(q^2)$. We know that this plane contains a Baer subplane. A Baer subline is any line of the Baer subplane contained in a given line. Let L be a line in Π and let ℓ be a Baer subline of Π. Let $\pi = \Pi^L$ be the affine plane formed by treating L as the line at infinity.

We shall construct $\tau = (\mathcal{A}, \mathcal{M}, \mathcal{J})$. Let $\mathcal{A}$ consist of the points of π, that is, the points of Π with the points of L removed. If m is a line of π such that the intersection of $\overline{m}$ and L, where $\overline{m}$ is the projective completion of m, is not a point of ℓ then $m \in \mathcal{M}$. A set E of $n+1$ points of L is a derivation set if every pair of distinct points q_1 and q_2 of π which determines a line meeting L in a point of E, there is a Baer subplane containing the two points and E. In this case, we say that a Baer subplane satisfying this condition belongs to E. Any Baer subplane that belongs to E restricted to the affine plane π is also in $\mathcal{M}$. The incidence relation $\mathcal{J}$ is given in the canonical manner. The structure τ is an affine plane and we call the plane τ the derived plane of Π. Its projective completion T is also called the derived plane. For $q > 2$, the derived planes are non-Desarguesian.

Given this construction we can state the following theorem.

Theorem 6.39 *There exists non-Desarguesian projective planes for all orders* q^2, *where* q *is a prime power and* $q > 2$.

For a detailed description of non-Dearguesian planes see [93].

Exercises

1. Verify that the structure given in (6.32) and (6.33) is a near field.

Designs

7

We shall now describe a more general incidence structure than a finite geometry, which is called a design. Affine and projective planes are examples of designs, but of course not all designs are planes, nor are they all finite geometries in the sense defined in the previous chapter. Essentially, the idea is that we wish to study a class of incidence structures, namely, points and sets of points, with some structure. The origin of the term comes from the fact that there were originally of interest as designs of experiments. In the enormously influential book by Ronald A. Fisher, The Design of Experiments [36], Fisher described how designs could be used to make various experiments, including the lady tasting tea experiment which sought to determine if Muriel Bristol could determine, by taste, whether the tea or the milk was added first to a cup. The book is foundational in the study of statistics and Fisher himself was largely a statistician, but the book also contains a chapter on Latin squares. Designs have many interesting applications inside and outside of mathematics. For example, they are still used extensively in the designs of experiments and in the construction of tournaments. Additionally, they also have many interesting connections to algebraic coding theory, graph theory, and group theory.

7.1 Designs

We begin with the standard definition of a design.

Definition 7.1 The structure $D = (\mathcal{P}, \mathcal{B}, \mathcal{I})$ is a t-(v, k, λ) design if $|\mathcal{P}| = v$, every block $B \in \mathcal{B}$ is incident with exactly k points, and every t distinct point is together incident with exactly λ blocks.

© The Editor(s) (if applicable) and The Author(s), under exclusive license
to Springer Nature Switzerland AG 2020
S. T. Dougherty, *Combinatorics and Finite Geometry*,
Springer Undergraduate Mathematics Series,
https://doi.org/10.1007/978_3_030_56395_0_7

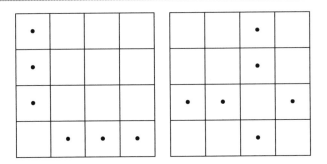

Fig. 7.1 Biplane of order 4

A projective plane of order n is a $2\text{-}(n^2 + n + 1, n + 1, 1)$ design and an affine plane of order n is a $2\text{-}(n^2, n, 1)$ design.

Later in the chapter we shall study k-nets of order n, which are a $1 - (n^2, n, k)$ design; balanced incomplete block design (BIBD) which is when $t = 2$; designs with the same number of points and blocks, that is, $v = b$, which are called symmetric designs; and Steiner triple systems which are when $k = 3$ and $\lambda = 1$.

As an example of a design that is not a plane, consider the design of points and planes in $AG_3(\mathbb{F}_2)$. Its incidence matrix is given in Table 6.1. Here the points of the design are the points in the space and the blocks in the design are the planes in the space. Therefore, we have that $v = 8$ and $k = 4$. We see that through any three points there is exactly one plane (since no three points are collinear in this space), so $t = 3$ and $\lambda = 1$. We see that there are $C(8, 3) = 56$ ways of choosing three points and each plane is counted $C(4, 3) = 4$ times so there are $\frac{56}{4} = 14$ planes. Therefore, the points and planes in $AG_3(\mathbb{F}_2)$ make a $3\text{-}(8, 4, 1)$ design with 14 blocks.

As another example of a design, we shall describe a biplane on 16 points. Consider the 16 points corresponding to the 16 places on a 4 by 4 grid. There are 16 blocks $\beta(i, j)$ where $\beta(i, j)$ corresponds to the 6 points that are in the i-th row and the j-th column distinct from the point (i, j). In Fig. 7.1, we show the blocks $\beta(1, 1)$ and $\beta(3, 2)$. Notice that these two blocks meet in two places. This design in Fig. 7.1 is a $2\text{-}(16, 6, 2)$ design.

The parameters $t\text{-}(v, k, \lambda)$ are not the only ones that describe the design. We denote the number of blocks by b and let r be the number of blocks through a point. There are, of course, restrictions on these values.

Theorem 7.1 *In a* $t\text{-}(v, k, \lambda)$ *design, we have*

$$vr = bk.$$

Proof To prove this we simply count the size of the set $\mathcal{I}$, that is, the incidence relation, in two different ways. That is we count the number of incidences. There are v points and r blocks through them so there are vr incidences. On the other hand, there are b blocks and k points on each block so there are bk incidences. Hence, $vr = bk$. ☐

Theorem 7.2 *In a 2-(v, k, λ) design, we have*

$$r(k - 1) = \lambda(v - 1).$$

Proof To prove this, we fix a point p_1 in the design, and hence there are $v - 1$ points distinct from p_1. For any other points p, there are exactly λ blocks through p and p_1 since $t = 2$. Each block is counted as many times as there are points on it other than p_1, that is, $k - 1$ times. This gives that the number of the blocks through p must be $\frac{(v-1)\lambda}{k-1}$. Of course, the number of blocks through p is r by definition which gives the result. □

Consider the 3-$(8, 4, 1)$ design formed by the points and planes in $AG_3(\mathbb{F}_2)$. We know $b = 14$. Using the equation in Theorem 7.1 we have $8(r) = 14(4)$ which gives that $r = 7$. This gives that through each point there are seven planes.

Let us consider the case of points and planes in $PG_3(\mathbb{F}_p)$. There are $\frac{p^4 - 1}{p - 1} = p^3 + p^2 + p + 1$ points and an equal number of planes. Hence $v = b = p^3 + p^2 + p + 1$. On any plane there are $p^2 + p + 1$ points, giving that $k = p^2 + p + 1$. Then we use $vr = bk$ which gives $(p^3 + p^2 + p + 1)r = (p^3 + p^2 + p + 1)(p^2 + p + 1)$ and then $r = p^2 + p + 1$. That is, through each point there are $p^2 + p + 1$ planes. If three points are collinear, then there are more planes through them than if there were not collinear so it is not a 3-design. Through any two points there are $p^3 + p^2 + p + 1 - (p + 1) = p^3 + p^2$ ways of choosing a non-collinear point. These three points give a plane since they generate a three-dimensional vector space. Doing this, each plane is picked $(p^2 + p + 1) - (p + 1) = p^2$ ways. Hence, through any two points there are $p + 1$ planes. Therefore, it is a 2-$(p^3 + p^2 + p + 1, p^2 + p + 1, p + 1)$ design.

If a set (v, b, r, k, λ) satisfies the equations $vr = bk$ and $r(k - 1) = \lambda(v - 1)$ then the parameters are said to be admissible. However, simply because they are admissible does not mean that a design exists. For example, there is an analog of the Bruck–Ryser theorem for designs which we state here without proof.

Theorem 7.3 (Bruck–Ryser–Chowla) *If a (v, k, λ) design of order n exists then either*

1. v is even and n is square or
2. v is odd and there exists integers a, b, c with $c^2 = na^2 + (-1)^{(v-1)/2}\lambda b^2$.

We denote by λ_s, $0 \le s \le t$, the number of blocks incident with s points. It follows that $\lambda_t = \lambda$.

Let $D = (\mathcal{P}, \mathcal{B}, \mathcal{I})$ be a t-(v, k, λ) design. Let S be a set of s points in $\mathcal{P}$, with $0 \le s < t$.

We shall count the number of sets of size t that contain S and the blocks containing these sets of size t. Specifically, consider the set

$$A = \{(T, B) \mid S \subseteq T \subseteq B, |T| = t, B \text{ a block in the design}\}.$$

Through any t points there are λ blocks and there are $\binom{v-s}{t-s}$ ways of choosing $t - s$ points to add to the s points of S to make t points. Hence, we have that

$$|A| = \lambda\binom{v-s}{t-s}. \tag{7.1}$$

Now through the s points of S there are λ_s blocks. In each of these blocks, there are $k - s$ points to choose $t - s$ from to make a set of size t, then we have that

$$|A| = \lambda_s\binom{k-s}{t-s}. \tag{7.2}$$

This gives the following.

Lemma 7.1 *Let* D *be a* t-(v, k, λ) *design, with* λ_s *be the number of blocks through* s *points with* $0 \le s \le t$. *Then we have*

$$\lambda\binom{v-s}{t-s} = \lambda_s\binom{k-s}{t-s}. \tag{7.3}$$

Proof This follows from counting the set A in two ways given in Eqs. 7.1 and 7.2.
□

We can now find a recurrence relation for λ_s.

Theorem 7.4 *Let* D *be a* t-(v, k, λ) *design, with* λ_s *the number of blocks through* s *points with* $0 \le s \le t$. *Then we have*

$$\lambda_s = \frac{v-s}{k-s}\lambda_{s+1}. \tag{7.4}$$

Proof The previous lemma implies

$$\lambda_s = \frac{\lambda\left(\frac{(v-s)!}{(t-s)!(v-s-(t-s))!}\right)}{\frac{(k-s)!}{(t-s)!(k-s-(t-s))!}} = \lambda\frac{(v-s)!(k-t)!}{(k-s)!(v-t)!}$$
$$= \lambda\frac{(v-s)(v-s-1)\cdots(v-t+1)}{(k-s)(k-s-1)\cdots(k-t+1)}.$$

It then follows immediately that

$$\lambda_s = \frac{(v-s)}{(k-s)}\lambda_{s+1}.$$

This gives the result.
□

Let I and J be sets of points with $|I| = i$, $|J| = j$, and $i + j \leq t$.

We define λ_i^j as the number of blocks containing I and disjoint from J. We notice from this definition that $\lambda_i^0 = \lambda_i$.

Let p be a point not in I nor J. Then each block through I, disjoint from J, is either incident with p or is not incident with p, and so is counted either in λ_{i+1}^j if it is incident with p or in λ_i^{j+1}. That is, either p is adjoined to I to make a set of size $i + 1$ or p is adjoined to J to make a set of size $j + 1$.

This gives the following theorem.

Theorem 7.5 *Let* D *be a design then*

$$\lambda_i^j = \lambda_{i+1}^j + \lambda_i^{j+1}. \tag{7.5}$$

This recursion can be used to make a type of Pascal's triangle. Notice that each of the λ_i^j must be a non-negative integer for a design to exist given a set of parameters.

Corollary 7.1 *Let* D *be a design then*

$$\lambda_i^j = \lambda \frac{\binom{v-i-j}{k-i}}{\binom{v-t}{k-t}}. \tag{7.6}$$

Proof We only need to show that it satisfies both the initial condition and the recursion.

First we examine the initial condition. If $j = 0$ then by Eq. 7.4 we have

$$\lambda_i^0 = \lambda_i = \lambda \frac{\binom{v-i}{t-i}}{\binom{k-i}{t-i}}$$
$$= \lambda \frac{\frac{(v-i)!}{(t-i)!(v-t)!}}{\frac{(k-i)!}{(t-i)!(k-t)!}}$$
$$= \lambda \frac{(v-i)!(k-t)!}{(v-t)!(k-i)!}.$$

By the formula we have

$$\lambda_i^0 = \lambda_i = \lambda \frac{\binom{v-i-0}{k-i}}{\binom{v-t}{k-t}}$$
$$= \lambda \frac{\frac{(v-i)!}{(k-i)!(v-k)!}}{\frac{(v-t)!}{(v-k)!(k-t)!}}$$
$$= \lambda \frac{(v-i)!(k-t)!}{(v-t)!(k-i)!}.$$

Next we have

$$\lambda \frac{\binom{v-i-1-j}{k-i-1}}{\binom{v-t}{k-t}} + \lambda \frac{\binom{v-i-j-1}{k-i}}{\binom{v-t}{k-t}} = \lambda \frac{\binom{v-i-j}{k-i}}{\binom{v-t}{k-t}} \tag{7.7}$$

and we have the result. □

We can now produce a Pascal type triangle for the values of λ_i^j. We write as follows (note that occasionally the mirror image of ours is given):

$$
\begin{array}{ccccccc}
& & & \lambda_0^0 & & & \\
& & \lambda_0^1 & & \lambda_1^0 & & \\
& \lambda_0^2 & & \lambda_1^1 & & \lambda_2^0 & \\
\lambda_0^3 & & \lambda_1^2 & & \lambda_2^1 & & \lambda_3^0 \\
& & & \vdots & & &
\end{array}
\tag{7.8}
$$

The minimum weight vectors in the extended Golay code form a very nice 5-$(24, 8, 1)$ design by the Assmus–Mattson theorem, which will be described in detail later in the text. It has the following triangle:

$$
\begin{array}{ccccccccc}
& & & & 759 & & & & \\
& & & 506 & & 253 & & & \\
& & 330 & & 176 & & 77 & & \\
& 210 & & 120 & & 56 & & 21 & \\
130 & & 80 & & 40 & & 16 & & 5 \\
78 & & 52 & & 28 & & 12 & & 4 & & 1
\end{array}
\tag{7.9}
$$

For the affine plane of order n, there are $n^2 + n$ lines and hence $\lambda_0 = n^2 + n$. We know that $\lambda_2 = 1$. Then, using Eq. 7.4, we get $\lambda_1 = n + 1$. Of course, we knew this since there are $n + 1$ lines through a point. Then using Eq. 7.5 we can fill in the table for the affine plane of order n and we have

$$
\begin{array}{ccccc}
& & n^2 + n & & \\
& n^2 - 1 & & n + 1 & \\
n^2 - n - 1 & & n & & 1
\end{array}
\tag{7.10}
$$

The design in Exercise 7 has $t = 5$, $v = 72$, $k = 16$, and 249849 blocks and is the design that would be formed by the minimum weight vectors of a $[72, 36, 16]$ Type II code. It is not known if such a code exists. If one were able to show that the design above does not exist then the code could not exist.

Let $D = (\mathcal{P}, \mathcal{B}, \mathcal{I})$ be a design. Let w be a block in D and define $D' = (\mathcal{P}', \mathcal{B}', \mathcal{I}')$ where $\mathcal{P}' = \mathcal{P} - \{p | (p, v) \in \mathcal{I}\}$, $\mathcal{B}' = \mathcal{B} - \{w\}$ and a point is incident with a block in D' if it was incident in D. The structure D' is called the residual design. We shall prove that it is a design in the following.

Theorem 7.6 *Let* D *be a* t-(v, k, λ) *design. Then* D$'$ *is a* t-$(v - k, k, \lambda)$ *design.*

Proof The fact that there are $v - k$ points in D$'$ follows from the fact that there are k points on each block of D.

Take any t points in D$'$. These are also points in D and in D there were exactly λ blocks. Each of these blocks has a restriction as a block in D$'$, since none of the points on the removed block are in D$'$. Thus, through these t points, there are λ blocks. If there were another block in D$'$ through these points then its extension would be a block through these points in D. Hence, D$'$ is a t-$(v - k, k, \lambda)$ design. $\square$

A parallel class in a design is a subset of the set of blocks that partitions the point set and has no points of intersection between any two blocks in the set. We say that a design is resolvable if the set of blocks can be partitioned into parallel classes. As an example an affine plane is a resolvable design.

Recall that a BIBD is a design with $t = 2$. This is perhaps the most widely studied subset of designs. The terms block and design are obvious. The reason for the incompleteness is that it is not simply the trivial design consisting of all possible subsets of size k and balanced refers to $t = 2$.

For this case, the triangle of λ_i^j is

$$
\begin{array}{ccccc}
 & & b & & \\
 & b - r & & r & \\
b - 2r + \lambda & & r - \lambda & & \lambda
\end{array}
\tag{7.11}
$$

Theorem 7.7 *If* D $= (\mathcal{P}, \mathcal{B}, \mathcal{I})$ *is a* (v, k, λ) *BIBD then if* $\mathcal{B}' = \{b' \mid b'$ *is the complement of a block in* $\mathcal{B}\}$ *then* D$' = (\mathcal{P}, \mathcal{B}', \mathcal{I})$ *is a* $(v, v - k, b - 2r + \lambda)$ *BIBD.*

Proof It is immediate that the number of points in D$'$ is v and that the number of points on a block is $v - k$. The number of complements of blocks containing two distinct points is the number of blocks that is disjoint from two points, i.e., λ_0^2. This can be read from the above table which gives that there are $b - 2r + \lambda$ complements of blocks through any two points. $\square$

The design given in this theorem is known as the complementary design.

Consider the complementary design of a finite projective plane. Any two blocks in this design have n^2 points. A point is in the intersection of two of these blocks if it was on neither line. Hence, any two blocks meet in exactly $n^2 - n$ points.

In the next few sections, we shall give several examples of BIBDs. The reader should consult [24] for an encyclopedic description of many of the designs we consider here.

1. Verify that the biplane on 16 points satisfies the equations in Theorems 7.1 and 7.2.
2. Verify that any two blocks meet exactly twice on the biplane on 16 points. Prove that the number of blocks through a point and the number of points on a block is 6.
3. Verify that finite affine and projective planes satisfy the equations in Theorems 7.1 and 7.2.
4. Produce the triangle of λ_i^j for the biplane of order 4.
5. Produce the triangle of λ_i^j for the projective plane of order n.
6. Let $\mathcal{P}$ be a set with m elements and let $\mathcal{B}$ consist of all subsets of $\mathcal{P}$ of size k. Show that $(\mathcal{P}, \mathcal{B}, \mathcal{I})$ is a design and determine its parameters.
7. Assume there exists a design with $t = 5$, $v = 72$, $k = 16$, and 249849 blocks. Construct the triangle of λ_i^j for this design. What is $\lambda_5 = \lambda$ for this design?
8. Determine the number of points in the intersection of two complements of lines in affine plane of order n.
9. Determine the parameters of the complementary design for affine and projective planes of order n.

7.2 Biplanes

We shall now examine a symmetric design similar to a projective plane, except that the number of times two lines meet is now two and there are two lines through any two points. This object is called a biplane. We begin with the definition.

Definition 7.2 (*Biplane*) A biplane is a set of points $\mathcal{P}$, a set of lines $\mathcal{L}$, and an incidence relation $\mathcal{I} \subseteq \mathcal{P} \times \mathcal{L}$ such that

1. Through any two points there are two lines incident with them. More precisely, given any two points $p, q \in \mathcal{P}$, there exists two lines $\ell, m \in \mathcal{L}$ with $(p, \ell) \in \mathcal{I}$, $(q, \ell) \in \mathcal{I}$, $(p, m) \in \mathcal{I}$ and $(q, m) \in \mathcal{I}$.
2. Any two lines intersect in two points. More precisely, given any two lines $\ell, m \in \mathcal{L}$, there exists two points p, q such that $p, q \in \mathcal{P}$ with $(p, \ell) \in \mathcal{I}, (p, m) \in \mathcal{I}, (q, \ell) \in \mathcal{I}, (q, m) \in \mathcal{I}$.

We shall now determine the number of points and lines in a biplane. Assume there are $n + 2$ points on a line. We shall call n the order of the plane. This is justified by the definition of order of a symmetric design since $\lambda = 2$. Given a point q off a line ℓ there are two lines connecting that point with each of the $n + 2$ points on ℓ. However, each line intersects the line ℓ exactly twice, and hence there are $\frac{(n+2)(2)}{2}$ lines through the point q. This gives the following.

Lemma 7.2 *In a biplane, there are* $n + 2$ *points on a line and* $n + 2$ *lines through a point.*

We can use this to get the number of points in a biplane.

Theorem 7.8 *Let* $\Pi = (\mathcal{P}, \mathcal{L}, \mathcal{I})$ *be a biplane of order* n, *then* $|\mathcal{P}| = |\mathcal{L}| = \frac{n^2 + 3n + 4}{2}$.

Proof Take a point q. There are $n + 2$ lines through q each of which has $n + 1$ points on it distinct from q. Each of these points is counted twice since any two lines through q intersect in a unique point distinct from q. Hence, there are $\frac{(n+2)(n+1)}{2}$ points distinct from q. Hence, there are $\frac{(n+2)(n+1)}{2} + 1 = \frac{n^2 + 3n + 4}{2}$ points in the biplane. By duality, the number of lines is the same. $\square$

Notice that $\frac{n^2 + 3n + 4}{2}$ is an integer (see Exercise 1) for all natural numbers n, and therefore no order is eliminated by this parameter not being an integer.

A biplane is a $2\text{-}(\frac{n^2 + 3n + 4}{2}, n + 2, 2)$ design. We have already given one example of a biplane in the previous section, namely, in Fig. 7.1. This is a biplane of order 4. We can give an example of a biplane of order 1 in the next example.

Example 7.1 Let $\mathcal{P} = \{A, B, C, D\}$ be a set of four points. Let the blocks be $\{A, B, C\}$, $\{A, B, D\}$, $\{A, C, D\}$, $\{B, C, D\}$. It is easy to see that through any two points there are exactly two blocks and that any two blocks have exactly two points in common. This is the biplane of order 1. We can represent it as follows (Fig. 7.2).

We note that there are $\frac{1^2 + 3(1) + 4}{2} = 4$ points and four blocks in this design as expected.

Example 7.2 We can now give an example of a biplane of order 2. Consider the projective plane of order 2. Recall that given the following diagram the set of hyperovals were exactly (Fig. 7.3)

Fig. 7.2 Biplane of order 1

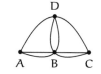

Fig. 7.3 Biplane of order 2

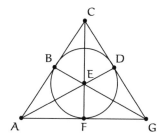

$$\{D, E, F, G\}$$
$$\{A, B, E, F\}$$
$$\{B, C, D, E\}$$
$$\{B, C, F, G\}$$
$$\{A, C, D, F\}$$
$$\{A, B, D, G\}$$
$$\{A, C, E, G\}$$

Notice that any two hyperovals intersect exactly twice and that there are seven of them. This gives the following. If Π is the projective plane of order 2 then the design is formed from the points of the plane and the blocks given by the hyperovals in a biplane of order 2. We note that there are $\frac{2^2+3(2)+4}{2} = 7$ points and blocks as expected.

We have just given examples of biplanes of orders 1, 2, and 4. At present, the only orders for which it is known that biplanes exist are $1, 2, 3, 4, 8, 9, 11$. The reader should be warned that sometimes in the literature the order of the biplane is given by $n + 2$ where n is the way order is defined here.

A nice description of the biplane of order 3 can be found in [14] "The fabulous $(11, 5, 2)$ Biplane". An interesting connection to coding theory is also presented. We shall construct this biplane later, see Theorem 10.15.

Exercises

1. Prove that $\frac{n^2+3n+4}{2}$ is always an integer.
2. Determine for which orders less than 100 the Bruck–Ryser–Chowla theorem rules out the existence of a biplane.

7.3   Symmetric Designs

One of the most interesting collections of designs is the symmetric designs. We have seen two examples of them already, namely, the projective planes ($\lambda = 1$) and the biplanes ($\lambda = 2$). We begin with the definition of a symmetric design.

Definition 7.3 A symmetric design is a 2-(v, k, λ) design where the number of points equals the number of blocks and the axioms are symmetric for points and blocks.

As we did for projective planes and for biplanes, it is a simple counting problem to determine the number of blocks and points in a symmetric design.

Theorem 7.9 *The number of blocks and the number of points in a symmetric 2-(v, k, λ) design of order n is $\frac{(n+\lambda-1)(n+\lambda)}{\lambda} + 1$.*

Proof Let there be $n + \lambda$ points on a block L. Through each point there are $n + \lambda - 1$ blocks other than L. Each of these blocks in the $(n + \lambda - 1)(n + \lambda)$ blocks is counted λ times. □

It is immediate from this that $(n + \lambda - 1)(n + \lambda)$ must be divisible by λ for a symmetric design to exist.

Theorem 7.10 *If a symmetric 2-(v, k, λ) design of order n exists, then $n^2 \equiv n$ (mod λ).*

Proof If the design exists then $\frac{(n+\lambda-1)(n+\lambda)}{\lambda}$ is an integer. Therefore, $(n + \lambda - 1)(n + \lambda)$ must be divisible by λ. This gives

$$(n + \lambda - 1)(n + \lambda) \equiv 0 \pmod{\lambda}$$
$$(n - 1)n \equiv 0 \pmod{\lambda}$$
$$n^2 \equiv n \pmod{\lambda},$$

which gives the result. □

Example 7.3 If $\lambda = 1$, namely, projective planes, then $n^2 \equiv n$ (mod 1) for all n and no orders are eliminated. If $\lambda = 2$, namely, biplanes, then $n^2 \equiv n$ (mod 2) for all n and no orders are eliminated. If $\lambda = 3$, then $n^2 \equiv n$ (mod 3) implies that n must be 0 or 1 (mod 3) which eliminates $n \equiv 2$ (mod 3).

Corollary 7.2 *If λ is a prime and a symmetric 2-(v, k, λ) design of order n exists, then $n \equiv 0$ or 1 (mod λ).*

Proof If λ is a prime, then $\mathbb{Z}_\lambda$ is a field and so there are only two solutions to $n^2 - n \equiv 0$ (mod λ), namely, 0 and 1. Then Theorem 7.10 gives the result. □

We shall show an example of an interesting symmetric design. Recall that $PG_m(\mathbb{F}_q)$ has points as one-dimensional subspaces of $\mathbb{F}_q^{m+1}$ and hyperplanes as m-dimensional subspaces of $\mathbb{F}_q^{m+1}$. We have seen before that the number of points and hyperplanes is that same and is $\frac{q^{m+1}-1}{q-1}$. There is a natural correlation between these two sets since the orthogonal of a one-dimensional subspace is an m-dimensional subspace in this setting. The number of points on a hyperplane is $\frac{q^m-1}{q-1}$ since it is simply the number of one-dimensional subspaces of a dimension m space.

To determine the parameter λ, consider two distinct points. We shall determine the number of blocks through these points. We need to count the number of m-dimensional subspaces of $\mathbb{F}_q^{m+1}$ containing two dimension one-dimensional subspaces.

Then we have

$$
\begin{aligned}
\lambda &= \frac{(q^{m+1} - q^2)(q^{m+1} - q^3) \cdots (q^{m+1} - q^{m-1})}{(q^m - q^2)(q^m - q^3) \cdots (q^m - q^{m-1})} \\
&= \frac{(q^{m-1} - 1)(q^{m-1} - q) \cdots (q^{m-1} - q^{m-3})}{(q^{m-2} - 1)(q^{m-2} - q) \cdots (q^{m-2} - q^{m-3})} \\
&= \frac{q^{m-1} - 1}{q - 1}.
\end{aligned}
$$

It is easy to determine n since $n + \lambda = k$. This means that

$$
n = \frac{q^m - 1}{q - 1} - \frac{q^{m-1} - 1}{q - 1} = \frac{q^m - q^{m-1}}{q - 1} = q^{m-1}. \tag{7.12}
$$

These results give the following.

Theorem 7.11 *The designs of points and hyperplanes in* $\mathrm{PG}_m(\mathbb{F}_q)$ *is a*

$$
2 - \left(\frac{q^{m+1} - 1}{q - 1}, \frac{q^m - 1}{q - 1}, \frac{q^{m-1} - 1}{q - 1} \right)
$$

symmetric design of order q^{m-1}.

We shall now consider the complement of a symmetric design and show that it is also a symmetric design. Consider a symmetric design with $n = \lambda$. It has

$$
\begin{aligned}
v &= \frac{(2n - 1)(2n)}{n} = 4n - 1, \\
k &= 2n, \\
\lambda &= n.
\end{aligned}
$$

Its complementary design has

$$
\begin{aligned}
v' &= 4n - 1, \\
k' &= 4n - 1 - 2n = 2n - 1, \\
\lambda' &= b - 2r + \lambda = 4n - 1 - 2(2n) + n = n - 1.
\end{aligned}
$$

This gives the following.

Theorem 7.12 *The complement of a symmetric* n-*design of order* n *is a symmetric* $(n-1)$-*design of order* n.

We have already seen an example of this theorem as the complementary design of the biplane of order 2 is the projective plane of order 2.

Exercises

1. Verify that the formulas in Theorem 7.9 are the same as were previously obtained for projective planes and biplanes.
2. Find all symmetric designs of orders 1, 2, and 3, by first finding all possible λ that satisfy the divisibility condition implicit in Theorem 7.9. Considering complementary designs should reduce the computation by half.
3. Verify that with the parameters in Theorem 7.11, we have that

$$v = \frac{(n+\lambda-1)(n+\lambda)}{\lambda} + 1.$$

7.4   Kirkman Schoolgirl Problem and Steiner Triple Systems

In [56], Kirkman asked the following question in a recreational publication:

Can 15 schoolgirls walk three abreast each day for a week such that each pair of girls walk abreast exactly once?

Kirkman found seven possible solutions and there are, in fact, exactly seven possible solutions to the problem up to isomorphism.

Essentially, the question asks to construct a resolvable design with $v = 15$ for the 15 girls, $k = 3$ since each group of girls walking abreast is a block, $r = 7$ since each day gives another block through each point, and $b = 35$ since there are 35 blocks. The value of λ is 1. The fact that the design is resolvable is vital since the girls are walking abreast.

As an example, if the girls are denoted by

$$\alpha, \beta, \gamma, \delta, \epsilon, \phi, \zeta, \eta, \theta, \iota, \kappa, \lambda, \mu, \nu, o$$

then a solution to the problem would be the following:

$$(\alpha\ \beta\ \gamma), (\delta\ \epsilon\ \phi), (\zeta\ \eta\ \theta), (\iota\ \kappa\ \lambda), (\mu\ \nu\ o)$$
$$(\alpha\ \delta\ \zeta), (\beta\ \epsilon\ \iota), (\gamma\ \phi\ \mu), (\eta\ \kappa\ \nu), (\theta\ \lambda\ o)$$
$$(\alpha\ \epsilon\ \nu), (\beta\ \delta\ o), (\gamma\ \eta\ \lambda), (\phi\ \theta\ \kappa), (\zeta\ \iota\ \mu)$$
$$(\alpha\ \theta\ \mu), (\beta\ \zeta\ \lambda), (\gamma\ \delta\ \kappa), (\epsilon\ \eta\ o), (\phi\ \iota\ \nu)$$
$$(\alpha\ \eta\ \iota), (\beta\ \kappa\ \mu), (\gamma\ \epsilon\ \theta), (\delta\ \lambda\ \nu), (\phi\ \zeta\ o)$$

$$(\alpha \phi \lambda), (\beta \ \theta \ v), (\gamma \ \iota \ o), (\delta \ \eta \ \mu), (\epsilon \ \zeta \ \kappa)$$
$$(\alpha \ \kappa \ o), (\beta \ \phi \ \eta), (\gamma \ \zeta \ v), (\delta \ \theta \ \iota), (\epsilon \ \lambda \ \mu).$$

This basic question generalizes to the idea of a Steiner triple system. A design where $k = 3$, that is, the block size is 3 is called a triple system. If $\lambda = 1$ and any two points are contained in a block, then the system is called a Steiner triple system. The parameters must satisfy the usual conditions, namely, $bk = vr$ and $(k - 1)r = \lambda(v - 1)$. Substituting $k = 3$ and $\lambda = 1$ gives

$$3b = vr$$
$$2r = v - 1.$$

It follows that

$$b = \frac{v(v - 1)}{6}$$
$$r = \frac{v - 1}{2}.$$

It follows that

$$v(v - 1) \equiv 0 \pmod 6$$
$$v - 1 \equiv 0 \pmod 2.$$

Therefore, a necessary condition for the existence of a Steiner triple system is that v must be either 1 or 3 (mod 6). Steiner first noticed this as a necessary condition and asked if it is also sufficient in [82]. It was proven that this was indeed the case by Reiss in [73]. Neither of these two men were aware of Kirkman's work years earlier. Of course, a solution to the Kirkman problem actually requires more, namely, that the blocks are resolvable. That is, they break up into classes which partition the space.

For $v = 13$, there are two non-isomorphic designs and for $v = 15$ there are 80. See [44] for details.

We shall now prove that these conditions are in fact sufficient as well as necessary. There are several ways of proceeding, we shall follow the path used by Hall in [44].

Theorem 7.13 *If there exists a Steiner triple system with v points and a Steiner triple system with v' points then there is a Steiner triple system with vv' points.*

Proof Let $D = (\mathcal{P}, \mathcal{B}, \mathcal{I})$ be a Steiner triple systems with v points and let $D' = (\mathcal{P}, \mathcal{B}, \mathcal{I})$ be a Steiner triple systems with v' points. Let $\mathcal{P}'' = \mathcal{P} \times \mathcal{P}'$. Define the set of blocks as $\mathcal{B}'' = \{(a_1, b_1), (a_2, b_2), (a_3, b_3) \mid (a_1 = a_2 = a_3$ and $(b_1, b_2, b_3) \in \mathcal{B}')$ or $(b_1 = b_2 = b_3$ and $(a_1, a_2, a_3) \in \mathcal{B})$ or $(a_1, a_2, a_3) \in \mathcal{B}$ and $(b_1, b_2, b_3) \in \mathcal{B}'\}$. Take any two points (a_1, b_1) and (a_2, b_2). If $a_1 = a_2$ there is a unique block in $\mathcal{B}'$ that contains b_1 and b_2, say the third point is b_3 then the two points are in the block

$$\{(a_1, b_1), (a_1, b_2), (a_1, b_3)\}.$$

If $b_1 = b_2$ there is a unique block in $\mathcal{B}$ that contains a_1 and a_2, say the third point is a_3 then the two points are in the block

$$\{(a_1, b_1), (a_2, b_1), (a_3, b_1)\}.$$

If $a_1 \neq a_2$ and $b_1 \neq b_2$ then let a_3 be the third point on the block containing a_1 and a_2 in $\mathcal{B}$ and b_3 be the third point on the block containing b_1 and b_2 in $\mathcal{B}'$, then the block

$$\{(a_1, b_1), (a_2, b_2), (a_3, b_3)\}.$$

Then these blocks make a Steiner triple system on vv' points. $\qquad\square$

Notice in the previous theorem by fixing the first or second coordinate there are subsystems with v_1 and v_2 points, respectively.

We shall verify that the number of points is $\frac{vv'(vv'-1)}{6}$ as it should be. The number of points of the first type is the number of blocks in $\mathcal{B}$ times the number of points in $\mathcal{P}'$ which is $\frac{v(v-1)v'}{6}$. The number of points of the second type is the number of blocks in $\mathcal{B}'$ times the number of points in $\mathcal{P}$ which is $\frac{v'(v'-1)v}{6}$. Let (a, b, c) be a block in $\mathcal{B}$ and (a', b', c') be a block in $\mathcal{B}'$. Then the following are blocks in $\mathcal{B}''$:

$$\{(a, a'), (b, b'), (c, c')\}, \{(a, a'), (b, c'), (c, b')\},$$
$$\{(a, b'), (b, a'), (c, c')\}, \{(a, b'), (b, c'), (c, a')\},$$
$$\{(a, c'), (b, a'), (c, b')\}, \{(a, c'), (b, b'), (c, a')\}.$$

Thus, the number of points of the third type is six times $|\mathcal{B}||\mathcal{B}'|$, which is $\frac{v(v-1)v'(v'-1)}{6}$. Then we add the three terms to get

$$\frac{v(v-1)v'}{6} + \frac{v'(v'-1)v}{6} + \frac{v(v-1)v'(v'-1)}{6} = \frac{vv'(v-1+v'-1+(v-1)(v'-1))}{6}$$

$$= \frac{vv'(vv'-1)}{6}$$

as desired.

Theorem 7.14 *If there exists a Steiner triple system with v_1 points and a Steiner triple system with v_2 points, where the system with v_2 points contains a subsystem with v_3 points, then there is a Steiner triple system with $v_3 + v_1(v_2 - v_3)$ points.*

Proof Let $D_1 = (\mathcal{P}_1, \mathcal{B}_1, \mathcal{I}_1)$ be a Steiner triple system with v_1 points, $D_2 = (\mathcal{P}_2, \mathcal{B}_2, \mathcal{I}_2)$ be a Steiner triple system with v_2 points, and $D_3 = (\mathcal{P}_3, \mathcal{B}_3, \mathcal{I}_3)$ be a subsystem of D_2 with v_3 points.

Define the point set $\mathcal{P}' = \mathcal{P}_3 \cup (\mathcal{P}_1 \times (\mathcal{P}_2 - \mathcal{P}_3))$. Note that $|\mathcal{P}'| = v + 3 + v_1(v_2 - v_3)$.

List the points of $\mathcal{P}_3$ by $p_1, p_2, \ldots, p_{v_3}$. The block set $\mathcal{B}'$ consists of the following types:

- Blocks in $\mathcal{B}_3$;
- Triples of the form $\{a, b, c\}$, where $a \in \mathcal{P}_3$, $b, c \in \mathcal{P}_2 - \mathcal{P}_3$, with $\{a, b, c\}$ a block in $\mathcal{B}_2$; and
- Triples of the form $\{(b, p_i), (c, p_j), (d, p_k) \mid i + j + k \equiv 0 \pmod{v_2 - v_3}$ and $(b, c, d) \in \mathcal{B}_1\}$.

If two points are in $\mathcal{P}_3$, then the block containing the two points in $\mathcal{B}_3$ is the block. If the point $a \in \mathcal{P}_3$ and another point $(b, p_i) \in \mathcal{P}_1 \times (\mathcal{P}_2 - \mathcal{P}_3)$, then we take the block $\{a, b, c\}$ in $\mathcal{B}_2$.

If two points (b, p_i) and (c, p_j) are in $\mathcal{P}_1 \times (\mathcal{P}_2 - \mathcal{P}_3)$ then we take a block of the third kind. Specifically, there is a unique d with $(b, c, d) \in \mathcal{B}_1$ and a unique k with $i + j + k \equiv 0 \pmod{v_2 - v_3}$. Then this is a Steiner triple system of the desired size. $\qquad\square$

It can be shown by using Theorems 7.13 and 7.14 that if $v \equiv 1$ or $3 \pmod 6$ then there exists a Steiner triple system with v points.

This does not answer the question of whether there exist a resolvable Steiner triple system, which is referred to as a Kirkman triple system. If the design is resolvable then v must be three $\pmod 6$ since three must divide the number of points. Hence, the question is when does there exist a $(v, 3, 1)$ resolvable design. The question remained open until 1970 when the following was shown by Ray-Chaudhuri and Wilson [71].

Theorem 7.15 *A resolvable $(v, 3, 1)$-design exists if and only if $v \equiv 3 \pmod 6$.*

Exercises

1. There is a unique Steiner triple system with $v = 3$ and a unique Steiner triple system with $v = 7$. Produce these two, the first is trivial and the second will be familiar to you.
2. Prove that the affine plane of order 3 is a Kirkman triple system with nine points.

7.5   **Nets and Transversal Designs**

In this section, we shall give another example of a resolvable design. Finite nets were introduced by Bruck in 1951, see [16]. Infinite nets had already been introduced under the German name *gewebe*, in 1928, by Reidemeister, see [72]. In some sense, one can think of a finite net as an incomplete affine plane. However, not all nets are extendable to an affine plane. In fact, it is a fundamental question of the theory, asking which nets have extensions to affine planes. They were introduced to help understand the structure of planes and to help determine for which orders planes exist. We shall see that nets are also canonically isomorphic to mutually orthogonal Latin squares as well, and are essentially a geometric way of looking at combinatorial structure.

Definition 7.4 A k-net of order n is an incidence structure consisting of n^2 points and nk lines satisfying the following four axioms:

1. Every line has n points.
2. Parallelism is an equivalence relation on lines, where two lines are said to be parallel if they are disjoint or identical.
3. There are k parallel classes each consisting of n lines.
4. Any two non-parallel lines meet exactly once.

It is immediate that a $(n + 1)$-net of order n is an affine plane of order n.

One of the most important tools in the study of nets is transversals. This is because transversals can be used to extend nets. We now give the definition of a transversal.

Definition 7.5 A transversal of a net is a set of n points having exactly one point in common with each line of the net.

Example 7.4 Consider the diagram of a 4-net of order 5. We mark a transversal with X. Notice that the transversal hits each line exactly once (Fig. 7.4).

Fig. 7.4 Transversal on a 4-net of order 5

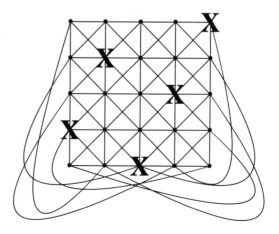

A k-net of order n is extendable if it is a subnet of a $(k + 1)$-net of order n. An extendable net must have n mutually exclusive transversals.

Proposition 7.1 *An $(n + 1)$-net of order n has no transversals.*

Proof Assume t is a transversal and let p and q be points on the transversal t. We know that there is a line through p and q since the net is an affine plane. This contradicts that t is incident with this line only once.   □

The following theorem is similar to Theorem 4.10.

Theorem 7.16 *A set of $(k - 2)$-MOLS of order n exists if and only if a k-net of order n exists.*

Proof Associate the n^2 points of the plane with the n^2 coordinates of the Latin squares. We can describe the points as (a, b) where $a, b \in \{0, 1, \dots, n - 1\} = \mathbb{Z}_n$.

Assume we have a set of $k - 2$ MOLS of order n on the alphabet $\{0, 1, \dots, n - 1\}$. As with the affine plane the first two parallel classes have the lines that are the horizontal and vertical lines of the grid. Let $L^s = (L_{ij})$ be the s-th Latin square, then the $(s + 2)$-nd parallel class has lines corresponding to the symbols of the square. That is, the c-th line is incident with the point (i, j) if and only if $L_{ij} = c$. The proof of Theorem 4.10 shows that this forms a k-net.

If we assume we have a k-net of order n then we use the first two parallel classes to describe the grid. That is, the point (i, j) is the point of intersection of the i-th line of the first parallel class and the j-th line of the second parallel class. Then we reverse the construction, $L^k_{i,j} = c$ if and only if the c-th line of the $(k + 2)$-nd parallel class is incident with the line (i, j). As in the proof of Theorem 4.10, it is easy to see that these form $k - 2$ MOLS of order n.   □

Example 7.5 As an example, consider the diagram in Fig. 7.5 of a 4-net of order 5 formed from the following Latin squares:

$$
\begin{array}{ccccc}
0\ 1\ 2\ 3\ 4 & \quad & 0\ 1\ 2\ 3\ 4 \\
1\ 2\ 3\ 4\ 0 & & 4\ 0\ 1\ 2\ 3 \\
2\ 3\ 4\ 0\ 1 & & 3\ 4\ 0\ 1\ 2 \\
3\ 4\ 0\ 1\ 2 & & 2\ 3\ 4\ 0\ 1 \\
4\ 0\ 1\ 2\ 3 & & 1\ 2\ 3\ 4\ 0
\end{array}
$$

Any Latin square of order n is equivalent to a 3-net of order n. For example, the Latin square

$$
\begin{pmatrix}
1 & 2 & 3 \\
3 & 1 & 2 \\
2 & 3 & 1
\end{pmatrix}
$$

Fig. 7.5 4-net of order 5

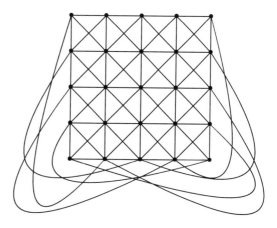

Fig. 7.6 3-net of order 3

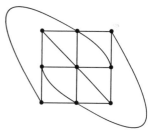

is equivalent to the 3-net in Fig. 7.6.

In terms of design parameters, a k-net of order n is a $1 - (n^2, n, k)$ design. The following is the standard definition of the direct product of nets.

Definition 7.6 Let N_k, N'_k be k-nets of order n and n', respectively. The direct product of nets $N_k \times N'_k$ is defined as follows:

1. the points $N_k \times N'_k$ are ordered pairs (q, q') with q a point of N_k and q' a point of N'_k;
2. the lines of $N_k \times N'_k$ are ordered pairs (m, m') with m, m' from the i-th parallel class of N_k and N'_k, respectively; and
3. the point (q, q') is incident with the line (m, m') if and only if q is incident with m and q' is incident with m'.

Example 7.6 The direct product of the 3-net of order 2 formed by the Latin square $\begin{pmatrix} 1 & 2 \\ 2 & 1 \end{pmatrix}$ and the 3-net of order 3 formed by the Latin square $\begin{pmatrix} 1 & 2 & 3 \\ 3 & 1 & 2 \\ 2 & 3 & 1 \end{pmatrix}$ is the 3-net formed by the Latin square:

$$\begin{pmatrix} A & B & C & D & E & F \\ C & A & B & F & D & E \\ B & C & A & E & F & D \\ D & E & F & A & B & C \\ F & D & E & C & A & B \\ E & F & D & B & C & A \end{pmatrix}.$$

The next lemma follows immediately from Exercise 2.

Lemma 7.3 *If there exists an* h-*net of order* n *and an* h-*net of order* n' *then there exists an* h-*net of order* nn'.

Theorem 7.17 *Let* $n = \prod p_i^{e_i}$, p_i *prime, with* $p_i > 2$ *or if* $p_i = 2$ *then* $e_i > 1$. *Then there exists a* 4-*net of order* n, *and therefore a Graeco-Latin square of order* n.

Proof We know that for any prime $p > 2$ there exists a $(p^e - 1)$-net of order p^e since there exists an affine plane of any prime power order. If $e > 1$, then $2^e - 1 > 2$ and there exists a $(2^e - 1)$-net of order 2^e. By repeatedly applying Lemma 7.3, we get that there exists at least a 4-net of order n. $\square$

This theorem means that the only orders for which there may not be a 4-net are 2 (mod 4). We have already stated that there is not a 4-net of order 2 and 6 and have exhibited a 2-net for $n = 10$. It is known that there exists a pair of MOLS for orders $n \equiv 2 \pmod 4$ for $n > 6$, and hence a 4-net, see [11,12]. Hence, the only orders for which there is no Graeco-Latin square are 2 and 6.

We shall now describe a design which is another way of examining MOLS which is dual to a net.

Definition 7.7 A transversal design D of order n, block size k, is a set of points V and blocks B such that

- V has kn points.
- V is partitioned into k classes, each of size n.
- Every pair of points is either contained in the same set of the partition or in a block.

There are more general definitions but this is the one we shall use.

Theorem 7.18 *The dual of a transversal design on* kn *points with block size* k *is a* k-*net of order* n.

Proof We notice that in the transversal design every pair of points is in the same set or in a block and so every pair of blocks in the dual either intersects once or is parallel. The number of blocks in the dual is kn and they are split into k parallel classes by the second axiom of transversal designs.   □

Exercises

1. Prove that any 2-net of order n has $n!$ transversals.
2. Prove that $N_k \times N'_k$ is a net of order nn'. Be sure to count the number of points, lines, and parallel classes are all correct.
3. Determine the parameters of a transversal design that would be a solution to the 36 officer problem.
4. Take the direct product of a 4-net (affine plane) of order 3 with itself to produce a Graeco-Latin square of order 9.
5. Determine the parameters of a transversal design that would be an affine plane.
6. Draw the net corresponding to 2 MOLS of order 10.
7. Prove that the 3-net corresponding to the Latin square:

$$\begin{pmatrix} 0\ 1\ 2 \\ 1\ 2\ 0 \\ 2\ 0\ 1 \end{pmatrix}$$

   is isomorphic to the 3-net corresponding to the Latin square:

$$\begin{pmatrix} 0\ 1\ 2 \\ 2\ 0\ 1 \\ 1\ 2\ 0 \end{pmatrix}.$$

8. Find all possible k-nets of order n, up to isomorphism, for $k \le n+1 \le 4$.
9. Prove that the 3-net formed from the following Latin square has no transversals:

$$\begin{pmatrix} 1\ 2\ 3\ 4 \\ 4\ 1\ 2\ 3 \\ 3\ 4\ 1\ 2 \\ 2\ 3\ 4\ 1 \end{pmatrix}.$$

10. Prove that the number of blocks through a point in a transversal design is n and use this to show there are n^2 blocks.

Combinatorial Objects

8.1   Introduction to Hadamard Matrices

Hadamard matrices were studied by Hadamard [42] in 1893 as solutions to an extremal problem in real analysis and they have numerous applications throughout mathematics. Sylvester [89] had already studied them.

Definition 8.1 A square n by n matrix H with $H_{ij} \in \{1, -1\}$ is said to be a Hadamard matrix if $HH^T = nI_n$, where I_n is the n by n identity matrix.

This implies that if $\mathbf{v}_1, \mathbf{v}_2, \ldots, \mathbf{v}_n$ are the rows of H then

$$\mathbf{v}_i \cdot \mathbf{v}_j = \begin{cases} 0 \text{ if } i \neq j \\ n \text{ if } i = j \end{cases},$$

where the operations are done in the integers.

Since $HH^T = nI_n$ we have that $H^TH = nI_n$ as well. This gives that any two distinct columns have dot product 0 as well.

Theorem 8.1 *If H is a Hadamard matrix of size n then H^T is a Hadamard matrix of size n.*

Proof We have seen that if the dot product of any distinct rows is 0 and any row with itself is n, then the same is true for the columns. This gives the result.   □

If $n = 1$ there are two matrices, namely, (1) and (-1). Notice that since there is only 1 row these matrices satisfy the definition trivially. For $n = 2$, there are eight different Hadamard matrices. They are

S. T. Dougherty, *Combinatorics and Finite Geometry*,
Springer Undergraduate Mathematics Series,
https://doi.org/10.1007/978-3-030-56395-0_8

$$\begin{pmatrix} x & x \\ x & -x \end{pmatrix}\begin{pmatrix} x & x \\ -x & x \end{pmatrix}\begin{pmatrix} x & -x \\ x & x \end{pmatrix}\begin{pmatrix} -x & x \\ x & x \end{pmatrix}, \qquad (8.1)$$

where $x = 1$ and $x = -1$.

Theorem 8.2 *Let H be an n by n Hadamard matrix with $n > 2$, then $n = 4k$ for some k.*

Proof Take two rows in a n by n Hadamard matrix, let $\chi_{i,j}$ denote the number of times the pair (i, j) occurs when overlapping the two rows. Then we have

$$\chi_{1,1} + \chi_{-1,-1} + \chi_{1,-1} + \chi_{-1,1} = n$$
$$2(\chi_{1,1} + \chi_{-1,-1}) = n$$

since $\chi_{1,1} + \chi_{-1,-1} = \chi_{1,-1} + \chi_{-1,1}$ because the dot product of the two rows must be 0. This gives that n is an even integer.

This gives that $n = 2m$ for some integer m and that any two rows of the matrix must agree in exactly m places. Now assume that n is at least 4 and let r_1, r_2, r_3 be three rows of the Hadamard matrix. If we assume that r_1 and r_2 agree in the coordinates of the set A and disagree in the coordinate set D. Then, let B be the subset of A where r_1 agrees with r_3 as well and E be the subset of D where r_3 disagrees with r_2. Notice that $|B| + |E| = m$ since this counts the number of places where r_3 agrees with r_1. Then, we have

$$|B| + |D - E| = m$$
$$|B| + |D| - |E| = m$$
$$|B| + m - |E| = m$$
$$|B| = |E|.$$

This gives that

$$|B| + |E| = m$$
$$|B| + |B| = m$$
$$2|B| = m$$

and m is even which makes n divisible by 4.   □

Given this theorem we see why usually one assumes that n is at least 4 when discussing Hadamard matrices. From this point on when referring to a Hadamard matrix of size $4n$ we shall say that n is the order of the matrix. It will become clear why we do this after we show how to construct a design from a Hadamard matrix.

Let $H = (h_{ij})$ and M be Hadamard matrices of size n and k, respectively. Define the Kronecker product of H and M to be the square matrix of size nk given by

$$H \otimes M = \begin{pmatrix} h_{11}M & h_{12}M & \dots & h_{1n}M \\ h_{21}M & h_{22}M & \dots & h_{2n}M \\ & & \vdots & \\ h_{n1}M & h_{n2}M & \dots & h_{nn}M \end{pmatrix}. \tag{8.2}$$

Theorem 8.3 *If H and M are Hadamard matrices of size n and k, respectively, then $H \otimes M$ is a Hadamard matrix of size nk.*

Proof Let x_i be the i-th row of M. We have

$$(h_{a1}x_i, h_{a2}x_i, \dots, h_{an}x_i) \cdot (h_{b1}x_j, h_{b2}x_j, \dots, h_{bn}x_j)$$
$$= (x_i \cdot x_j) \sum_{\alpha=1}^{n} h_{a\alpha}h_{b\alpha}.$$

If $i = j$ and $a = b$ then this value is nk. If $i \neq j$ then $x_i \cdot x_j = 0$ and so the value is 0. If $i = j$ and $a \neq b$ then the second sum is 0. Hence, the dot product of a row with itself is the size of the matrix and the dot product of any two distinct rows is 0. Therefore, the Kronecker product is a Hadamard matrix. $\square$

As an example, let

$$H = \begin{pmatrix} 1 & 1 \\ 1 & -1 \end{pmatrix}, \quad M = \begin{pmatrix} 1 & 1 & 1 & 1 \\ -1 & -1 & 1 & 1 \\ 1 & -1 & -1 & 1 \\ -1 & 1 & -1 & 1 \end{pmatrix}. \tag{8.3}$$

Then

$$H \otimes M = \begin{pmatrix} M & M \\ M & -M \end{pmatrix} = \begin{pmatrix} 1 & 1 & 1 & 1 & 1 & 1 & 1 & 1 \\ -1 & -1 & 1 & 1 & -1 & -1 & 1 & 1 \\ 1 & -1 & -1 & 1 & 1 & -1 & -1 & 1 \\ -1 & 1 & -1 & 1 & -1 & 1 & -1 & 1 \\ 1 & 1 & 1 & 1 & -1 & -1 & -1 & -1 \\ -1 & -1 & 1 & 1 & 1 & 1 & -1 & -1 \\ 1 & -1 & -1 & 1 & -1 & 1 & 1 & -1 \\ -1 & 1 & -1 & 1 & -1 & 1 & -1 & 1 \end{pmatrix}. \tag{8.4}$$

Corollary 8.1 *There exists Hadamard matrices of arbitrarily large order.*

Proof Let H be a Hadamard matrix of size 2. Then taking the repeated Kronecker product with itself t times gives a Hadamard matrix of size 2^t. This gives that there exists a Hadamard matrix of all orders that are a power of 2. Hence, there exists Hadamard matrices of arbitrarily large size.   □

This does not prove that there are Hadamard matrices of all possible orders. In fact, it is an open question as to whether there exists Hadamard matrices of all possible orders. The smallest unknown case is $4(167) = 668$. See [55] for details and a construction of the previous largest unknown case, which was 428. However, we do have the following conjecture.

Conjecture 8.1 There exists Hadamard matrices for all sizes of the form $4n$.

We shall describe another construction of Hadamard matrices. It is known as the Payley construction. To give the construction we shall require some more information about finite fields. If F is a field, then a is a square in F if there exists $b \in F$ with $b^2 = a$. We usually consider 0 to be a special case so we shall only consider non-zero elements in the next examples. Consider the field $\mathbb{F}_3$ then $1 = 1^2 = 2^2$ and so 1 is a square and 2 is a non-square. For $\mathbb{F}_5$, $1 = 1^2 = 4^2$ and $4 = 2^2 = 3^2$ and so 1 and 4 are squares and 2 and 3 are non-squares.

Lemma 8.1 *In a finite field of odd order, exactly half of the non-zero elements are squares.*

Proof For each x we have $x^2 = (-x)^2$. Since, the order is odd, we have $x \neq -x$. Therefore, the cardinality of the set of squares is at most half the cardinality of the non-zero elements. Given a square, say $a = b^2$, the equation $x^2 - a = 0$ factors into $(x - b)(x + b) = 0$. Since a finite field has no zero divisors, this equation has at most two solutions. Therefore, precisely half of the non-zero elements are squares.   □

Example 8.1 Consider the field of order 7. Here $1^2 = 6^2 = 1$, $2^2 = 5^2 = 4$, $3^2 = 4^2 = 2$. Therefore, the squares are $1, 2$, and 4 and the non-squares are $3, 5$, and 6.

Example 8.2 For fields of even order Lemma 8.1 is not true. Consider the field of order 4, $\{0, 1, \omega, \omega^2\}$. Then $1^2 = 1$, $\omega^2 = \omega^2$, $(\omega^2)^2 = \omega$. Therefore, every element is a square.

Define the following function $\chi : \mathbb{F}_q \to \{0, 1, -1\}$:

$$\chi(x) = \begin{cases} 0 & x = 0, \\ 1 & x \text{ is a square,} \\ -1 & x \text{ is not a square.} \end{cases} \tag{8.5}$$

Lemma 8.2 *If* $a, b \in \mathbb{F}_q$ *then* $\chi(ab) = \chi(a)\chi(b)$.

Proof If a or b is 0 then the result is trivial so we assume that neither a nor b is 0. First we note that if a and b are both squares, then their product is a square, that is, if $a = c^2$ and $b = d^2$ then $ab = (cd)^2$. Hence, multiplication on the set of squares is closed. Let S be the set of squares then, if e is a non-square, the set $\{fe \mid fe \in S\}$ is the set of all non-squares since the number of non-squares is equal to the number of squares and the map is injective. Then, if neither a nor b are squares then $a = ef$ and $b = eg$ where $f = k^2$ and $g = h^2$. Then $ab = efeg = (ehk)^2$.

Now consider the case if one of a and b is square and the other is not. Without loss of generality assume $a = c^2$ and b is not a square. If their product were m^2 then $c^2 b = m^2$ and $b = \frac{m}{c}^2$ which is a contradiction. $\square$

The previous proof could be simplified a great deal if we had the theory of groups which we discuss in the next chapter.

We shall let I denote the identity matrix, that is, the matrix that is 0 off the diagonal and 1 along the diagonal and we shall let J denote the matrix where every element is 1. Let M be a matrix indexed by the elements of $\mathbb{F}_q$ with

$$M_{a,b} = \chi(b - a). \tag{8.6}$$

Notice that M has 0 on the diagonal.

We form the matrix

$$H = \begin{pmatrix} 1\,1\,1 & \cdots & 1 \\ 1 & & \\ \vdots & M - I & \\ \vdots & & \\ 1 & & \end{pmatrix}. \tag{8.7}$$

Theorem 8.4 *If* $q \equiv 3 \pmod 4$ *then the matrix* H *defined in Eq. 8.7 is a Hadamard matrix of size* $q + 1$.

Proof First we note that every element of H is either 1 or -1. We shall prove that $HH^\mathsf{T} = (q + 1)I$.

Using Exercise 6 we see that

$$H^\mathsf{T} = \begin{pmatrix} 1\,1\,1 & \cdots & 1 \\ 1 & & \\ \vdots & -M - I & \\ \vdots & & \\ 1 & & \end{pmatrix}.$$

It is straightforward to see that

$$
HH^T = \begin{pmatrix} q+1\,0\,0 & \cdots & 0 \\ 0 & & \\ \vdots & J + (M-I)(-M-I) & \\ \vdots & & \\ 0 & & \end{pmatrix}.
$$

Now we shall consider the value of $(M-I)(-M-I) = -M^2 + M - M + I = -M^2 + I$. All that is needed now is to show that $-M^2 = MM^T = qI - J$. If the a-th row in M is $(m_0, m_1, \ldots, m_{q-1})$, then $m_a = 0$ and the a-th column of M^T is the a-th row of M and so we have $MM^T_{a,a} = q - 1$. This is as we expected, i.e., $q - 1$ down the diagonal. Now consider the a-th row of M and the b-th column of M^T which is the b-th row of M. The inner product of these two rows is

$$
(MM^T)_{a,b} = \sum_{c \in \mathbb{F}_q} \chi(c-a)\chi(c-b)
$$

$$
= \sum_{d \in \mathbb{F}_q} \chi(d)\chi(d+e) \text{ with } d = c - a, e = a - b
$$

$$
= \sum_{d \in \mathbb{F}_q} \chi(d)\chi(df) \text{ with } df = d + e \text{ notice } f \neq 1, d = 0 \text{ drops out}
$$

$$
= \sum_{d \in \mathbb{F}_q} \chi(d)\chi(d)\chi(f)
$$

$$
= \sum_{f \neq 1} \chi(f) = -1.
$$

This gives the result.   □

As an example, consider the finite field of order 3. Then indexing by $0, 1, 2$ we have

$$
M = \begin{pmatrix} 0 & 1 & -1 \\ -1 & 0 & 1 \\ 1 & -1 & 0 \end{pmatrix}.
$$

Then the Hadamard matrix is

$$
H = \begin{pmatrix} 1 & 1 & 1 & 1 \\ 1 & -1 & 1 & -1 \\ 1 & -1 & -1 & 1 \\ 1 & 1 & -1 & -1 \end{pmatrix}.
$$

1. Show that multiplying a column or a row by -1 in a Hadamard matrix produces a Hadamard matrix. Use this to show that each Hadamard matrix can be transformed into a Hadamard matrix where the first row and column are all one.
2. Give a complete proof of Theorem 8.1.
3. Show that there are no Hadamard matrices with $n = 3$.
4. Prove that -1 is a square in the finite field $\mathbb{Z}_q$, q a prime, if and only if $q \equiv 1 \pmod 4$.
5. Construct a 4 by 4 Hadamard matrix, i.e., a Hadamard matrix of order 1.
6. Use Exercise 4 and Lemma 8.2 to prove that if $q \equiv 3 \pmod 4$ then $M_{a,b} = (-1)M_{b,a}$.
7. Find all the squares and non-squares in the field of order 13.
8. Construct the Hadamard matrix from the Paley construction for the finite field of order 7.
9. Determine where the proof of Theorem 8.4 fails when $q \equiv 1 \pmod 4$. Hint: consider MM^{T} in this case.

8.2 Hadamard Designs

We shall show how Hadamard matrices relate to designs.

Let $\{p_1, p_2, \ldots, p_{4n}\} = \mathcal{P}$ be a set of points corresponding to the coordinates of a Hadamard matrix H_i of size $4n$. Denote by H_i the i-th row of H and let $H_i = \{x_1, x_2, \ldots, x_{4n}\}$. For any j different from i with $H_j = \{y_1, y_2, \ldots, y_{4n}\}$ let

$$B_j = \{p_\alpha \mid y_\alpha = x_\alpha\} \tag{8.8}$$

and

$$\overline{B_j} = \mathcal{P} - B_j. \tag{8.9}$$

Let the block set be

$$\mathcal{B} = \{B_j \mid j \neq i\} \cup \{\overline{B}_j \mid j \neq i\}. \tag{8.10}$$

The incidence between a point and a block is given by set-theoretic containment.

Lemma 8.3 *Given a Hadamard matrix of size $4n$ the incidence structure defined above is a $3 - (4n, 2n, n - 1)$ design.*

Proof The fact that $v = 4n$ is obvious. Given any two rows H_i and H_j, we know that the number of coordinates in which they agree is equal to the number in which they disagree since if they agree in q_1 places and disagree in q_2 places then the dot product $H_i \cdot H_j = q_1 - q_2$. However, we know that the dot product must be 0 and hence $q_1 = q_2 = 2n$ since $q_1 + q_2 = 4n$. This gives that $k = 2n$.

Take any three points from $\mathcal{P}$, that is, any three coordinates of the matrix H. It was shown in the proof of Theorem 8.2 that any three rows of a Hadamard matrix of size $4n$ coincide in exactly n places. Now we can assume that the row H_i is all ones by multiplying coordinates by -1 if necessary. Hence, the number of blocks through any three points is $n - 1$, corresponding to the $n - 1$ coordinates (not including the coordinates in H_i) where the three columns coincide. Hence, we have $\lambda = n - 1$ and $t = 3$. $\square$

It follows from the construction that there are $8n - 2 = 2(4n - 1)$ blocks and hence we can compute the following triangle of λ_i^j for the Hadamard $3 - (4n, 2n, n - 1)$ design:

$$
\begin{array}{ccccccc}
 & & & 8n - 2 & & & \\
 & & 4n - 1 & & 4n - 1 & & \\
 & 2n - 1 & & 2n & & 2n - 1 & \\
n - 1 & & n & & n & & n - 1
\end{array}
\qquad (8.11)
$$

Theorem 8.5 *Let* $D = (\mathcal{P}, \mathcal{B})$ *be a* $3 - (4n, 2n, n - 1)$ *design. Then if* D' *is formed by removing a point from* $\mathcal{P}$ *and the blocks not incident with it, we have that* D' *is a symmetric* $2 - (4n - 1, 2n - 1, n - 1)$ *design.*

Proof The facts that $v = 4n - 1$ and $k = 2n - 1$ are obvious since we removed the point p. Let q_1 and q_2 be two points. Through q_1, q_2, and p there were $n - 1$ blocks in D, and hence these $n - 1$ blocks are incident with q_1 and q_2 in D'. If a block in D was incident with q_1 and q_2 but not incident with p, then its restriction would not be a block in D'. Hence D' is a $2 - (4n - 1, 2n - 1, n - 1)$ design. It is a simple computation then to compute that the number of blocks is equal to the number of points which is $4n - 1$. $\square$

Its triangle of λ_i^j is given by

$$
\begin{array}{ccccc}
 & & 4n - 1 & & \\
 & 2n & & 2n - 1 & \\
n & & n & & n - 1
\end{array}
\qquad (8.12)
$$

Lemma 8.4 *If there exists a* $3 - (4n, 2n, n - 1)$ *design, then there exists a Hadamard matrix of size* $4n$.

Proof First we need to show that any two blocks are either disjoint or meet in n places. Assume that two blocks do meet and take any point p in their intersection. Using Theorem 8.5 we see that D' formed from the $3 - (4n, 2n, n - 1)$ by removing the point p is a symmetric $2 - (4n - 1, 2n - 1, n - 1)$ design. Hence, any two blocks in D' meet in $n - 1$ places. However, since they are blocks in D' they also intersect at p and hence meet at n places.

Next we need to show that the complement of a block is indeed a block. Let B be a block. Through each point on the block there are $4n - 2$ blocks different from B since there are $4n - 1$ blocks through each point. There are $2n$ points on the block. Hence, the number of blocks that intersects B is $\frac{(4n-2)(2n)}{n} = 8n - 4$. Each block is counted n times, since it intersects B in n places. However, there are $8n - 2$ blocks, which gives that there is B and $8n - 4$ blocks that intersect B making $8n - 3$ blocks which implies there must be a block that does not intersect B.

Use the $4n$ points of the design to make n coordinates for a matrix H. Then for each pair of disjoint blocks B and $\overline{B}$ define a row v with $v_\alpha = 1$ if p_α is incident with B and $v_\alpha = -1$ if p_α is incident with $\overline{B}$. Then add a row of ones as the first row of the matrix. This gives that every row (except the first) has n coordinates with a 1 and n coordinates with a -1. If v is a row corresponding to the pair B, $\overline{B}$ and w is a row corresponding to the pair B', $\overline{B}'$ then B shares n coordinates with B' and n coordinates with $\overline{B}'$ and the same is true for $\overline{B}$. Hence, the dot product of v with w is 0. Hence, the matrix H is a Hadamard matrix.   □

Theorem 8.6 *A Hadamard matrix of size $4n$ exists if and only if a $3 - (4n, 2n, n - 1)$ design exists.*

Proof Follows from Lemmas 8.3 and 8.4.   □

This design and the corresponding 2-design given in Theorem 8.5 are called Hadamard designs.

As an example consider the following Hadamard matrix:

$$\begin{pmatrix} 1 & 1 & 1 & 1 \\ 1 & 1 & -1 & -1 \\ 1 & -1 & 1 & -1 \\ 1 & -1 & -1 & 1 \end{pmatrix}.$$

Denote the four points by p_1, p_2, p_3, p_4 corresponding to the four coordinates and let H_i in the construction be H_1. Then the blocks of the 3-design are

$$\{p_1, p_2\}, \{p_3, p_4\}, \{p_1, p_3\}, \{p_2, p_4\}, \{p_1, p_4\}, \{p_2, p_3\}.$$

Using p_1 we construct a 2-design with blocks

$$\{p_2\}, \{p_3\}, \{p_4\}.$$

Then the incidence matrix of the 2-design is given by

$$\begin{pmatrix} 1 & 0 & 0 \\ 0 & 1 & 0 \\ 0 & 0 & 1 \end{pmatrix}.$$

Notice that if every 0 is replaced by -1 and a first row and column of all ones are added then the incidence matrix produces the Hadamard matrix.

1. Construct the 3-design and 2-design formed from the Hadamard matrix given by the product of H and M in (8.3).

8.3   Generalized Hadamard Matrices

In this section, we shall give a generalization of Hadamard matrices. We begin with the definition.

Definition 8.2 A generalized Hadamard matrix $H(q, \lambda) = (h_{ij})$ of order $n = q\lambda$ over $\mathbb{F}_q$ is a $q\lambda \times q\lambda$ matrix with entries from $\mathbb{F}_q$ such that for every i, j, $1 \leq i < j \leq q\lambda$, each of the multisets $\{h_{is} - h_{js} : 1 \leq s \leq q\lambda\}$ contains every element of $\mathbb{F}_q$ exactly λ times.

Example 8.3 The following is an example of an $H(3, 2)$ generalized Hadamard matrix. Note that each element of $\mathbb{F}_3$ appears twice in the difference of any two rows.

$$H(3, 2) = \begin{pmatrix} 0\,0\,0\,0\,0\,0 \\ 0\,0\,1\,1\,2\,2 \\ 0\,1\,2\,0\,1\,2 \\ 0\,1\,0\,2\,2\,1 \\ 0\,2\,1\,2\,1\,0 \\ 0\,2\,2\,1\,0\,1 \end{pmatrix}. \tag{8.13}$$

Two generalized Hadamard matrices H_1 and H_2 of order n are said to be equivalent if one can be obtained from the other by a permutation of the rows and columns and a series of adding the same element of $\mathbb{F}_q$ to all the coordinates in a row or in a column. It is always possible to change the first row and column of a generalized Hadamard matrix into zeros. This generalized Hadamard matrix is called normalized.

Theorem 8.7 *The multiplication table of* $\mathbb{F}_q$ *is an* $H(q, 1)$ *generalized Hadamard matrix.*

Proof The α and β rows ($\alpha \neq \beta$) of the multiplication table are of the form

$$\alpha a_1, \alpha a_2, \ldots, \alpha a_q$$

and

$$\beta a_1, \beta a_2, \ldots, \beta a_q$$

where $a_1, a_2, \ldots, a_q$ are the elements of $\mathbb{F}_q$. Then the difference is the multiset $\{\alpha a_i - \beta a_i \mid 1 \leq i \leq n\}$. If $\alpha a_i - \beta a_i = \alpha a_j - \beta a_j$, then $(\alpha - \beta)a_i = (\alpha - \beta)a_j$. Since $\alpha \neq \beta$ we have $a_i = a_j$ and $i = j$. Therefore, the n elements in the multiset are distinct giving that the matrix is a generalized Hadamard matrix. $\square$

Example 8.4 Consider the following multiplication table of $\mathbb{F}_5$:

$$\begin{pmatrix} 0 & 0 & 0 & 0 & 0 \\ 0 & 1 & 2 & 3 & 4 \\ 0 & 2 & 4 & 1 & 3 \\ 0 & 3 & 1 & 4 & 2 \\ 0 & 4 & 3 & 2 & 1 \end{pmatrix}. \tag{8.14}$$

Here each element of $\mathbb{F}_5$ appears once in the difference between any two rows. Therefore, $\lambda = 1$. Hence, this is a $H(5,1)$ generalized Hadamard matrix.

Theorem 8.8 *If* $H(q, \lambda)$ *is a generalized Hadamard matrix then* $H(q, \lambda)^\top$ *is a generalized Hadamard matrix, where* $H(q, \lambda)^\top$ *denotes the transpose of* $H(q, \lambda)$.

Proof It follows from the fact that since $(\mathbb{F}_q, +)$ is an abelian group then $H(q, \lambda)^\top$ is also a generalized Hadamard matrix. $\square$

Theorem 8.9 *An ordinary Hadamard matrix of order* 4λ *exists if and only if a generalized Hadamard matrix* $H(2, 2\lambda)$ *over* $\mathbb{F}_2$ *exists.*

Proof Let H be an ordinary Hadamard matrix of order 4λ. Any two rows of this matrix consists of 1s and -1s such that any two rows have equally many occurrences of $(1,1), (-1,-1)$ and $(1,-1), (-1,1)$. Form H' by sending each 1 to 0 and each -1 to 1. Now in any two rows of H' there are equally many occurrences of $(0,0), (1,1)$ and $(0,1), (1,0)$. This gives that in the multiset formed from the differences of any two rows there are an equal number $(2\lambda$ in particular$)$ of 0 and 1. Hence, H' is a generalized Hadamard matrix $H(2, 2\lambda)$ over $\mathbb{F}_2$.

The other direction is identical reversing the map. $\square$

We shall now give a standard method to construct generalized Hadamard matrices from existing generalized Hadamard matrices. This technique is known as the Kronecker sum construction. Let $H(q, \lambda) = (h_{ij})$ be any $q\lambda \times q\lambda$ generalized Hadamard matrix over $\mathbb{F}_q$, and let $B_1, B_2, \ldots, B_{q\lambda}$ be any $q\mu \times q\mu$ generalized Hadamard matrices over $\mathbb{F}_q$.

The matrix in Eq. 8.15 gives a $q^2\lambda\mu \times q^2\lambda\mu$ generalized Hadamard matrix over $\mathcal{F}_q$, denoted by $H \oplus [B_1, B_2, \ldots, B_n]$, where $n = q\lambda$. If $B_1 = B_2 = \cdots = B_n = B$, then we write $H \oplus [B_1, B_2, \ldots, B_n] = H \oplus B$.

$$H \oplus [B_1, B_2, \ldots, B_n] = \begin{pmatrix} h_{11} + B_1 & h_{12} + B_1 & \cdots & h_{1n} + B_1 \\ h_{21} + B_2 & h_{22} + B_2 & \cdots & h_{2n} + B_2 \\ \vdots & \vdots & \vdots & \vdots \\ h_{n1} + B_n & h_{n2} + B_n & \cdots & h_{nn} + B_n \end{pmatrix}. \quad (8.15)$$

We leave the proof that the Kronecker sum construction gives a generalized Hadamard matrix as Exercise 2

Let S_q be the normalized generalized Hadamard matrix $H(q, 1)$ given by the multiplicative table of $\mathbb{F}_q$. As for ordinary Hadamard matrices over $\mathbb{F}_2$, starting from a generalized Hadamard matrix $S^1 = S_q$, we can recursively define S^T as a generalized Hadamard matrix $H(q, q^{t-1})$, constructed as $S^T = S_q \oplus [S^{t-1}, S^{t-1}, \ldots, S^{t-1}] = S_q \oplus S^{t-1}$ for $t > 1$, which is called a Sylvester generalized Hadamard matrix.

Exercises

1. Verify that the matrix in Eq. 8.13 is a generalized Hadamard matrix.
2. Prove that the matrix given in Eq. 8.15 is a generalized Hadamard matrix.

8.4   Latin Hypercubes

In this section, we shall describe a generalization of the Latin squares called Latin hypercubes. Essentially, these objects are multidimensional Latin squares. In other words, the same condition on rows and columns that makes a matrix a Latin square is applied to a k-dimensional matrix to make a Latin hypercube. The word hypercube simply means a cube in (possibly) more than three dimensions.

To be specific we have the following definition. Note that we use the alphabet $\{0, 1, \ldots, n-1\}$ right from the beginning of the discussion for simplicity.

Definition 8.3 A Latin k-hypercube $L_{j_1, \ldots, j_k}$ of order n is a k-dimensional array of size n, whose elements come from the set $\{0, 1, 2, \ldots, n-1\}$. If $j_i = j_{i'}$ for all $i \neq \alpha$ and $j_\alpha \neq j_{\alpha'}$ then $L_{j_1, j_2, \ldots, j_k} \neq L_{j_1', j_2', \ldots, j_k'}$. That is, in any direction, each "row" has each element from the set exactly once.

If L is a hypercube of order n then eliminating any set of dimensions will result in a hypercube. Specifically, we have that the subhypercube of a Latin k-hypercube of order n formed by fixing $k - s$ coordinates is a Latin s-hypercube of order n. This is because each line in the hypercube still contains each element exactly once.

Notice that a 1-hypercube, which we shall call a Latin line is simply a vector with each of the elements from 0 to $n - 1$ listed once and 2-hypercube is a Latin square.

We can count the possible number of subhypercubes in the following theorem.

Theorem 8.10 *For* $s \leq k$, *a Latin* k-*hypercube of order* n *contains* $\binom{k}{k-s} n^{k-s} = \binom{k}{s} n^{k-s}$ *s-hypercubes of order* n.

Proof There are exactly $\binom{k}{k-s}$ different ways of picking $k - s$ coordinates out of the k to fix. For each coordinate, there are n choices of which element to fix in that coordinate. □

We shall make a definition of orthogonality that mimics the definition of orthogonality for Latin squares.

Definition 8.4 Two Latin k-hypercubes are said to be orthogonal if each corresponding pair of Latin subsquares is orthogonal. A set of s Latin k-hypercubes of order n is said to be mutually orthogonal if each pair is orthogonal. In this case, we say we have a set of s mutually orthogonal Latin k-hypercubes (MOLkC).

Theorem 8.11 *There do not exist* n *mutually orthogonal Latin* k-*hypercubes of order* n.

Proof If there were n MOLkCs of order n then fixing any $k - 2$ dimension in every hypercube we would get n Latin squares. These corresponding Latin squares would be n MOLS of order n, which we have proven do not exist. □

We shall now show how to construct a multidimensional magic hypercube in a manner that is similar to the construction of a magic square.

Let $L_1, \ldots, L_k$ be a set of k-MOLkCs of order n. Define a k-dimensional hypercube of order n as follows:

$$\begin{aligned}
\mathrm{Mag}(L_1, \ldots, L_k)_{i_1, i_2, \ldots, i_k} &= (L_1)_{i_1, i_2, \ldots, i_k} + (L_2)_{i_1, i_2, \ldots, i_k} n \\
&\quad + \cdots + (L_k)_{i_1, i_2, \ldots, i_k} n^{k-1} \\
&= \sum_{j=1}^{k} (L_j)_{i_1, i_2, \ldots, i_k} n^{j-1}.
\end{aligned}$$

We shall show that the sum along any line in this hypercube is a constant and that elements in the hypercube each appear exactly once.

Theorem 8.12 *Let* $L_1, L_2, \ldots, L_k$ *be a set of* k- *MOLkCs, then* $\mathrm{Mag}(L_1, L_2, \ldots, L_k)$ *is a* k-*hypercube of order* n *such that the numbers* $0, 1, 2, \ldots, n^k - 1$ *can be placed in the hypercube so that in each* i-*line the digits in any place of each* k-*digit number written in base* n *are a permutation of* $0, 1, \ldots, n - 1$.

Proof We leave the proof as an exercise, see Exercise 1   □

Theorem 8.13 *Let* $\text{Mag}(L_1, L_2, \ldots, L_k)$ *be a k-hypercube of order* n *formed from MOLkCs* $L_1, \ldots, L_k$ *then the sum of every* i-*line is* $(\frac{n(n^k-1)}{2})$.

Proof Every number in $\{0, 1, 2, \ldots, n-1\}$ occurs once in each digit place. When summing the numbers in each digit's place in an i-line we have $(0 + 1 + 2 + \cdots + n - 1)(1 + n + n^2 + \cdots + n^{k-1}) = (\frac{(n-1)n}{2})(\frac{n^k-1}{n-1}) = (\frac{n(n^k-1)}{2})$.   □

Thus, in this hypercube, the sum in each i-line gives the same sum, namely, $(\frac{n(n^k-1)}{2})$.

The $[6, 3, 4]$ MDS code over $\mathbb{F}_5$ gives the following magic hypercube of order 5, formed by stacking the following squares:

Using 3 MOLkCs of order 5 we can form the following magic hypercube. Simply place the squares on top of each other.

$$\begin{pmatrix} 0 & 83 & 36 & 119 & 72 \\ 64 & 17 & 95 & 28 & 106 \\ 123 & 51 & 9 & 87 & 40 \\ 32 & 110 & 68 & 21 & 79 \\ 91 & 49 & 102 & 55 & 13 \end{pmatrix}, \begin{pmatrix} 43 & 121 & 54 & 7 & 85 \\ 77 & 30 & 113 & 66 & 24 \\ 11 & 94 & 47 & 100 & 58 \\ 70 & 3 & 81 & 39 & 117 \\ 109 & 62 & 15 & 98 & 26 \end{pmatrix}$$

$$\begin{pmatrix} 56 & 14 & 92 & 45 & 103 \\ 115 & 73 & 1 & 84 & 37 \\ 29 & 107 & 60 & 18 & 96 \\ 88 & 41 & 124 & 52 & 5 \\ 22 & 75 & 33 & 111 & 69 \end{pmatrix}, \begin{pmatrix} 99 & 27 & 105 & 63 & 16 \\ 8 & 86 & 44 & 122 & 50 \\ 67 & 20 & 78 & 31 & 114 \\ 101 & 59 & 12 & 90 & 48 \\ 35 & 118 & 71 & 4 & 82 \end{pmatrix}$$

$$\begin{pmatrix} 112 & 65 & 23 & 76 & 34 \\ 46 & 104 & 57 & 10 & 93 \\ 80 & 38 & 116 & 74 & 2 \\ 19 & 97 & 25 & 108 & 61 \\ 53 & 6 & 89 & 42 & 120 \end{pmatrix}.$$

Exercises

1. Produce a proof of Theorem 8.12 modeling the proof done for Latin squares.
2. Produce a magic hypercube that is 3 by 3 by 3.

8.5 Partially Ordered Sets

Next, we shall study the combinatorial object partially ordered sets which has numerous applications in mathematics and computer science. Recall, that a relation R on

Fig. 8.1 Partially ordered set with four elements

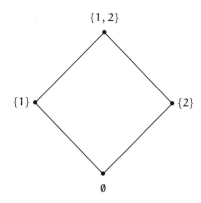

a set A is a subset of $A \times A$. A relation $R \subseteq A \times A$ is reflexive if for all $a \in A$, $(a, a) \in R$, it is antisymmetric if $(a, b) \in R$ and $(b, a) \in R$ implies $a = b$, and it is transitive if $(a, b) \in R$ and $(b, c) \in R$ implies $(a, c) \in R$.

Definition 8.5 A partially ordered set is a set A together with a relation R on A such that R is reflexive, antisymmetric, and transitive.

For example, $(\mathbb{Z}, \leq)$ is a partially ordered set since $\leq$ is reflexive, antisymmetric, and transitive. In general, since R is reflexive, antisymmetric, and transitive, we often denote the relation by $\leq$ when no confusion will arise. If there is a standard notation for the relation we shall use the standard notation.

Example 8.5 The relation $<$ on $\mathbb{Z}$ does not give a partially ordered set since $<$ is not reflexive. The relation $\geq$ on $\mathbb{Z}$ does give a partially ordered set since the relation is reflexive, antisymmetric, and transitive.

As with any relation we can construct the directed graph corresponding to the relation. Specifically, if $(a, b) \in R$ then (a, b) is a directed edge in the corresponding graph. We can simplify the graph since it is reflexive and transitive. Namely, since it is reflexive, we can eliminate all loops on an element, since each element has such a loop. Moreover, since it is transitive, we can take a minimal set of edges which gives all edges by completing the transitivity. Put simply, if vertex a has a path going up to vertex b then we have that $a \leq b$. This graph is known as the Hasse diagram of the partially ordered set. We begin with a simple example.

Example 8.6 Consider the partially ordered set $(\mathcal{P}(\{1, 2\}), \subseteq)$ shown in Fig. 8.1. The set has four elements, namely, $\emptyset$, $\{1\}$, $\{2\}$, and $\{1, 2\}$.

We can generalize this result in the following theorem.

Theorem 8.14 *Let A be a set, then $(\mathcal{P}(A), \subseteq)$ is a partially ordered set.*

Fig. 8.2 The poset
$\mathcal{P}(\{1,2,3\})$

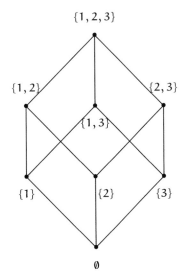

Proof It is immediate that $B \subseteq B$ for all sets B. Then if $B \subseteq C$ and $C \subseteq B$ then $C = B$. Finally, if $B \subseteq C \subseteq D$, then every $x \in B$ must be in C and then must be in D giving that $B \subseteq D$. Therefore, the relation is reflexive, antisymmetric, and transitive. $\square$

Example 8.7 Consider the partially ordered set $\mathcal{P}(\{1,2,3\})$ in Fig. 8.2. The set has eight elements, namely, $\emptyset, \{1\}, \{2\}, \{3\}, \{1,2\}, \{1,3\}, \{2,3\}, \{1,2,3\}$. Its Hasse diagram is given as follows.

Definition 8.6 An element a in a partially ordered set $(A, \leq)$ is maximal if there exists no b with $a < b$. An element c in a partially ordered set $(A, \leq)$ is minimal if there exists no d with $d < c$.

Example 8.8 Consider the partially ordered set given in Example 8.6, then $\emptyset$ is a minimal element and $\{1, 2\}$ is a maximal element. Consider the partially ordered set given in Example 8.7, then $\emptyset$ is a minimal set and $\{1, 2, 3\}$ is a maximal element.

Example 8.9 Consider the partially ordered set in Fig. 8.3.
    Here d, e, and f are all maximal elements and a and b are both minimal elements. Notice that these elements do not have the property that they are greater than or less than all elements in the partially ordered set. They simply have nothing greater than or less than themselves.

    For any relation R on a set A, that is, $R \subseteq A \times A$, we can define the inverse relation to be $R^{-1} = \{(b, a) \mid (a, b) \in R\}$.

Fig. 8.3 Poset on six points

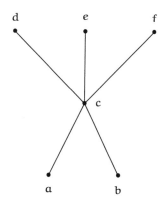

Theorem 8.15 *Let* R *be a relation on a set* A. *If* (A, R) *is a partially ordered set then* (A, R^{-1}) *is a partially ordered set.*

Proof If a is any element in A, then $(a, a) \in R$ which implies $(a, a) \in R^{-1}$. Therefore, R^{-1} is reflexive.

If $(a, b) \in R^{-1}$ and $(b, a) \in R^{-1}$ then $(b, a) \in R$ and $(a, b) \in R$ which implies $a = b$ since R is antisymmetric. Therefore, R^{-1} is antisymmetric.

If $(a, b) \in R^{-1}$ and $(b, c) \in R^{-1}$, then $(b, a) \in R$ and $(c, b) \in R$ which implies $(c, a) \in R$ since R is transitive. This gives that $(a, c) \in R^{-1}$ which gives that R^{-1} is transitive. Therefore, (A, R^{-1}) is a partially ordered set. □

Example 8.10 We have that $(\mathbb{Z}, \leq)$ is a partially ordered set. Then by Theorem 8.15, we have that $(\mathbb{Z}, \geq)$ is a partially ordered set.

Example 8.11 In Example 8.7, we considered the partially ordered set $\mathcal{P}(\{1, 2, 3\})$ with set containment as the relation. Using Theorem 8.15, we can use the same set with reverse containment to get a partially ordered set with the Hasse diagram given in Fig. 8.4. The set has eight elements, namely, $\emptyset, \{1\}, \{2\}, \{3\}, \{1, 2\}, \{1, 3\}, \{2, 3\}, \{1, 2, 3\}$.

We can now look at specific elements satisfying certain properties in a partially ordered set.

Definition 8.7 An element **I** in a partially ordered set $(A, \leq)$ is the greatest element if for all $a \in A$, we have $a \leq I$. An element **0** in a partially ordered set $(A, \leq)$ is the least element if for all $a \in A$, we have $0 \leq a$.

Example 8.12 In the partially ordered set $(\mathcal{P}(A), \subseteq)$, the greatest element is A and the least element is $\emptyset$.

We shall show the relationship between maximal elements and greatest elements and between minimal elements and least elements.

Fig. 8.4 Reverse poset of
$\mathcal{P}(\{1,2,3\})$

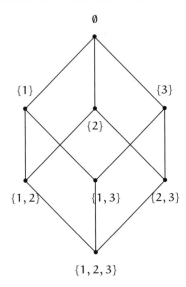

Theorem 8.16 *Let* $(A, \leq)$ *be a partially ordered set. If* a *is the greatest element then* a *is a maximal element. If* b *is the least element then* b *is a minimal element.*

Proof Let a be a greatest element, then for any c in A we have $c \leq a$ by definition. Therefore, there is no c with $a \leq c, c \neq a$. Let b be a least element, then for any c in A we have $b \leq c$ by definition. Therefore, there is no c with $c \leq b, c \neq b$.   □

The converse of this theorem is not true as we have seen in Example 8.9.
For finite lattices we can say something more.

Theorem 8.17 *Let* $(A, \leq)$ *be a partially ordered set with* $|A|$ *finite. Then* A *must have a maximal and minimal element.*

Proof Let a_1 be an element in A. If there is no a_2 with $a_1 < a_2$ then a_1 is maximal. Similarly, if there is no a_{i+1} with $a_i < a_{i+1}$ then a_i is maximal. Given that the set is finite, the sequence a_i must be finite, that is, there is an a_n that is maximal.

A similar proof works for minimal elements by reversing the order of the relation.   □

We note that the maximal and minimal elements guaranteed in the previous theorem are not necessarily unique. That is, a partially ordered set may have numerous maximal and minimal elements. This is not true for the greatest and least elements as we see in the following theorem.

Theorem 8.18 *Let* $(A, \leq)$ *be a partially ordered set, the greatest and least elements, if they exist, are unique.*

Fig. 8.5 Poset on 6 points

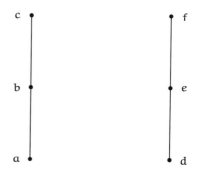

Proof Assume I and I' are two greatest elements of the partially ordered set. Then I ≤ I' since I' is greatest and I' ≤ I since I is greatest. Then since the relation is antisymmetric, we have I = I'.

Assume 0 and 0' are two least elements of the partially ordered set. Then 0 ≤ 0' since 0 is least and 0' ≤ 0 since 0' is least. Then since the relation is antisymmetric, we have 0 = 0'. □

Theorem 8.19 *Let* I *be the greatest element and let* 0 *be the least element in a partially ordered set* (A, R), *then* 0 *be the greatest element and* I *is the least element in a partially ordered set* (A, R^{-1}).

Proof If (a, I) ∈ R for all a then (I, a) ∈ R^{-1} and if (0, a) ∈ R for all a then (a, 0) ∈ R^{-1}. This gives the result. □

We now want to examine specific kinds of partially ordered sets. We begin with a necessary definition.

Definition 8.8 Let (A, ≤) be a partially ordered set and let B ⊆ A. The element a ∈ A is a least upper bound of B if b ≤ a for all b ∈ B and if b ≤ c for all b ∈ B then a ≤ c. The element d ∈ A is a greatest lower bound of B if d ≤ b for all b ∈ B and if e ≤ b for all b ∈ B then e ≤ d.

Example 8.13 A subset of a partially ordered set need not have a least upper bound nor a greatest lower bound. Consider the partially ordered set in Fig. 8.5.

The set {b, e} has neither a greatest lower bound nor a least upper bound.

For a set {a, b} we denote the least upper bound of the set {a, b} by l.u.b.(a, b) and greatest lower bound of the set {a, b} by g.l.b.(a, b).

Definition 8.9 A partially ordered set is a lattice if and only if for every two elements a and b in the partially ordered set, the least upper bound l.u.b.(a, b) and the greatest lower bound g.l.b.(a, b) always exist.

In a lattice, we denote $l.u.b.(a,b)$ as $a \vee b$ and $g.l.b.(a,b)$ as $a \wedge b$. Given a lattice with greatest element I and least element 0 the following rules are immediate for all a in the lattice:

1. $I \wedge a = a$;
2. $I \vee a = I$;
3. $0 \wedge a = 0$;
4. $0 \vee a = a$.

Theorem 8.20 *Let* D_n *be the set of positive integers dividing* n. *For* $a, b \in D_n$, *let* $a \leq b$ *if and only if* a *divides* b. *Then* D_n *is a lattice.*

Proof First, we note that if a is a divisor of n then a divides a so the relation is reflexive. If a and b divide n where a divides b and b divides a, then $a = b$ making the relation antisymmetric. If a, b, and c are divisors of n, where a divides b and b divides c, then $b = ka$ for some k and $c = gb$ for some b giving $c = gk(a)$ and a divides c making the relation transitive. Therefore, D_n is a partially ordered set.

Given a and b divisors of n, the $g.l.b\{a, b\}$ is the largest number d such that d divides a and d divides b, that is, the greatest common divisor of a and b. Then, if $n = ka$, we have $n = kda'$, where $a = da'$ so d is a divisor of n. Therefore, the greatest lower bound exists. The $l.u.b.\{a, b\}$ is the smallest number m such that a divides m and b divides m, which is the least common multiple of a and b which is $\frac{ab}{d} = a'b$. Then $a'b$ divides n since b divides n and a' divides n and they are relatively prime. Therefore, m is a divisor of n. Therefore, the least upper bound exists and D_n is a lattice. $\qquad\square$

The proof of this theorem gives the following corollary.

Corollary 8.2 *Let* D_n *be the set of positive integers dividing* n. *Then* $a \vee b = \text{lcm}(a, b)$ *and* $a \wedge b = \text{gcd}(a, b)$.

Theorem 8.21 *The partially ordered set* $(\mathcal{P}(\{1, 2, 3, \ldots, k\}), \subseteq)$ *is a lattice.*

Proof Given any two subsets A and B of $\{1, 2, 3, \ldots, k\}$, $l.u.b.(A, B)$ is the smallest set C such that $A \subseteq C$ and $B \subseteq C$. This set C is $A \cup B$ by definition. Additionally, $g.l.b.(A, B)$ is the largest set D such that $D \subseteq A$ and $D \subseteq B$. This set D is $A \cap B$ by definition. Therefore, it is a lattice. $\qquad\square$

We can make the standard definition for the equivalence of partially ordered sets.

Definition 8.10 A partially ordered set $(A, \leq)$ is linearly ordered if for all $a, b \in A$, either $a \leq b$ or $b < a$.

Fig. 8.6 Linearly ordered set
with five points

Colloquially, this definition can be rephrased by saying that any pair of elements is comparable.

Theorem 8.22 *A linearly ordered partially ordered set* $(A, \leq)$ *is a lattice.*

Proof If a, b are any two elements in A, then either $a \leq b$ or $b \leq a$. If $a = b$ then $g.l.b.\{a, b\} = l.u.b.\{a, b\} = a$. If $a < b$ then $g.l.b.\{a, b\} = a$ and $l.u.b.\{a, b\} = b$. If $b < a$ then $g.l.b.\{a, b\} = b$ and $l.u.b.\{a, b\} = a$. Therefore, $(A, \leq)$ is a lattice. □

Example 8.14 The Hasse diagram of a linearly ordered set is simply a vertical line as shown in Fig. 8.6 for a linearly ordered set with five elements.

Definition 8.11 Let $(A, \leq)$ and $(A', \leq')$ be partially ordered sets. Then $(A, \leq)$ and $(A', \leq')$ are isomorphic if there is a bijection $\Phi : A \rightarrow A'$ such that for $a, b \in A$, $a \leq b$ if and only if $\Phi(a) \leq' \Phi(b)$.

We give a theorem which shows two families of lattices are isomorphic.

Theorem 8.23 *Let* $p_1, p_2, \ldots, p_k$ *be distinct primes with* $n = \prod_{i=1}^{k} p_i$. *Then* D_n *is isomorphic to* $\mathcal{P}(\{1, 2, 3, \ldots, k\})$.

Proof Make a map between the elements of D_n with $\mathcal{P}(\{1, 2, 3, \ldots, k\})$ as follows:

$$\Phi\left(\prod_{i \in A} p_i\right) = A. \tag{8.16}$$

Fig. 8.7 Poset D_{30}

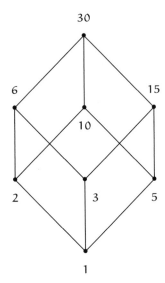

Since the p_i are distinct and every divisor of n is a unique product of the p_i by the Fundamental Theorem of Arithmetic, we have that the map Φ is a bijection. Then it is immediate that $\prod_{i \in A} p_i$ divides $\prod_{i \in B} p_i$ if and only if $A \subseteq B$. Therefore, the partially ordered sets are isomorphic. $\qquad\square$

Example 8.15 Consider the partially ordered set D_{30} in Fig. 8.7. The set has eight elements, namely, $1, 2, 3, 5, 6, 10, 15, 30$.

Notice that the vertices and lines of this Hasse diagram are a two-dimensional representation of a cube in three dimensions. It is clear from the picture that D_{30} is isomorphic to $\mathcal{P}(\{1, 2, 3\})$ by the mapping:

$$1 \to \emptyset$$
$$2 \to \{1\}$$
$$3 \to \{2\}$$
$$5 \to \{3\}$$
$$6 \to \{1, 2\}$$
$$10 \to \{1, 3\}$$
$$15 \to \{2, 3\}$$
$$30 \to \{1, 2, 3\}$$

The following idea was first put forth by George Boole in [9]. We are only going to study finite Boolean algebras. It is possible to also discuss infinite Boolean algebras, but we are going to restrict ourselves to the finite case. As such, we take the definition that is easiest and most natural for the finite case.

Definition 8.12 A Boolean algebra is a lattice that is isomorphic to $(\mathcal{P}(A), \subseteq)$ for some set A.

Theorem 8.24 *Consider a Boolean algebra isomorphic to* $(\mathcal{P}(A), \subseteq)$ *for some set A. Then* $\text{l.u.b.}\{B, C\} = B \cup C$ *and* $g.\text{l.b.}\{B, C\} = B \cap C$.

Proof Any set that contains B and C must contain $B \cup C$, moreover $B \cup C$ is the smallest set containing both B and C by definition. Therefore, $\text{l.u.b.}\{B, C\} = B \cup C$.

Any set that is contained in B and C must be in $B \cap C$, moreover $B \cap C$ is the largest set contained in both B and C by definition. Therefore, $g.\text{l.b.}\{B, C\} = B \cap C$. $\square$

This result shows why the notation is so natural, that is, in the language of lattices, we have

$$B \wedge C = B \cap C$$

and

$$B \vee C = B \cup C.$$

It naturally follows from this proof that any Boolean algebra is necessarily a lattice.

Example 8.16 Consider the partially ordered set D_{210} given in Fig. 8.8. The set has 16 elements, namely, 1, 2, 3, 5, 6, 7, 10, 14, 15, 21, 30, 35, 42, 70, 105, 210.

Notice that the vertices and lines of this Hasse diagram are a two-dimensional representation of a tesseract, which is a cube in four dimensions. It is clear from the picture that D_{210} is isomorphic to $\mathcal{P}(\{1, 2, 3, 4\})$ by the mapping:

$$1 \to \emptyset$$
$$2 \to \{1\}$$
$$3 \to \{2\}$$
$$5 \to \{3\}$$
$$7 \to \{4\}$$
$$6 \to \{1, 2\}$$
$$10 \to \{1, 3\}$$
$$14 \to \{1, 4\}$$
$$15 \to \{2, 3\}$$
$$21 \to \{2, 4\}$$
$$35 \to \{3, 4\}$$
$$30 \to \{1, 2, 3\}$$
$$42 \to \{1, 2, 4\}$$

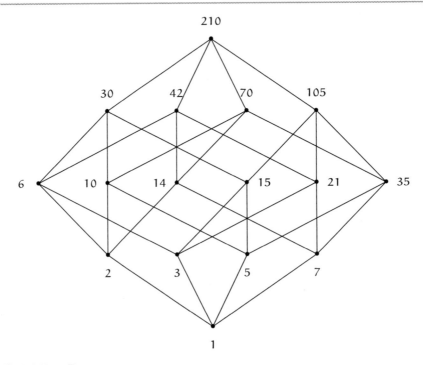

Fig. 8.8 Poset D_{210}

$$70 \to \{1,3,4\}$$
$$105 \to \{2,3,4\}$$
$$210 \to \{1,2,3,4\}$$

Therefore, this partially ordered set is a Boolean algebra.

Theorem 8.25 *Let* $(A, \leq)$ *be a finite Boolean algebra then* $|A| = 2^n$ *for some* n.

Proof If $(A, \leq)$ is a Boolean algebra then there is a bijection between A and $\mathcal{P}(B)$ for some finite set B. Then $\mathcal{P}(B)$ has cardinality $2^{|B|}$. This gives the result. $\square$

Theorem 8.26 *If* (S, R) *is a Boolean algebra, then* (S, R^{-1}) *is a Boolean algebra.*

Proof If (A, R) is a Boolean algebra then it is isomorphic to $\mathcal{P}(S)$ for some set S. Define the map $\Phi : \mathcal{P}(S) \to \mathcal{P}(S)$, by $\Phi(C) = C'$, where C' is the complement of the set C. Then we have $C \subset D$ if and only if $D' \subset C'$ which gives that Φ is a lattice isomorphism. The lattice (S, R^{-1}) is canonically isomorphic to $(\mathcal{P}(S), \supseteq)$, which gives the result. $\square$

It is immediate that the Hasse diagram for (S, R^{-1}) is formed by inverting the Hasse diagram for (S, R) as in Example 8.11.

Theorem 8.27 *Let D_n be the set of positive integers dividing n. For $a, b \in D_n$, let $a \leq b$ if and only if a divides b. If n is the product of distinct primes, then D_n is a Boolean algebra.*

Proof Assume n is the product of distinct primes, i.e., $n = \prod_{i=1}^{s} p_i$ where $p_i \neq p_j$ if $i \neq j$. Define a map $\psi : D_n \to (\{1, 2, 3, \ldots, s\})$ by

$$\psi : \prod_{i \in C} p_i = C,$$

where $C \subseteq \{1, 2, 3, \ldots, s\}$. It is immediate that if $k, m \in D_n$, then k divides m if and only if $\psi(k) \subseteq \psi(m)$.   □

The converse is left as an exercise, see Exercise 2

We can give an additional operation on a Boolean algebra for any element a, namely, a', which is defined as the unique element such that $a' \vee a = I$ and $a' \wedge a = 0$ where I is the greatest element and 0 is the least element in the Boolean algebra. We notice that this corresponds to the complement function for sets. Given the canonical isomorphism between any Boolean algebra and $\mathcal{P}(A)$ the following rules are immediate for all a, b in the Boolean algebra:

- $a \wedge a = a$;
- $a \vee a = a$;
- $a \wedge b = b \wedge a$;
- $a \vee b = b \vee a$;
- $(a \wedge b) \wedge c = a \wedge (b \wedge c)$;
- $(a \vee b) \vee c = a \vee (b \vee c)$;
- $a \wedge (b \vee c) = (a \wedge b) \vee (a \wedge c)$;
- $a \vee (b \wedge c) = (a \vee b) \wedge (a \vee c)$;
- $(a \wedge b)' = a' \vee b'$;
- $(a \vee b)' = a' \wedge b'$;
- $I' = 0$;
- $0' = I$;
- $(a')' = a$.

We shall show a canonical way of representing a Boolean algebra. Let A be a finite set and let B be a subset of A. Define the following characteristic function:

$$\chi_B(a) = \begin{cases} 1 & a \in B \\ 0 & a \notin B. \end{cases}$$

Then if $|A| = n$, list the elements as $a_1, a_2, \ldots, a_n$. Then each subset of A can be defined as a vector of length n, where the ith coordinate is given by $\chi_B(a_i)$.

Fig. 8.9 Boolean algebra on $\{0,1\}^2$

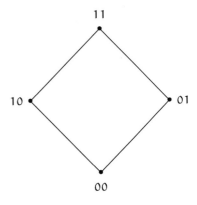

In this sense, each Boolean algebra can be thought of as a partially ordered set on $\{0,1\}^n$. The lattice operations can be done coordinate-wise with the following functions:

$$
\begin{array}{c|cc}
\wedge & 0 & 1 \\
\hline
0 & 0 & 0 \\
1 & 0 & 1
\end{array}
\qquad
\begin{array}{c|cc}
\vee & 0 & 1 \\
\hline
0 & 0 & 1 \\
1 & 1 & 1
\end{array}.
$$

Applying these functions coordinate-wise, we have for $\mathbf{v}, \mathbf{w} \in \{0,1\}^n$,

$$
\mathrm{l.u.b.}\{\mathbf{v}, \mathbf{w}\} = \mathbf{v} \vee \mathbf{w}
$$

and

$$
g.\mathrm{l.b}\{\mathbf{v}, \mathbf{w}\} = \mathbf{v} \wedge \mathbf{w}.
$$

For example, $0011 \wedge 1110 = 0010$ and $0011 \vee 1110 = 1111$.

We note then that the compliment function is given as $0' = 1$ and $1' = 0$.

Example 8.17 The Hasse diagram for the Boolean algebra on $\{0,1\}^2$ is given in Fig. 8.9:

The Hasse diagram for the Boolean algebra on $\{0,1\}^3$ is given in Fig. 8.10:

The Hasse diagram for the Boolean algebra on $\{0,1\}^4$ is given in Fig. 8.11:

Exercises

1. In D_{72} compute the following:

   a. $8 \vee 6$
   b. $8 \wedge 6$
   c. $(2 \vee 9) \wedge 4$
   d. $(2 \wedge 9) \vee 4$

2. Prove that if n is not square-free then D_n is not a Boolean algebra.

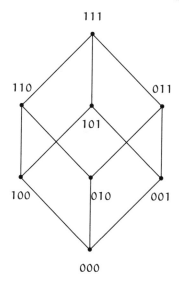

Fig. 8.10 Boolean algebra on $\{0,1\}^3$

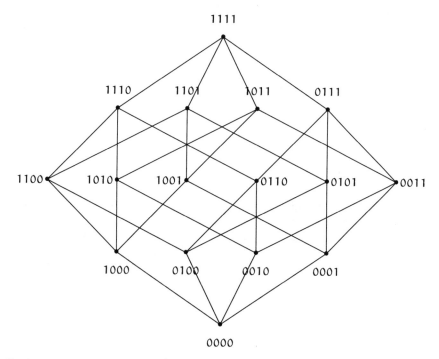

Fig. 8.11 Boolean algebra on $\{0,1\}^4$

3. Compute the following in $\{0,1\}^5$:

    a. $11011 \wedge 01011$
    b. $10011 \vee 11100$
    c. $(10100 \wedge 00111)'$
    d. $(00111 \vee 11101)'$

4. Draw the Hasse diagram for Boolean algebras on $\{0,1\}^n$ for $n = 2,3,4$ using the inverse relation.
5. Prove that any two Boolean algebras with the same cardinality are isomorphic.
6. Give an example of the following:

    a. A partially ordered set that has no greatest element.
    b. A partially ordered set that has no least element.
    c. A maximal element that is not a greatest element.
    d. Two non-isomorphic partially ordered sets with four elements.
    e. A partially ordered set with no maximal element.
    f. A partially ordered set with no minimal element.
    g. A partially ordered set that is not a lattice.
    h. A lattice that is not a Boolean algebra.

7. Prove that if (A, R) is a lattice, then (A, R^{-1}) is a lattice. Describe its Hasse diagram.
8. Draw the Hasse diagram for D_{2310}.
9. Prove that the if $n > 2$, then a Boolean algebra with n elements is not linearly ordered.

8.6   Association Schemes

Association schemes have their origin in statistics in terms of the design of experiments. They have found numerous applications in combinatorics and algebra and have found particular application in coding theory. Association schemes were introduced in [10]. For further study, see [7]. We begin with the standard definition.

Definition 8.13 Let X be a finite set, with $|X| = v$. Let R_i be a subset of $X \times X$, for all $i \in I = \{i \mid i \in \mathbb{Z}, 0 \leq i \leq d\}$ with $d > 0$. We define $\mathfrak{R} = \{R_i\}_{i \in I}$. We say that $(X, \mathfrak{R})$ is a d-class association scheme if the following properties are satisfied:

1. The relation $R_0 = \{(x, x) : x \in X\}$ is the identity relation.
2. For every $x, y \in X$, $(x, y) \in R_i$ for exactly one i.
3. For every $i \in I$, there exists $i' \in I$ such that $R_i^T = R_{i'}$, that is, we have $R_i^T = \{(x, y) \mid (y, x) \in R_i\}$.

4. If $(x, y) \in R_k$, the number of $z \in X$ such that $(x, z) \in R_i$ and $(z, y) \in R_j$ is a constant p_{ij}^k.

The values p_{ij}^k are called the intersection numbers of the association scheme. The elements $x, y \in X$ are called i-th associates if $(x, y) \in R_i$. If $i = i'$ for all i, namely, $R_i^T = R_i$ for all i, then the association scheme is said to be symmetric. An association scheme that does not satisfy this condition is said to be non-symmetric. The association scheme $(X, \mathfrak{R})$ is said to be commutative if $p_{ij}^k = p_{ji}^k$, for all $i, j, k \in I$. Note that a symmetric association scheme is always commutative but that the converse of this statement is not true.

Relations are often described by their adjacency matrices. Namely, given a relation R on a set X, the elements of X are the coordinates of the matrix M_R, and

$$(M_R)_{a,b} = \begin{cases} 1 & (a, b) \in R \\ 0 & (a, b) \notin R. \end{cases}$$

In this setting, we shall denote the adjacency matrix for the relation R_i by A_i. Then A_i is the $v \times v$ matrix whose rows and columns are labeled by the points of X and defined by

$$(A_i)_{x,y} = \begin{cases} 1 & \text{if } (x, y) \in R_i \\ 0 & \text{otherwise.} \end{cases}$$

Conditions $1 - -4$ in the definition of $(X, \mathfrak{R})$ are equivalent to the following conditions:

1. The matrix $A_0 = I_v$, where I_v is the v by v identity matrix.
2. Summing the matrices, we have $\sum_{i \in I} A_i = J_v$, where J_v is the v by v matrix where all entries are 1.
3. For all $i \in I$, there exists $i' \in I$, such that $A_i = A_{i'}^T$.
4. For all $i, j \in I$ we have $A_i A_j = \sum_{k \in I} p_{ij}^k A_k$.

If the association scheme is symmetric, then it follows that $A_i = A_i^T$, for all $i \in I$. If the association scheme is commutative, then matrix multiplication on the A_i commutes, namely, $A_i A_j = A_j A_i$, for all $i, j \in I$. The adjacency matrices generate a $(n + 1)$-dimensional algebra $\mathbf{A}$ of symmetric matrices. This algebra is called the Bose-Mesner algebra.

Given the description of association schemes it is easy to see that an association scheme can be thought of as the complete graph on v vertices where an edge (a, b) is marked by i if $(a, b) \in R_i$. Similarly, we can think of it as a v by v matrix where the entry in (a, b) is i if $(a, b) \in R_i$.

We shall describe one of the most useful and interesting association schemes. Consider the space $\mathbb{F}_2^n$. It consists of 2^n vectors of length n with entries either 0 or 1. Given two such vectors $\mathbf{v}$ and $\mathbf{w} \in \mathbb{F}_2^n$, the Hamming distance is defined as $d_H(\mathbf{v}, \mathbf{w}) = |\{i \mid v_i \neq w_i\}|$.

Theorem 8.28 *Let R_i be the relation on $\mathbb{F}_2^n$, where $(\mathbf{v}, \mathbf{w}) \in R_i$ if and only if $d_H(\mathbf{v}, \mathbf{w}) = i$. Then $(\mathbb{F}_2^n, \{R_0, R_1, \ldots, R_n\})$ is an n-class association scheme.*

Proof First, we have that a vector $\mathbf{v}$ has Hamming distance 0 from a vector $\mathbf{w}$ if and only if $\mathbf{v} = \mathbf{w}$. Therefore, R_0 is the identity relation.

Secondly, any two vectors $\mathbf{v}$ and $\mathbf{w}$ have a unique Hamming distance; therefore, for all $\mathbf{v}, \mathbf{w} \in \mathbb{F}_2^n$, we have that $(\mathbf{v}, \mathbf{w}) \in R_i$ for exactly one i.

Thirdly, since $d_H(\mathbf{v}, \mathbf{w}) = d_H(\mathbf{w}, \mathbf{v})$ we have that each relation R_i is symmetric.

Fourthly, let $(\mathbf{v}, \mathbf{w}) \in R_k$, that is, $d_H(\mathbf{v}, \mathbf{w}) = k$. We need to determine the number of $\mathbf{u}$ such that $d_H(\mathbf{v}, \mathbf{u}) = i$ and $d_H(\mathbf{u}, \mathbf{w}) = j$. We are changing i coordinates of the vector $\mathbf{v}$ to change to get to $\mathbf{u}$. Then we have j coordinates of $\mathbf{u}$ to change to get to $\mathbf{w}$. Noting that we must change in total k coordinates to get from $\mathbf{v}$ to $\mathbf{w}$. We get that

$$p_{ij}^k = \begin{cases} C(k, \frac{i-j+k}{2})C(n-k, \frac{i+j-k}{2}) & i+j-k \equiv 0 \pmod 2 \\ 0 & i+j-k \equiv 1 \pmod 2. \end{cases}$$

We note that $\frac{1-j+k}{2} + \frac{1+j-k}{2} = i$ which is the number of coordinates of $\mathbf{v}$ that are changed.

Therefore, this is an association scheme. $\square$

This association scheme is known as the Hamming scheme. We note immediately that the scheme is symmetric.

Example 8.18 Consider the Hamming scheme on $\mathbb{F}_2^2$. It is a 2-class association scheme with the following matrices:

A_0	$(0,0)$	$(0,1)$	$(1,0)$	$(1,1)$
$(0,0)$	1	0	0	0
$(0,1)$	0	1	0	0
$(1,0)$	0	0	1	0
$(1,1)$	0	0	0	1

A_1	$(0,0)$	$(0,1)$	$(1,0)$	$(1,1)$
$(0,0)$	0	1	1	0
$(0,1)$	1	0	0	1
$(1,0)$	1	0	0	1
$(1,1)$	0	1	1	0

A_2	$(0,0)$	$(0,1)$	$(1,0)$	$(1,1)$
$(0,0)$	0	0	0	1
$(0,1)$	0	0	1	0
$(1,0)$	0	1	0	0
$(1,1)$	1	0	0	0

We note that the sum of the matrices is the all one matrix J_4.

We can realize the association scheme in the following complete graph. Note that since the scheme is symmetric, the edges do not need to be directed.

$$
\begin{array}{c}
\text{(diagram)}
\end{array}
$$

```
        1
(0,1) ───────── (1,1)
   |  \    2   /  |
   1    \    /    1
   |      \/      |
   |      /\      |
   |    /    \    |
   |  / 2      \  |
(0,0) ───────── (1,0)
        1
```

Finally, we can use a matrix to represent this graph as follows:

	(0,0)	(0,1)	(1,0)	(1,1)
(0,0)	0	1	1	2
(0,1)	1	0	2	1
(1,0)	1	2	0	1
(1,1)	2	1	1	0

Example 8.19 Geometrically, we can think of the elements in the Hamming scheme as the vertices on an n-dimensional hypercube and the distance as the distance, in edges, in that hypercube. For $n = 2$, we have the following 3-dimensional cube.

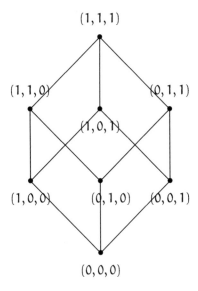

The matrix for this scheme is as follows:

	(0,0,0)	(0,0,1)	(0,1,0)	(0,1,1)	(1,0,0)	(1,0,1)	(1,1,0)	(1,1,1)
(0,0,0)	0	1	1	2	1	2	2	3
(0,0,1)	1	0	2	1	2	1	3	2
(0,1,0)	1	2	0	1	2	3	1	2
(0,1,1)	2	1	1	0	3	2	2	1
(1,0,0)	1	2	2	3	0	1	1	2
(1,0,1)	2	1	3	2	1	0	2	1
(1,1,0)	2	3	1	2	1	2	0	1
(1,1,1)	3	2	2	1	2	1	1	0

In each row and column, there are 3 ones, 3 twos, 1 three, and 1 zero. This corresponds to the line in Pascal's triangle 1 3 3 1 giving the binomial coefficients for $n = 3$.

We shall now show how to make an association scheme from any finite group. Groups are defined in Chap. 10 in Definition 10.1.

Theorem 8.29 *Let* G *be a finite group with elements* $g_0, g_1, \ldots, g_{n-1}$ *where* g_0 *is the identity element of the group. Define* $R_i = \{(x, y) \mid x = g_i y\}$. *Then* $(G, \{R_0, R_1, \ldots, R_{n-1}\})$ *is an association scheme.*

Proof Firstly, the relation R_0 consists of all pairs (x, y) such that $x = g_0 y$ where g_0 is the identity. Therefore, R_0 is the identity relation by definition.

Secondly, given a pair (x, y), we have that if $xg = y$, then $g = x^{-1} y$. Hence, there is a unique g_i with $(x, y) \in R_i$.

Thirdly, if $(x, y) \in R_i$ then $x = g_i y$. Then $g_i^{-1} x = y$, giving that $(y, x) \in R_{i'}$ where $g_{i'} = g_i^{-1}$. Therefore, $R_i^T = R_{i'}$.

Fourthly, if $(x, y) \in R_k$ this gives that $xy^{-1} = g_k$. If $(x, z) \in R_i$ and $(z, y) \in R_j$, then we have $xz^{-1} = g_i$ and $zy^{-1} = g_j$. Then we have $xz^{-1} zy^{-1} = xy^{-1}$ so we we need $g_i g_j = g_k$. Given any g_i there is a unique g_j satisfying this condition. For this pair i, j, there is a unique z satisfying $xz^{-1} = g_i$. Therefore, the fourth condition is satisfied.   □

It is clear that the association scheme is commutative if and only if the group G is commutative. The scheme is symmetric if and only if each element is its own inverse.

Example 8.20 Consider the symmetric group of 3 letters, which consists of all permutations of three elements. The group consists of the following elements: $\{g_0 =$

$(1), g_1 = (1, 2), g_2 = (1, 3), g_3 = (2, 3), g_4 = (1, 2, 3), g_5 = (1, 3, 2)\}$. We shall give the matrix coming from the complete graph associated with this association scheme.

	g_0	g_1	g_2	g_3	g_4	g_5
g_0	0	1	2	3	4	5
g_1	1	0	5	4	2	3
g_2	2	4	0	5	3	1
g_3	3	5	4	0	1	2
g_4	5	2	3	1	0	4
g_5	4	3	1	2	5	0

We note that in this case the matrix of the association scheme is a Latin square which is not symmetric. Each matrix A_i for this scheme is actually a permutation matrix, with precisely six ones in each A_i.

Exercises

1. Construct the matrix for the association scheme on the group $(\mathbb{F}_2 \times \mathbb{F}_2, +)$. Prove that this scheme is symmetric.
2. Let S be a set with n elements. For the Johnson scheme, the points are the $C(n, k)$ subsets of S with k elements. Two subsets x, y satisfy $(x, y) \in R_i$ if and only if $|x \cap y| = k - i$. Prove that this is an association scheme.
3. Construct the A_i for the relation given in Example 8.20.
4. Draw the labeled complete graph for the relation given in Example 8.20.
5. Construct an infinite family of association schemes, based on groups, that is symmetric.

Discrete Probability—A Return to Counting

9

In this chapter, we return to counting problems, but we study them to understand the probability of a given event. While most branches of mathematics have fairly lofty origins, probability has its origins in gambling. Specifically, in determining when a game of chance is fair to all involved. One of the earliest texts on the subject is Christaan Huygens' book [51].

The probability of an event is a number between 0 and 1 where 0 indicates that event is impossible and 1 indicates that the event must occur. While probability has been extended to infinite sets as well, we shall restrict ourselves to combinatorial probability, meaning probability applied to finite sets. There are many important differences between probability with finite sets as opposed to probability with infinite sets. For example, with infinite sets a probability of 0 indicates something will almost never happen and a probability of 1 indicates that an event will almost always happen. For example, the probability of picking an irrational number out of the real numbers is 1, even though rational numbers do exist. However, with finite sets the adjective *almost* is done away with and we are talking about events never occurring or always occurring.

9.1 Definitions and Elementary Probability

We begin with the standard definition of probability.

Definition 9.1 Given a sample space S and an event $E \subset S$, the probability of event E is

$$P(E) = \frac{|E|}{|S|}. \tag{9.1}$$

© The Editor(s) (if applicable) and The Author(s), under exclusive license
to Springer Nature Switzerland AG 2020
S. T. Dougherty, *Combinatorics and Finite Geometry*,
Springer Undergraduate Mathematics Series,
https://doi.org/10.1007/978-3-030-56395-0_9

Example 9.1 Assume $S = \{1, 2, 3, 4, 5, 6, 7, 8, 9, 10\}$. The probability of picking an even number from this set is $\frac{5}{10}$, and the probability of picking an odd number is $\frac{5}{10}$. The probability of picking a number greater than 1 is $\frac{9}{10}$. The probability of picking 17 is 0.

Example 9.2 In a well-known lottery game, a player is asked to pick a 3-digit number ranging from 000 to 999. If the number chosen matches the number chosen for that day, the state running the game will pay 500 dollars on a one dollar bet. There are 1000 possible outcomes and 1 that wins so the probability of winning is $P(W) = \frac{1}{1000}$. Hence, the game is highly unfair since a payoff of 1000 dollars would be required to make the game fair.

Example 9.3 In a common lottery game, a player picks six numbers from 1 to 40. The state running the game also picks 6 for the day. If the player's 6 matches the state's 6 then the player wins. There are $C(40, 6) = 3,838,380$ possible ways of picking 6 from 40. Therefore, the probability of winning is $P(W) = \frac{1}{3,838,380}$. For the game to be fair, a payoff of $3,838,380$ dollars on a 1 dollar bet would be required. In general, the payoff is determined by how many people have bet and when the last winner was. In other words, if there is no winner on a play, the money collected in the pot continues on to the next play. Given this scenario, if a player only wanted to play a fair game, they should only play the game when the pot hits $3,838,380$ dollars. Of course, even though the game would be fair, the probability of actually winning would still be quite low.

Example 9.4 In American roulette, a small ball is placed on a wheel which is spun. The ball can land in any of 38 positions corresponding to $00, 0, 1, 2, \ldots, 36$. Of these 38 positions, 18 are red, 18 are black, and 2 (0 and 00) are green. A player can choose a variety of ways of betting. For example, one can bet that the number will be red, black, even, odd, one of four numbers (for certain groups), one of two numbers (for certain pairs). If there are n ways a player can win, then the probability of winning on that bet $\frac{n}{38}$. For example, the probability of winning on a bet of red is $P(R) = \frac{18}{38} = \frac{9}{19}$, and the probability of winning on a bet of single number is $\frac{1}{38}$. However, in the first case, the payoff on a bet of 1 dollar would be 2 dollars and in the second case the payout on a bet of one dollar is 36 dollars (including the dollar bet for both instances). In essence, the casino pays off as if there were 36 possibilities rather than 38 possibilities. Therefore, the game is not fair.

Theorem 9.1 *Let S be a finite sample space and let* A *and* B *be subsets of* S.

1. $0 \leq P(A) \leq 1$.
2. $P(\emptyset) = 0$, $P(S) = 1$.
3. $P(A \cup B) = P(A) + P(B) - P(A \cap B)$.
4. $P(A) + P(\overline{A}) = 1$.
5. If $A \subseteq B$ *then* $P(A) \leq P(B)$.

Proof Statements 1, 2, and 5 are immediate from the definition.
For Statement 3, we know that $|A \cup B| = |A| + |B| - |A \cap B|$. Therefore

$$P(A \cup B) = \frac{A \cup B|}{|S|} = \frac{|A| + |B| - |A \cap B|}{|S|}$$
$$= \frac{|A|}{|S|} + \frac{|B|}{|S|} - \frac{|A \cap B|}{|S|}$$
$$= P(A) + P(B) - P(A \cap B).$$

$\square$

For Statement 4, we have $\overline{A} = S \setminus A$, so $|A| = |S| - |A|$. Then

$$P(A) + P(\overline{A}) = \frac{|A|}{|S|} + \frac{|S| - |A|}{|S|} = \frac{|S|}{|S|} = 1.$$

Definition 9.2 Two events A and B in a sample space S are mutually exclusive if $A \cap B = \emptyset$.

It follows from the definition and Statement 3 in Theorem 9.1 that if two events A and B are mutually exclusive then $P(A \cup B) = P(A) + P(B)$.

Assume you have a standard deck of 52 cards. The probability of picking either an ace or a spade is $\frac{16}{52} = \frac{4}{13}$ since there are 13 spades one of which is an ace and 3 other aces. However, the probability of picking an ace is $\frac{1}{13}$, and the probability of picking a spade is $\frac{1}{4}$ and $\frac{1}{13} + \frac{1}{4} = \frac{17}{52} \neq \frac{16}{52}$, since these events are not mutually exclusive.

Example 9.5 The probabilities for rolling a pair of dice are given below:

Dice sum	Probability
2	$\frac{1}{36}$
3	$\frac{2}{36}$
4	$\frac{3}{36}$
5	$\frac{4}{36}$
6	$\frac{5}{36}$
7	$\frac{6}{36}$
8	$\frac{5}{36}$
9	$\frac{4}{36}$
10	$\frac{3}{36}$
11	$\frac{2}{36}$
12	$\frac{1}{36}$

These events are all mutually exclusive. As an example, the probability of getting a 3, 4, 7, or 12 is $\frac{2}{36} + \frac{3}{36} + \frac{6}{36} + \frac{1}{36} = \frac{12}{36} = \frac{1}{3}$.

Definition 9.3 Two events A and B in a sample space S are independent if $P(A \cap B) = P(A)P(B)$.

Intuitively, one thinks of this definition as A and B are independent if the occurrence of A has no bearing on the occurrence of B. For example, assume you pick a ball out of a bin with a green ball, a red ball, and a yellow ball, then you roll a die. There are 18 possible outcomes. The probability of picking a green ball followed by rolling a 2, 4, or 6 is $\frac{3}{18} = \frac{1}{6}$ as the three events are $(green, 2), (green, 4)$, and $(green, 6)$. Now the probability of picking a green ball out of the bin is $\frac{1}{3}$, and the probability of picking either a $2, 4$, or 6 is $\frac{1}{2}$. Then the product of these probabilities is $\frac{1}{3}(\frac{1}{2}) = \frac{1}{6}$. Therefore, the events are independent as we would suspect.

Example 9.6 The probability of flipping a coin ten times in a row and getting a head each time is $(\frac{1}{2})^{10} = \frac{1}{1024}$ since the events (each flip of the coin) are independent. One might also think of this as there are 2^{10} possible outcomes, and HHHHHHHHHH is the only one with all heads. Therefore, the probability is $\frac{1}{1024}$.

Example 9.7 The probability that flower A is blooming is 0.9, and the probability that flower B is blooming is 0.7. Then the probability that both flowers are blooming is $(0.9)(0.7) = 0.63$. The probability that neither is blooming is $(0.1)(0.3) = 0.03$. The probability that flower A is blooming and flower B is not is $(0.9)(0.3) = 0.27$, and the probability that flower A is not blooming and flower B is blooming is $(0.1)(0.7) = 0.07$. We note that $0.63 + 0.03 + 0.27 + 0.07 = 1$.

Theorem 9.2 *Assume that the probability of an event occurring is P. The probability that the event will occur k times in n trials is $C(n, k)P^k(1 - P)^{n-k}$.*

Proof There are $C(n, k)$ ways that the event will occur k times in n trials. In each of these, there are k locations for the event to occur and $n - k$ locations for the event not to occur. Therefore, the probability of this occurring is $P^k(1 - P)^{n-k}$.   □

We notice that in this theorem, the value is the coefficient in the binomial theorem with P and $1 - P$ substituted for x and y. Specifically

$$(x + y)^n = \sum_{k=0}^{n} C(n, k)x^k y^{n-k}$$

$$(P + (1 - P))^n = \sum_{k=0}^{n} C(n, k)P^k(1 - P)^{n-k}$$

$$1 = \sum_{k=0}^{n} C(n, k)P^k(1 - P)^{n-k}.$$

This shows that the sum of the probabilities of the event occurring is $0, 1, 2, \ldots, n$ times is 1 as we would have expected.

Example 9.8 Assume there is a box with a red ball, a green ball, and a blue ball. Assume a ball is extracted 4 times replacing it each time. Then, the probability that a green ball is picked exactly twice is $C(4, 2)(\frac{1}{3})^2(\frac{2}{3})^2 = 6(\frac{1}{9}\frac{4}{9}) = \frac{24}{81}$.

Example 9.9 Assume a coin is flipped 7 times. To determine the probability that we get at least 5 heads, we sum the probability that we get for 5 heads, 6 heads, and 7 heads. Namely, the probability is

$$C(7,5)(\frac{1}{2})^5(\frac{1}{2})^2 + C(7,6)(\frac{1}{2})^6(\frac{1}{2})^1 + C(7,7)(\frac{1}{2})^7 = (21 + 7 + 1)(\frac{1}{128}) = \frac{29}{128}.$$

Exercises

1. Determine the probability of rolling a pair of dice and getting a number greater than or equal to 9.
2. Determine the probability of flipping a coin 10 times in a row and getting no heads.
3. Determine the probability of flipping a coin 8 times and getting exactly 3 heads.
4. Determine the probability of flipping a coin 8 times and getting at least 6 heads.
5. Determine the probability of rolling a pair of dice twice and having a sum greater than or equal to 6 for each throw.
6. Determine the probability of shuffling a deck of cards and having spades in the first 13 spaces.
7. Determine the probability that the first and second numbers are different when rolling a single die twice.
8. Determine the probability that the first and second numbers are equal when rolling a single die twice.
9. Determine the probability that a coin is flipped 9 times and you get exactly 4 tails.
10. Determine the probability that a coin is flipped 7 times and you get exactly 5 heads.
11. Determine the probability that a coin is flipped 10 times and you get at least 9 heads.
12. Determine the probability that if six people, all with different ages, are arranged in a row, that they will be arranged in descending order.
13. Assume there are six married couples in a room. Determine the probability that if two people are chosen at random that they are married.
14. The hands in 5-card poker are two or kind, three of a kind, straight, flush, full house, four of a kind, straight flush, royal flush. Determine the probabilities of

getting these hands and show that this is the proper order for the hands given that a hand with a lesser probability should beat a hand with a greater probability.

15. Assume a standard deck of cards is shuffled. Determine the probability that it is arranged in the same manner as a given deck.

16. Assume there are 10 questions on a multiple-choice test, where each question has three possible answers and that a student answers the questions randomly. Determine the probability that the student passes the test (where a pass is a 70%).

17. Assume p, q, and r are distinct primes. Determine the probability that a number is chosen between 1 and pqr is relatively prime to pqr.

18. Assume an event has probability P. Prove that there are a number of trials such that the probability of the event occurring at least once is greater than or equal to $\frac{2}{3}$.

19. Determine the probability that three points on a projective plane of order n chosen randomly are on a line.

20. Determine the probability that three points on an affine plane of order n chosen randomly are on a line.

9.2 Conditional Probability

We shall now discuss conditional probability.

Definition 9.4 In a sample space S, with events A and B, the conditional probability is defined as

$$P(A|B) = \frac{P(A \cap B)}{P(B)}.$$

We see immediately from the definition that if A and B are independent, then $P(A \cap B) = P(A)P(B)$ and so $P(A|B) = P(A)$ and $P(B|A) = P(B)$.

Example 9.10 Given a standard deck of 52 cards. To determine the probability of choosing a queen given that you have chosen a card, that is, a heart is $P(A|B) = \frac{1}{52}/\frac{1}{4} = \frac{1}{13}$. To determine the probability of choosing a heart given that the card chosen is a queen is $P(B|A) = \frac{1}{52}/\frac{1}{13} = \frac{1}{4}$.

We shall now discuss one of the most interesting examples of conditional probability. It is known as the Monty Hall problem. It is called this because of its similarity to events on an American television show hosted by Monty Hall. The scenario is as follows. There are three curtains which hide prizes. Label these 1, 2, and 3. Behind two of the curtains are worthless prizes, for the purpose of this description let us say there is a can of corn behind these curtains. Behind one of the curtains is a valuable prize, for the purpose of this description let us say there is a car behind this curtain. The person playing the game wants to win the car. We assume that the prizes are randomly assigned to the curtains.

Fig. 9.1 Monty Hall
problem

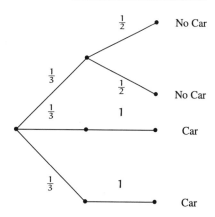

The host offers the player a choice of the three curtains. Let us suppose the player picks curtain 1. The host then reveals one of the remaining two curtains as hiding the can of corn. Note that we assume that the host chooses randomly if both contain a can of corn. He then turns to the player and asks if the player would like to change their answer. The question is should the player now change their answer to the remaining curtain or stick with the curtain they originally chose. It is tempting to assume that you should not change since the probability that the car is behind any curtain is $\frac{1}{3}$. However, this is not the correct decision to make.

An intuitive way to think about the problem is that if a player does invoke the switching strategy the only way they could lose is if they choose the correct curtain initially, which has probability $\frac{1}{3}$. Therefore, the probability of winning with this strategy is $1 - \frac{1}{3} = \frac{2}{3}$.

A more rigorous description uses the following argument. The diagram in Fig. 9.1 indicates the probability of getting a car by initially choosing a door and then switching after the host raises a curtain that does not contain the car. Each of the first three probabilities is $\frac{1}{3}$ for the possibilities of the car being behind each curtain. The top one is if the player chooses the curtain where the car is initially. Then the next two probabilities are $\frac{1}{2}$ for the possibilities for which curtain the host opens. In the bottom two are the possibilities when the player chooses a curtain that does not contain the car. Then the host has only one curtain which can be opened.

The probability for getting the car is $(\frac{1}{3})(1) + (\frac{1}{3})(1) = \frac{2}{3}$.

Finally, we can use conditional probability to determine the probability of success using the switching technique. We take the probability of event A (ending on the right curtain) given that a curtain unequal to the original choice is revealed and we have

$$\frac{\frac{1}{3}}{\frac{1}{3} + \frac{1}{6}} = \frac{2}{3}.$$

For a complete description of the Monty Hall problem and its implications, see [76].

Theorem 9.3 (Bayes' Theorem) *In a sample space* S, *with events* A, B,

$$P(A|B) = \frac{P(B|A)P(A)}{P(B)}. \tag{9.2}$$

Proof By the definition of conditional probability we have

$$P(A|B) = \frac{P(A \cap B)}{P(B)}$$

and

$$P(B|A) = \frac{P(B \cap A)}{P(A)}.$$

This gives $P(A)P(B|A) = P(B)P(A|B)$ and finally $P(A|B) = \frac{P(B|A)P(A)}{P(B)}.$ □

Theorem 9.4 *Let* S *be a sample space with events* A, B.

1. *Events* A *and* B *are independent if and only if* $P(A) = P(A|B)$.
2. *Events* A *and* B *are independent if and only if* $P(B) = P(B|A)$.

Proof We shall prove the first statement; the proof of the second is identical. Recall by Definition 9.3 that A and B are independent if $P(A \cap B) = P(A)P(B)$. Then

$$P(A)P(B) = P(A \cap B) \Leftrightarrow P(A) = \frac{P(A \cap B)}{P(B)} = P(A|B).$$

This gives the result. □

Exercises

1. Determine the probability that a card chosen from a standard deck is an ace given that it is a spade.
2. The probability that a train departs on time is 0.9 and the probability that a train both departs and arrives on time is 0.75. Given that the train departs on time determine the probability that it arrives on time.
3. Determine the probability that a card chosen from a standard deck of cards is a face card given that it is red.
4. Determine the probability that a card chosen from a standard deck of cards is a king given that it is a spade.

Automorphism Groups

10

In this chapter, we shall use one of the most important structures in abstract algebra as a tool to study finite incidence structures. The algebraic structure is a group. It is generally the first structure one encounters in studying abstract algebra. We shall begin with a very elementary study of finite groups, and then we shall study the groups associated with various combinatorial structures.

10.1   Groups

The theory of groups is one of the largest and most important branches in all of abstract algebra. It would be almost impossible to exhaust the study of groups in a lifetime, let alone in a text. Here, we shall only require the most elementary facts about groups in order to understand a bit about how to use them to study combinatorial objects. The study of finite groups has many connections to the study of combinatorial objects, and important results in both areas have come from this connection.

We begin with the definition of a group.

Definition 10.1 A group $(G, *)$ is a set G with an operation $*$ such that

1. (Closure) If $a, b \in G$ then $a * b \in G$.
2. (Associativity) For all $a, b, c \in G$, $(a * b) * c = a * (b * c)$.
3. (Identity) There exists an $e \in G$ with $e * a = a * e = a$ for all $a \in G$.
4. (Inverses) For all $a \in G$ there exists $b \in G$ with $a * b = e$.

The order of the group $(G, *)$ is $|G|$. If $H \subseteq G$ and $(H, *)$ is a group, then H is said to be a subgroup of G.

S. T. Dougherty, *Combinatorics and Finite Geometry*, Springer Undergraduate Mathematics Series

Example 10.1 The following are examples of groups: $(\mathbb{Z}, +), (\mathbb{Q}, +), (\mathbb{R}, +), (\mathbb{Q} - \{0\}, *), (\mathbb{R} - \{0\}, *), (\mathbb{Z}_n, +)$. Notice that $(\mathbb{Z} - \{0\}, *)$ is not a group since, for example, 2 has no multiplicative inverse in $\mathbb{Z} - \{0\}$.

We note that all of the groups in the previous example have a commutative operation. These are given a special designation.

Definition 10.2 If $(G, *)$ is a group such that $a * b = b * a$ for all $a, b \in G$, then the group is said to be an Abelian group.

The term Abelian comes from the name of the Norwegian mathematician Niels Henrik Abel.

Example 10.2 Let S_n denote the set of all $n!$ permutations on a set of size n. Let the operation on this set be functional composition. Then $(S_n, \circ)$ is a group of order $n!$ that is non-Abelian. This group is called the symmetric group on n letters. The group S_n is an extremely important group since it can be shown that all groups can be viewed as subgroups of this group.

Theorem 10.1 *The operation table of a finite group is a Latin square.*

Proof Let $(G, *)$ be a finite group, and let L be a matrix indexed by the elements of the finite group $G = \{g_1, g_2, \ldots, g_n\}$ with $L_{g_i, g_j} = g_i * g_j$.
Consider the i-th row of L. This consists of the elements

$$g_i * g_1 \quad g_i * g_2 \quad \cdots \quad g_i * g_n.$$

If $g_i * g_j = g_i * g_k$ then if g_i^{-1} is the inverse guaranteed by axiom 4 then

$$g_i^{-1} * g_i * g_j = g_i^{-1} * g_i * g_k$$
$$g_j = g_k.$$

Hence, each element of the group appears exactly once in each row. The j-th column of L consists of the elements

$$g_1 * g_j \quad g_2 * g_j \quad \cdots \quad g_n * g_j.$$

If $g_i * g_j = g_k * g_j$ then if g_j^{-1} is the inverse guaranteed by axiom 4 then

$$g_i * g_j * g_j^{-1} = g_k * g_j * g_j^{-1}$$
$$g_j = g_k.$$

Hence, each element of the group appears exactly once in each column.   □

Example 10.3 We give the operation table of the Klein-4 group, where e denotes the identity.

$$
\begin{array}{c|cccc}
* & e & a & b & c \\
\hline
e & e & a & b & c \\
a & a & e & c & b \\
b & b & c & e & a \\
c & c & b & a & e \\
\end{array}
$$

We note that the 4 by 4 table gives a non-circulant Latin square of order 4. We can also give the operation table of the group $(\mathbb{Z}_4, +)$.

$$
\begin{array}{c|cccc}
+ & 0 & 1 & 2 & 3 \\
\hline
0 & 0 & 1 & 2 & 3 \\
1 & 1 & 2 & 3 & 0 \\
2 & 2 & 3 & 0 & 1 \\
3 & 3 & 0 & 1 & 2 \\
\end{array}
$$

Here, the 4 by 4 table gives a circulant Latin square of order 4.

For the rest of this section, we shall refer to a group $(G, *)$ simply as G whenever the operation is understood. Additionally, the operation will be denoted by juxtaposition.

The order of an element g in a group G with identity e is the smallest natural number $n > 0$ such that $g^n = e$.

Let G and H be groups; then, the cross product of G and H is defined by

$$G \times H = \{(g, h) \mid g \in G, h \in H\}, \tag{10.1}$$

where

$$(g, h) * (g', h') = (g *_G g', h *_H h') \tag{10.2}$$

and $*_G$ is the operation in G and $*_H$ is the operation in H.

From this point, we can use juxtaposition to indicate the operation of the group.

Theorem 10.2 *Let G and H be groups. Then $G \times H$ is a group.*

Proof If $(g, h), (g', h') \in G \times H$, then $gg' \in G, hh' \in H$ since they are groups which gives $(g, h)(g', h') \in G \times H$. Therefore it is closed.

If $(g, h), (g', h'), (g'', h'') \in G \times H$ then

$$((g, h)(g', h'))(g'', h'') = (gg', hh')(g'', h'') = ((gg')g'', (hh')h'') = $$
$$(g(g'g''), h(h'h'')) = (g, h)(g'g'', h'h'') = (g, h)((g', h')(g'', h'')).$$

Therefore, the operation is associative.

Let e_G and e_H be the identity elements of G and H, respectively. Then (e_G, e_H) $(g, h) = (e_g g, e_H h) = (g, h) = (g e_g, h e_H) = (g, h)(e_G, e_H)$. Therefore, $G \times H$ has an identity element.

Let g^{-1} be the inverse of g in G and let h^{-1} be the inverse of h in H. Then

$$(g, h)(g^{-1}, h^{-1}) = (g g^{-1}, h h^{-1}) = (e_G, e_H)$$
$$= (g^{-1} g, h^{-1} h) = (g^{-1}, h^{-1})(g, h).$$

Therefore, every element has an inverse. □

Example 10.4 Consider the group $\mathbb{Z}_2 \times \mathbb{Z}_2$. It has the following table:

+	(0,0)	(0,1)	(1,0)	(1,1)
(0,0)	(0,0)	(0,1)	(1,0)	(1,1)
(0,1)	(0,1)	(0,0)	(1,1)	(1,0)
(1,0)	(1,0)	(1,1)	(0,0)	(0,1)
(1,1)	(1,1)	(1,0)	(0,1)	(0,0)

Let $(G, *)$ be a group and H a subgroup of G, that is, $H \subseteq G$. A (left) coset of H is $aH = \{a * h \mid h \in H\}$.

Lemma 10.1 *Let H be a subgroup of a finite group G then* $|aH| = |H|$ *for all* $a \in G$.

Proof Let $\psi : H \to aH$ by $\psi(g) = ag$. We shall show that ψ is a bijection. If $\psi(g) = \psi(g')$ then $ag = ag'$ which implies $a^{-1}ag = a^{-1}ag'$ since we are in a group and we know that a has an inverse with respect to the operation of G. This gives that $g = g'$ and ψ is injective. Any element in aH is of the form ag for some $g \in H$ so ψ is naturally surjective and hence ψ is a bijection. Therefore, we have that $|aH| = |H|$. □

Example 10.5 Consider the group $(\mathbb{Z}_8, +)$. The group $\{0, 4\}$ is a subgroup. The cosets all have two elements, namely, $1 + \{0, 4\} = \{1, 5\}$, $2 + \{0, 4\} = \{2, 6\}$, and $3 + \{0, 4\} = \{3, 7\}$.

Lemma 10.2 *The cosets of a subgroup H partition G.*

Proof We shall show that membership in a coset is an equivalence relation. That is, a is related to b if $b \in aH$.

- Reflexive: Any element a is related to itself since $ae \in aH$ where e is the identity element of G.

- Symmetric: If $b \in aH$ then $b = ag$ for some $g \in H$. Then multiplying by g^{-1} on the right, which must also be in H since H is a group gives that $a = bg^{-1}$ and $a \in bH$.
- Transitive: If $b \in aH$ and $c \in bH$ then $b = ag$ and $c = bg'$ for some $g, g' \in H$. Then $c = a(gg') \in aH$ and the relation is transitive.

The equivalence classes partition the group G. □

Example 10.6 Consider the cosets given in Example 10.5. The set $\{0, 1, 2, 3, 4, 5, 6, 7\}$ is partitioned by $\{0, 4\}, \{1, 5\}, \{2, 6\},$ and $\{3, 7\}$.

For a group G and a subgroup H, we define the index of H in G as the number of cosets of H in G and denote it by $[G : H]$.

Theorem 10.3 (Lagrange's Theorem) For a finite group G and a subgroup H, we have

$$|G| = |H|[G : H]. \tag{10.3}$$

Proof Let G be a group and H a subgroup. Then by Lemma 10.2, we have

$$G = a_1H \cup a_2H \cup \cdots \cup a_{[G:H]}H, \tag{10.4}$$

where $a_1 = e$ and $a_iH \cap a_jH = \emptyset$ when $i \neq j$. Then since $|a_iH| = |H|$ by Lemma 10.1 we have that $|G| = |H|[G : H]$, which gives the result. □

Corollary 10.1 Let G be a finite group G and let H be a subgroup of G. Then the order of H must divide the order of G.

Proof By the previous theorem, we have $|G| = |H|[G : H]$. Then, since $[G : H]$ is an integer we have the result. □

Example 10.7 Consider the group $(\mathbb{Z}_{10}, +)$, then the non-trivial subgroups are $\{0, 5\}$ and $\{0, 2, 4, 6, 8\}$. These are the only non-trivial subgroups since 2 and 5 are the only integers dividing 10.

We shall now describe the action of a group on a set. We have seen an example of this in Sect. 1.4 as S_n acted on any set with n elements. Let G be a group and S a set. Then we say G acts on the set S if $gs \in S$ for all $g \in G$ and $s \in S$ and $g(g's) = (gg')s$ for $g, g' \in G$.

Define the orbit of an element $a \in S$ as

$$\mathrm{Orb}(a) = \{b \mid \text{there exits } \sigma \in G \text{ with } \sigma(a) = b\}. \tag{10.5}$$

Theorem 10.4 *Let G be a group acting on a set S. For* $a \in S$ *the set of elements of G that fix* a, *i.e.,* $H = \{\sigma \mid \sigma \in G, \sigma(a) = a\}$ *is a subgroup of G.*

Proof If σ_e is the identity of G, then $\sigma_e a = a$ so $a \in H$ and it is non-empty. If $\sigma, \tau \in H$, then $(\sigma\tau)(a) = \sigma(\tau(a)) = \sigma(a) = a$ and H is closed. Also $\sigma^{-1}\sigma(a) = e_\sigma(a) = a$ but $\sigma^{-1}(\sigma(a)) = \sigma^{-1}(a)$ which gives $\sigma^{-1}(a) = a$ and therefore H is a subgroup of G. □

Let G be a group acting on a set S. Let $a \in S$ and let H be the subgroup of G of elements that fix a. If $\tau \in \sigma H$ then $\tau(a) = \sigma(a)$. This is easy to see since $\tau = \sigma \circ \sigma'$ where σ' fixes a so $\tau(a) = \sigma(\sigma'(a)) = \sigma(a)$. Hence, the number of elements in the orbit is precisely the number of cosets of the fixed subgroup in the ambient group.

This gives the following theorem.

Theorem 10.5 *Let G be a group acting on a set S. For* $a \in S$ *we have*

$$|\text{Orb}(a)| = \frac{|G|}{|H(a)|}, \tag{10.6}$$

where $H(a)$ *is the subgroup of G that fixes* a.

Consider the unit circle in the complex plane given in Fig. 10.1. Let $\xi_n = e^{\frac{2\pi i}{n}}$ which corresponds to the complex number on the unit circle $\frac{2\pi}{n}$ radians around the circle in the counterclockwise direction. We have that ξ_n is a primitive root of unity, meaning $\xi_n^n = 1$ and $\xi_n^i \neq 1$ for $1 \leq i < n$.

It is immediate that the set $\{1, \xi_n, \xi_n^2, \ldots, \xi_n^{n-1}\}$ forms a group of order n. This group is the cyclic group of order n, denoted C_n. More abstractly, it can be defined as

$$C_n = \langle a \mid a^n = 1 \rangle = \{a^i \mid 0 \leq i \leq n - 1\}. \tag{10.7}$$

Theorem 10.6 *The cyclic group of order* n *is an abelian group.*

Proof If $x, y \in C_n$ then $x = a^i$ and $y = a^j$ for some integers i and j. Then $xy = a^i a^j = a^{i+j} = a^{j+i} = a^j a^i = yx$ and so the group is abelian. □

Consider the following triangle. We shall consider the group of symmetries of the triangle on the plane. A rigid motion of the plane is an injective function from the plane to itself that preserves distance. A symmetry of an object in the plane is a rigid motion of the plane that maps the object to itself.

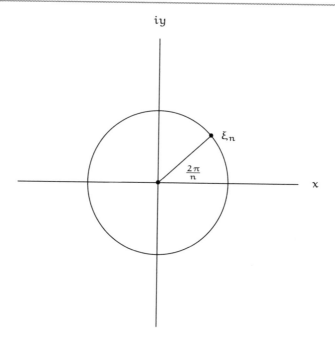

Fig. 10.1 Primitive root of unity

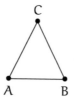

$$(10.8)$$

Consider rotating the triangle $\frac{2\pi}{3}$ radians counterclockwise. This moves A to B's original location, B to C's original location, and C to A's original location. We shall call this action α. Note that α^2 is also a symmetry and α^3 is the identity map, we shall denote the identity by ι.

Consider a line from C to midpoint between A and B. Rotating the triangle π around this line interchanges A and B and leaves C fixed. We shall call this symmetry β. This symmetry is of order 2.

Consider the element $\alpha\beta$. This element is the same as rotating the triangle π around the line from B to the midpoint of A and C. In fact, these two elements generate all the symmetries of the triangle. The group of symmetries of an equilateral triangle on the plane is a non-abelian group of order 6.

We note that $\alpha\beta \neq \beta\alpha$. We can build a multiplication table of the group generated by α and β.

$$
\begin{array}{c|cccccc}
* & \iota & \alpha & \alpha^2 & \beta & \alpha\beta & \beta\alpha \\
\hline
\iota & \iota & \alpha & \alpha^2 & \beta & \alpha\beta & \beta\alpha \\
\alpha & \alpha & \alpha^2 & \iota & \alpha\beta & \beta\alpha & \beta \\
\alpha^2 & \alpha^2 & \iota & \alpha & \beta\alpha & \beta & \alpha\beta \\
\beta & \beta & \beta\alpha & \alpha\beta & \iota & \alpha^2 & \alpha \\
\alpha\beta & \alpha\beta & \beta & \beta\alpha & \alpha & \iota & \alpha^2 \\
\beta\alpha & \beta\alpha & \alpha\beta & \beta & \alpha^2 & \alpha & \iota
\end{array}
\qquad (10.9)
$$

This group is the first non-trivial example of the dihedral group. The dihedral group is defined as

$$D_{2t} = \langle a, b \mid b^t = 1, a^2 = 1, aba^{-1} = b^{-1}\rangle. \qquad (10.10)$$

This group can be seen as the automorphism group of a regular t sided figure in a Euclidean plane. The element b corresponds to shifting the object $\frac{2\pi}{t}$ radians, and the element a corresponds to flipping the object over.

Definition 10.3 We say that two groups $(G, *)$ and $(G', *')$ are isomorphic if there is a bijection $\phi : G \to G'$ such that $\phi(g * h) = \phi(g) *' \phi(h)$ for all $g, h \in G$.

Example 10.8 Consider the Klein-4 group given in Example 10.3 and the group given in Example 10.4. The following bijection is an isomorphism:

$$
\begin{aligned}
e &\to (0,0) \\
a &\to (0,1) \\
b &\to (1,0) \\
c &\to (1,1).
\end{aligned}
$$

Therefore, the two groups are isomorphic.

Exercises

1. Prove that $(Z_n, +)$ is a group.
2. Prove that the table of a subgroup $(H, *)$ of the group $(G, *)$ forms a Latin subsquare of the square formed from the group $(G, *)$.
3. Prove that the set of rational numbers with addition is a group.
4. Prove that the set of rational numbers with multiplication is not a group but that the set of non-zero rational numbers with multiplication is a group.
5. Prove that $(Z_n - \{0\}, *)$, where $*$ is multiplication, is a group if and only if p is prime.
6. Prove that the set of permutations on n letters S_n defined in Sect. 1.4 with the operation of functional composition is a group.

7. Prove that the identity in a group is unique.
8. Determine the group of symmetries of a square, D_8, in the plane. Write its multiplication table.
9. Prove that the subset of the symmetries of a geometric object in the plane that fixes a point on the object is a subgroup.
10. Determine the group of symmetries of a cube in three-dimensional Euclidean space.
11. Prove that $(\mathbb{Z}_n, +)$ is isomorphic to the cyclic group of order n.
12. Prove that the dihedral group D_{2t} is non-commutative for $t \geq 3$.
13. Prove that if G is a cyclic group of prime order then G has no non-trivial subgroups.
14. Determine the group of symmetries of a regular tetrahedron.

10.2 Automorphisms of a Design

One of the most useful ways of examining the structure of any mathematical object is to consider the group of automorphisms of an object.

Definition 10.4 An automorphism of an incidence structure $D = (\mathcal{P}, \mathcal{B}, I)$ is a bijection $\psi : D \to D$, where ψ is a bijection from the set of points to the set of points, a bijection from the set of blocks to the set of blocks and

$$(p, B) \in I \iff (\psi(p), \psi(B)) \in I.$$

Notice that an automorphism is simply an isomorphism of a design with itself. An automorphism is also called a collineation.

If ϕ and ψ are two automorphisms of a design D, then the composition of these two functions is defined as $\phi \circ \psi(x) = \phi(\psi(x))$.

Theorem 10.7 *The set of automorphisms of a design* $D = (\mathcal{P}, \mathcal{L}, I)$ *with functional composition is a group.*

Proof It is well known that the composition of bijections is a bijection. We know $(p, \ell) \in I$ if and only if $(\phi(p), \phi(\ell)) \in I$ and $(p, \ell) \in I$ if and only if $(\psi(p), \psi(\ell)) \in I$. Therefore $(p, \ell) \in I$ if and only if $(\phi(p), \phi(\ell)) \in I$ if and only if $(\psi(\phi(p)), \psi(\phi(\ell))) \in I$. Therefore, the composition of two automorphisms is an automorphism and the set is closed under functional composition.

It is also well known that functional composition is associative.

The identity map serves as the identity of the group. Since if $\iota(x) = x$ for all $x \in D$ then $\iota \circ \phi = \phi \circ \iota = \phi$.

If ϕ is an isomorphism then define the inverse ϕ^{-1} by $\phi(x) = y$ if and only if $\phi^{-1}(y) = x$. Then $\phi \circ \phi^{-1} = \iota$ and there are inverses. Hence, it is a group. $\square$

Fig. 10.2 Projective plane of
order 2

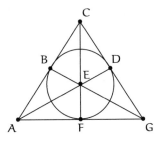

Table 10.1 The points and
lines under the automorphism
ϕ

p	$\phi(p)$	L	$\phi(L)$
A	F	L_1	L_3
B	G	L_2	L_5
C	A	L_3	L_7
D	E	L_4	L_1
E	C	L_5	L_4
F	B	L_6	L_2
G	D	L_7	L_6

Recall the following description of the projective plane of order 2 given in
Fig. 10.2.

List the lines $L_1, \ldots, L_7$ as follows:

$$L_1 \leftrightarrow \{A, B, C\}$$
$$L_2 \leftrightarrow \{C, D, G\}$$
$$L_3 \leftrightarrow \{A, F, G\}$$
$$L_4 \leftrightarrow \{C, E, F\}$$
$$L_5 \leftrightarrow \{A, D, E\}$$
$$L_6 \leftrightarrow \{B, E, G\}$$
$$L_7 \leftrightarrow \{B, D, F\}$$

Consider the following automorphism ϕ (Table 10.1).

As an example, we can compute ϕ^2 as well (Table 10.2).

Consider the automorphism ϕ as a permutation of the points. It can be written as
(A, F, B, G, D, E, C). In the same way, the automorphism can be viewed as a permutation of the lines as $(L_1, L_3, L_7, L_6, L_2, L_5, L_4)$. It is clear that both permutations
have order 7, that is, applying them seven times results in the identity. This gives the
following.

Theorem 10.8 *The group* $G_7 = \langle \phi \rangle$ *is a commutative group of order 7.*

This group corresponds to a Singer cycle, that is, the points of the plane are viewed
as non-zero elements in the field of order 8, and the automorphisms correspond to
multiplication by a non-zero element.

Table 10.2 The points and lines under the automorphism ϕ^2

p	$\phi^2(p)$	L	$\phi^2(L)$
A	B	L_1	L_7
B	D	L_2	L_4
C	F	L_3	L_6
D	C	L_4	L_3
E	A	L_5	L_1
F	G	L_6	L_5
G	E	L_7	L_2

Table 10.3 The points and lines under the automorphism ψ_1

p	$\psi_1(p)$	L	$\psi_1(L)$
A	A	L_1	L_5
B	D	L_2	L_6
C	E	L_3	L_3
D	B	L_4	L_4
E	C	L_5	L_1
F	F	L_6	L_2
G	G	L_7	L_7

Definition 10.5 A group G acts transitively on S if given $a, b \in S$ there exists a $\sigma \in G$ with $\sigma(a) = b$.

Theorem 10.9 *The group G_7 acts transitively on the points and lines of the projective plane of order 2.*

Proof This can be seen directly by examining the tables constructed in Exercise 3. More elegantly, consider points p and q as elements in the field of order 8. Then the automorphism corresponding to $p^{-1}q$ sends p to q. In other words, it is a solution to the equation $px = q$ in the field. $\qquad\square$

The group G_7 is not the full automorphism of the projective plane of order 2. Notice that in each of the automorphisms ϕ^i that there are no fixed points. Let $p_1, \ldots, p_7$ be the points of the plane of order 2. Construct an automorphism τ with no fixed points by mapping p_1 to p_2, p_2 to p_3, and p_3 to the third point on the line through p_1 and p_2 and continuing in this manner. Construct another automorphism by mapping p_1 to p_2, p_2 to p_3, and p_3 to the third point on the line through p_1 and p_3 and continuing in this manner. There are 48 automorphisms constructed in this manner. We leave verification of this facts as an exercise, see Exercise 4.

Consider the following automorphism which we call ψ_1 given in Table 10.3. We construct this by fixing the points on line L_1, and the lines through point F.

Table 10.4 The points and lines under the automorphism ψ_2

p	$\psi_2(p)$	L	$\psi_2(L)$
A	A	L_1	L_5
B	E	L_2	L_2
C	D	L_3	L_1
D	G	L_4	L_7
E	F	L_5	L_3
F	B	L_6	L_4
G	G	L_7	L_6

Table 10.5 The points and lines under the automorphism ψ_3

p	$\phi(p)$	L	$\phi(L)$
A	A	L_1	L_5
B	D	L_2	L_4
C	E	L_3	L_3
D	C	L_4	L_6
E	B	L_5	L_1
F	G	L_6	L_7
G	F	L_7	L_2

The map ψ_1 is an automorphism, and there are 21 automorphisms formed by fixing a line and the remaining two lines through that point and then switching the other four points in pairs. We leave this verification as an exercise, see Exercise 1.

Consider the following automorphism which we call ψ_2 given in Table 10.4. We construct this by fixing the point A and cycling the three points on line L_2.

There are 56 automorphisms constructed in the same manner as ψ_2. We leave this verification as an exercise, see Exercise 6.

Consider the following automorphism which we call ψ_3 given in Table 10.5. We construct this by fixing a point and permuting the other two points on a line and cycling the remaining four points.

There are 42 automorphisms constructed in the same manner as ψ_3. We leave this verification as an exercise, see Exercise 4.

Combining the 21 automorphisms from Exercise 1, 48 automorphisms from Exercise 4, the 56 automorphisms from Exercise 6, the 42 automorphisms from Exercise 8, and the identity, we have 168 automorphisms. These automorphisms form the full automorphism group of the projective plane of order 2. We state this as follows.

Theorem 10.10 *The automorphism group of the plane of order 2 has order 168.*

The automorphism group of the affine plane of order 2 is easier to construct. We describe it in the following theorem.

Theorem 10.11 *The automorphism group of the affine plane of order 2 has 24 elements.*

Proof There are four points in the affine plane of order 2. If σ is a permutation of the points, then σ induces an automorphism since any two points form a line. That is, if a, b are points then the line through a and b is mapped to the line through $\sigma(a)$ and $\sigma(b)$. There are $4! = 24$ automorphisms. □

Exercises

1. Verify that ψ_1 is an automorphism and prove that there are 21 automorphisms formed by fixing a line and the remaining two lines through that point and then switching the other four points in pairs.
2. Verify that ϕ is an automorphism.
3. Compute the automorphisms ϕ^i for $i = 3, 4, 5, 6$.
4. Verify that map τ is an automorphism, and that there are 48 automorphisms constructed in this manner.
5. Verify that ψ_2 is an automorphism.
6. Prove that there are 56 automorphisms constructed in the same manner as ψ_2.
7. Verify that ψ_3 is an automorphism.
8. Prove that there are 42 automorphisms constructed in the same manner as ψ_3.

10.3 Quasigroups

We have seen that the multiplication table of a group is a Latin square. Yet, it is clear that not all Latin squares can be seen as the multiplication table of a group nor are they equivalent in some way to a multiplication table of a group. We shall examine an algebraic structure which does correspond to a Latin square. We begin with the definition.

Definition 10.6 A quasigroup $(G, *)$ is a set G with an operation $*$ such that

1. (Closure) If $a, b \in G$ then $a * b \in G$.
2. (Left cancelation) For all $x, y \in G$, $a * x = a * y$ implies $x = y$.
3. (Right cancelation) For all $x, y \in G$, $x * a = y * a$ implies $x = y$.

For example, the following gives a quasigroup:

*	A	B	C	D	E	F
A	A	B	C	D	E	F
B	C	A	B	F	D	E
C	B	C	A	E	F	D
D	D	E	F	A	B	C
E	F	D	E	C	A	B
F	E	F	D	B	C	A

Note that there is no identity element nor is it associative. For example, $B * (D * E) = B * B = A$, whereas $(B * D) * E = F * E = C$. Also without an identity there is no way to describe an inverse. Since a group has left and right cancelation, it is easy to see that a group is a quasigroup but as we have shown that a quasigroup is not necessarily a group.

Theorem 10.12 *A table is a Latin square if and only if it is the multiplication table of a quasigroup.*

Proof Let $(Q, *)$ be a quasigroup. Let $T_{a,b} = a * b$. We shall show that T is a Latin square. If $T_{a,b} = T_{a,c}$ then $a * b = a * c$ which implies that $b = c$ by left cancelation. Therefore, each row contains each element exactly once. If $T_{a,b} = T_{c,b}$ then $a * b = c * b$ and by right cancelation $a = c$; hence, each element appears exactly once in each column. Then, T is a Latin square.

Let L be a Latin square and define $a * b = T_{a,b}$. If $a * x = b$ then $T_{a,x} = b$ and since each row has each element exactly once there is a unique x satisfying this equation. If $y * a = b$ then $T_{y,a} = b$ and since each column has each element exactly once there is a unique y satisfying this equation. This means that the operation $*$ on the alphabet of L is a quasigroup. □

Let $(Q, *)$ be a quasigroup. A set R of Q is a subquasigroup of Q if $R \subseteq Q$ and R is a quasigroup. In the table given above, the set $\{A, B, C\}$ is a subquasigroup.

Definition 10.7 A loop is a quasigroup that has an identity element. That is, there exists an element e such that for all a we have $e * a = a * e = a$.

Theorem 10.13 *Every Latin square is equivalent to a Latin square that is a loop.*

Proof Let L be a Latin square and assume that the elements in the square are from the set $\{0, 1, 2, \ldots, n - 1\}$. Permute the columns so that the column's first elements are in standard order, i.e., $0, 1, 2, \ldots, n - 1$. Then permute the rows so that the first elements are in standard order to form the Latin square L'. Then 0 is the identity element of the quasigroup corresponding to L'. □

Example 10.9 Let L be the Latin square:

$$\begin{pmatrix} 2\ 3\ 1\ 0 \\ 1\ 0\ 2\ 3 \\ 3\ 1\ 0\ 2 \\ 0\ 2\ 3\ 1 \end{pmatrix}.$$

Permute the columns to obtain

$$\begin{pmatrix} 0\ 1\ 2\ 3 \\ 3\ 2\ 1\ 0 \\ 2\ 0\ 3\ 1 \\ 1\ 3\ 0\ 2 \end{pmatrix}.$$

Permute the rows to obtain the Latin square L':

$$\begin{pmatrix} 0\ 1\ 2\ 3 \\ 1\ 3\ 0\ 2 \\ 2\ 0\ 3\ 1 \\ 3\ 2\ 1\ 0 \end{pmatrix}.$$

Then L' is a loop with 0 as the identity.

It is immediate then that if a loop is associative then it is a group.

Exercises

1. Prove that in a loop each element has a left inverse.
2. Construct a loop that is not a group.
3. Let $(G, *)$ and $(G', *')$ be two quasigroups. Define the direct product of quasigroups to be $(G \times G', *'')$ where $(a, b) *'' (c, d) = (a * c, b *' d)$. Prove that the direct product of quasigroups is a quasigroup.

10.4 Difference Sets

We shall describe a combinatorial object which uses groups and is useful in constructing designs.

Definition 10.8 Let G be a group, where the operation is denoted by $+$ and the inverse of g is denoted by $-g$. A difference set in G is a subset D of G such that each non-identity element g of G can be written in exactly λ different ways of the form $x - y$ with $x, y \in D$.

Theorem 10.14 *Let* D *be a difference set in a group* G *with* $|G| = v$ *and* $|D| = k$. *Let* G *be the point set* $\mathcal{P}$ *and define the set of blocks* $\mathcal{B}$ *by* $\{D + g \mid g \in G\}$. *Then* $(\mathcal{P}, \mathcal{B})$ *is a* $2 - (v, k, \lambda)$ *design.*

Proof Let a and b be two elements of G. Let $c = a - b$. We know c appears exactly λ times in the difference set D. That is, $a - b = d_i - d_j$ for λ pairs of i, j and $d_i, d_j \in D$. This means that $a - d_i = b - d_j$. Let $g_i = a - d_i$. Then a and b are both elements of $D + g_i$. Since there are exactly λ elements g_i, we have proven that any two points are together incident with λ blocks.   □

Example 10.10 Let us consider the following example. Consider the group formed by addition modulo 7. Let $D = \{1, 2, 4\}$.

Each non-identity element can be written as a difference from this set exactly once, namely,

$$2 - 1 = 1$$
$$4 - 2 = 2$$
$$4 - 1 = 3$$
$$1 - 4 = 4$$
$$2 - 4 = 5$$
$$1 - 2 = 6.$$

Consider the sets $D + g$:

$$\{1, 2, 4\}, \{2, 3, 5\}, \{3, 4, 6\}, \{4, 5, 0\}, \{5, 6, 1\}, \{6, 0, 2\}, \{0, 1, 3\}.$$

This is a design we have seen before. Consider the diagram in Fig. 10.3.

Consider the group formed by addition modulo 11. Let $D = \{1, 3, 4, 5, 9\}$.

We invite the reader to verify that each non-identity element in D can be written as a difference of elements of D exactly twice. For example $2 = 3 - 1$ and $2 = 5 - 3$.

Fig. 10.3 Projective plane of order 2

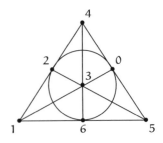

Table 10.6 Incidence matrix

	0	1	2	3	4	5	6	7	8	9	10
B_1	0	1	0	1	1	1	0	0	0	1	0
B_2	0	0	1	0	1	1	1	0	0	0	1
B_3	1	0	0	1	0	1	1	1	0	0	0
B_4	0	1	0	0	1	0	1	1	1	0	0
B_5	0	0	1	0	0	1	0	1	1	1	0
B_6	0	0	0	1	0	0	1	0	1	1	1
B_7	1	0	0	0	1	0	0	1	0	1	1
B_8	1	1	0	0	0	1	0	0	1	0	1
B_9	1	1	1	0	0	0	1	0	0	1	0
B_{10}	0	1	1	1	0	0	0	1	0	0	1
B_{11}	1	0	1	1	1	0	0	0	1	0	0

The cosets are

$$B_1 = \{1, 3, 4, 5, 9\}$$
$$B_2 = \{2, 4, 5, 6, 10\}$$
$$B_3 = \{3, 5, 6, 7, 0\}$$
$$B_4 = \{4, 6, 7, 8, 1\}$$
$$B_5 = \{5, 7, 8, 9, 2\}$$
$$B_6 = \{6, 8, 9, 10, 3\}$$
$$B_7 = \{7, 9, 10, 0, 4\}$$
$$B_8 = \{8, 10, 0, 1, 5\}$$
$$B_9 = \{9, 0, 1, 2, 6\}$$
$$B_{10} = \{10, 1, 2, 3, 7\}$$
$$B_{11} = \{0, 2, 3, 4, 8\}$$

We shall show the incidence matrix in Table 10.6.

We notice that each block has five points, and each point is on five blocks and that any two points are incident with exactly two blocks. Theorem 10.14 gives that it is an $(11, 5, 2)$ design. Therefore, we have the following.

Theorem 10.15 *The design formed from the difference set $\{1, 3, 4, 5, 9\}$ in the group formed from addition modulo 11 is a biplane of order 3.*

These two constructions generalize. We shall state the theorem but the proof requires more number theory than we have assumed. First, we need a definition. A non-zero element a of $\mathbb{Z}_p$ is a square if there exists a $b \in \mathbb{Z}_p$, with $a = b^2$.

Theorem 10.16 *Let* $p \equiv 3 \pmod 4$, *with* $p = 4t - 1$. *Then the set of squares in* $\mathbb{Z}_p$ *is a difference set and produces a* $2 - (4t - 1, 2t - 1, t - 2)$ *design.*

For proof of the theorem, see [44].

Exercises

1. Verify that any two elements are in exactly one set together in Example 10.10.
2. Verify that the diagram in Fig. 10.3 corresponds to the blocks given in Example 10.10.
3. Use the incidence matrix to draw a diagram of the biplane of order 3.

Codes

11

11.1 Introduction

Algebraic coding theory arose in the last half of the twentieth century to answer applications in the field of electronic communication. Specifically, the idea is to transmit information over a channel such that, when it is received, whatever errors that were made can be corrected with a high degree of certainty. Since its initial study of this application, coding theory has grown into an interesting branch of both pure and applied mathematics. From the very beginning of the study of coding theory, numerous interesting connections between codes, combinatorics, and finite geometry were found. Codes were used to study designs, and designs were used to construct some of the best codes. We shall give some examples of the connection between codes and combinatorics in this chapter. The reader interested in an in-depth description of coding theory that can consult the standard texts of coding theory, namely, MacWilliams and Sloane [62] or Huffman and Pless [49]. For more elementary introductions to this interesting branch of mathematics, see the text by Hill [48]. For an advanced description of algebraic coding theory over finite commutative rings, see [29]. Coding theory is related to, but distinct from, cryptography which is the study of secret communication, namely, how to communicate so that no one can intercept your message. In coding theory, the idea is to communicate so that it can be decoded correctly, whereas in cryptography the idea is to keep the information secret.

From our point of view, we will not be overly concerned with applications of coding theory to electronic communication but rather with the study of codes as algebraic structures and their application to combinatorics and finite geometry. We take the view of coding theory as a branch of pure mathematics alongside algebra, number theory, and combinatorics. Much of the research done in coding theory by mathematicians is far removed from its application. However, the application of this theory benefits electronic communication greatly and moreover gives fascinating

S. T. Dougherty, *Combinatorics and Finite Geometry*,
Springer Undergraduate Mathematics Series,
https://doi.org/10.1007/978-3-030-56395-0_11

open problems to mathematicians. In many ways, coding theory is in an ideal situation as it is studied for its own sake and benefits from its close relationship to other branches of mathematics and with its applications in electronic communication.

We shall begin with a description of the origin of coding theory and its basic principles. For a machine, there are only two pieces of information: on or off, yes or no, or from a mathematical point of view 1 or 0. All the information that is transmitted over the phone, stored in your computer, or bounced off satellites is simply a set of ones and zeros. It is in the way that the machines interpret these ones and zeros that give them meaning to the users of these machines.

In the later half of the twentieth century, it was realized that techniques must be developed to insure that the information could be communicated accurately. Shannon wrote the landmark paper [78] in 1948 and gave birth to the subjects of Information Theory and Coding Theory. The early work of Shannon and Hamming [47] set the stage for a tremendous amount of research that followed. This research was very important in the revolution of the information age that occurred in the late twentieth century.

Consider the following situation. Someone is standing across a crowded room and you want to convey the message "yes" to this person. If you simply said "yes" once, there is a very good chance that he will not hear you and may assume that you said no. There are two things that you will usually do in this situation. One is to say yes a few times, so that he will assume you said the answer he heard most often. In other words, if you say it four times and twice he distinctly heard yes, and once he could not tell and once he thought you said no, he would still assume the correct answer is yes. The other thing is that he would nod his head and say yes at the same time. In that manner, if the person thought he heard no but saw your head nodding in the affirmative he would know that it did not make sense and so motion to you to repeat.

In a simplistic manner, these are two of the most commonly used methods to insure correct information. The first is the repetition of information. The second is a check to be sure that the message makes sense. The first technique hopes that the correct information can be gleaned from the message and the second that a message that is not received correctly can be discarded and new information will be requested.

Techniques not much more complicated than this were used at the beginning of electronic communication. For example, instead of sending 1, the message 11111 might be sent. If it was received 11011 it would still be decoded correctly as 1. The trouble with this technique is that it requires a good deal of information, in this case 5 times as much, to be sent over the communication channel.

As an example of the second technique, where some other information is sent as a check, we will show how a parity check is used. Assume the desired message was 111; in this case, the message 1111 is sent and if 110 was to be sent then 1100 is sent. In other words, you put a 1 at the end if there are oddly many ones and a 0 at the end if there are evenly many ones. More mathematically, the last coordinate is simply the sum of the other coordinates using the addition of the finite field of order 2. This is known as a parity check. If the message 1110 is received, then you know that it is not a possible message and so it is rejected. Although you can tell that an error has occurred, you cannot tell in which place the error occurred. Any of 1111, 1100, 1010, and 0110 could have been the desired message. Being able to tell if an error is made is known as error detecting.

Amazingly, the techniques which are now in use not only can tell if an error has been made but can determine where it was made and correct it! The mathematics is based on nineteenth-century linear algebra over finite fields and ideas from combinatorics and finite geometry.

11.2 Basics of Coding Theory

We shall now give the basic definitions in a mathematical setting.

A code of length n over an alphabet A is simply a subset of

$$A^n = \{(a_1, a_2, \ldots, a_n) \mid a_i \in A\}.$$

One can think of this as all possible words that can be *spelled* with the letters of the alphabet. Most usually, the alphabet used is either a finite field or the finite ring $\mathbb{Z}_m$. However, in the twenty-first century, the collection of acceptable alphabets has grown significantly to include all finite Frobenius rings and even some infinite rings as well. An element of a code is called a codeword. For most applications in the field of electronic communications, the alphabet is the binary field $\mathbb{F}_2 = \{0, 1\}$.

To the ambient space A^n, we attach a distance. The Hamming distance between any two codewords is defined to be the number of places where they differ, specifically

$$d(\mathbf{v}, \mathbf{w}) = |\{i \mid v_i \neq w_i\}|.$$

For example, the distance between $(0, 0, 1, 1, 0)$ and $(0, 0, 0, 0, 1)$ is 3 because they differ in three places, namely, the last three coordinates.

Theorem 11.1 *The Hamming distance is a metric on the space A^n. That is, it satisfies the following three conditions:*

1. *We have $d(\mathbf{v}, \mathbf{w}) = 0$ if and only if $\mathbf{v} = \mathbf{w}$ and $d(\mathbf{v}, \mathbf{w}) \geq 0$ for all $\mathbf{v}$ and $\mathbf{w}$.*
2. *For all $\mathbf{v}$ and $\mathbf{w}$, $d(\mathbf{v}, \mathbf{w}) = d(\mathbf{w}, \mathbf{v})$.*
3. *For all $\mathbf{u}, \mathbf{v}, \mathbf{w}$ we have $d(\mathbf{u}, \mathbf{w}) \leq d(\mathbf{u}, \mathbf{v}) + d(\mathbf{v}, \mathbf{w})$. This is known as the triangle inequality.*

Proof To prove the first statement, we simply notice that the function is determined as a cardinality which is always greater than or equal to 0. If it is 0, then the vectors agree in all coordinates and so are equal.

The second statement follows from the fact that $\{i \mid v_i \neq w_i\} = \{i \mid w_i \neq v_i\}$.

Consider the three vectors $\mathbf{u}, \mathbf{v}, \mathbf{w}$. We can change $\mathbf{u}$ into $\mathbf{v}$ by changing $d(\mathbf{u}, \mathbf{v})$ coordinates. We can change $\mathbf{v}$ into $\mathbf{w}$ by changing $d(\mathbf{v}, \mathbf{w})$ coordinates. Therefore, the maximum number of coordinates in which $\mathbf{u}$ and $\mathbf{w}$ differ is $d(\mathbf{u}, \mathbf{v}) + d(\mathbf{v}, \mathbf{w})$. This gives the third statement. $\qquad\square$

Given a code $C \subseteq A^n$, the minimum Hamming distance d_C is the smallest distance between any two distinct vectors, that is,

$$d_C = \min\{d(\mathbf{v}, \mathbf{w}) \mid \mathbf{v}, \mathbf{w} \in C, \mathbf{v} \neq \mathbf{w}\}.$$

In general, we look at codes that have some sort of algebraic structure, rather than those that are simply a collection of vectors. We define a linear code over the alphabet F, where F is a field, to be a vector space over the field F. In other words, C is a linear code if the following conditions hold:

- If $\mathbf{v}, \mathbf{w} \in C$ then $\mathbf{v} + \mathbf{w} \in C$.
- If $\mathbf{v} \in C$ and $\alpha \in F$ then $\alpha\mathbf{v} \in C$.

Example 11.1 Let p be a prime and n a non-zero integer. Then the code of length n consisting of the codewords

$$\{(0,0,\ldots,0),(1,1,\ldots,1),\ldots,(p-1,p-1,\ldots,p-1)\}$$

is a linear code over $\mathbb{F}_p$ with p elements. This code is known as the repetition code.

Example 11.2 Consider the binary code $C = \{(0,0,0),(1,1,0),(0,1,0)\}$. This code is not a linear code since $(1,1,0) + (0,1,0) = (1,0,0) \notin C$.

The weight of a vector is the number of non-zero elements in a vector. The weight of vector $\mathbf{v}$ is denoted by $wt(\mathbf{v})$. The minimum weight of a code is the smallest of all non-zero weights in a code.

Theorem 11.2 *For a linear code, the minimum weight of the code is the minimum distance.*

Proof If $\mathbf{v}$ and $\mathbf{w}$ are vectors in a linear code C, then $d(\mathbf{v}, \mathbf{w}) = d(\mathbf{v} - \mathbf{w}, \mathbf{0}) = wt(\mathbf{v} - \mathbf{w})$ where $\mathbf{0}$ is the all 0 vector. So if d is the minimum distance of a code then there is a vector of weight d in the code. If there is a vector of weight d, then the distance between that vector and the all-zero vector is d. Therefore, for a linear code the minimum distance and the minimum weight are equal.   □

Lemma 11.1 *Let V be a vector space of dimension k over the finite field $\mathbb{F}_q$, then $|V| = q^k$.*

Proof Let $\mathbf{v}_1, \mathbf{v}_2, \ldots, \mathbf{v}_k$ be the basis for V. Then since these vectors are linearly independent, each of the following sums

$$\alpha_1\mathbf{v}_1 + \alpha_2\mathbf{v}_2 + \ldots, \alpha_k\mathbf{v}_k$$

must be distinct. Therefore, there are q choices for each α_i and hence q^k vectors in V.   □

This leads immediately to the following.

Theorem 11.3 *If C is a linear code over the field $\mathbb{F}_q$, then $|C| = q^k$ for some k.*

Proof By definition, a linear code is a vector space. Then by Lemma 11.1, we have that $|C| = q^k$ where k is the dimension of C as a vector space. □

Definition 11.1 Two codes C and C' of length n over $\mathbb{F}_q$ are said to be permutation equivalent if C' can be obtained from C by a permutation of the coordinates.

Since any linear code is a vector space of dimension k in $\mathbb{F}_q^n$, it has a basis of k vectors. Using the standard technique of Gaussian elimination, we have that any linear code over a finite field $\mathbb{F}_q$ is permutation equivalent to a code that has a generator matrix of the form:

$$(I_k \mid A),$$

where k is the dimension of the code and I_k is the k by k identity matrix.

Example 11.3 Consider the code generated by vectors $(1, 1, 1, 1)$, $(1, 1, 0, 0)$, and $(0, 0, 0, 1)$. Applying Gaussian elimination, we get the following matrix:

$$\begin{pmatrix} 1 & 1 & 0 & 0 \\ 0 & 0 & 1 & 0 \\ 0 & 0 & 0 & 1 \end{pmatrix}.$$

Then using the permutation $(2, 4, 3)$ on the coordinates, we obtain

$$\begin{pmatrix} 1 & 0 & 0 & 1 \\ 0 & 1 & 0 & 0 \\ 0 & 0 & 1 & 0 \end{pmatrix},$$

which is of the form $(I_k \mid A)$. This code has cardinality 8.

If a code is of length n, has M vectors, and minimum distance d, then the code is said to be an (n, M, d) code. If the code is linear and its dimension is k, then it is said to be a $[n, k, d]$ code. Of course, in this case we have that $M = |\mathbb{F}_q|^k$. If the minimum weight is not known, then we may simply call it an $[n, k]$ code.

The sphere of radius r around a vector $\mathbf{v} \in C$, where C is a code, consisting of all vectors in $\mathbb{F}_q^n$ that are Hamming distance less than or equal to r. When a vector $\mathbf{w}$ is received, it is decoded to the vector $\mathbf{v} \in C$ that is closest to it (if such a vector exists). This notion of nearest neighbor decoding leads to the following theorem.

Theorem 11.4 *Let C be an $[n, k, d]$ code, then C can detect $d - 1$ errors. If $d = 2t + 1$ then C can correct t errors.*

Proof Given a vector $\mathbf{v} \in C$, if $d - 1$ or less errors are made, then the resulting vector cannot be in C since the minimum distance of C is d. Hence, the code can correct up to $d - 1$ errors.

If $d = 2t + 1$ and t errors are made to a vector $\mathbf{v} \in C$, then by the triangle inequality, any other vector in C must have distance at least $t + 1$ from the resultant vector. Therefore, there is a unique vector that is closest to this vector, namely, $\mathbf{v}$, and the errors are corrected.   □

Example 11.4 Consider the code C of length n over $\mathbb{F}_q$ generated by $(1, 1, \ldots, 1)$. The minimum distance of this code is n; hence, it can detect $n - 1$ errors and correct $\lfloor \frac{n-1}{2} \rfloor$ errors.

Exercises

1. Prove that there exists a linear $[n, k]$ code for all k with $1 \leq k \leq n$.
2. Produce the generating matrix in standard form by row reducing the following over $\mathbb{F}_2$:

$$\begin{pmatrix} 1 & 1 & 0 & 1 \\ 0 & 0 & 1 & 1 \\ 1 & 1 & 1 & 1 \\ 1 & 0 & 0 & 0 \\ 0 & 1 & 1 & 0 \end{pmatrix}.$$

3. Produce the generating matrix in standard form by row reducing the following over $\mathbb{F}_3$:

$$\begin{pmatrix} 1 & 2 & 2 & 1 \\ 2 & 1 & 1 & 1 \\ 1 & 1 & 1 & 2 \\ 0 & 1 & 0 & 1 \end{pmatrix}.$$

4. Find all vectors in $\mathbb{F}_2^4$ that are at distance 2 or less from the vector $(1, 0, 0, 1)$.
5. Produce a non-linear code whose minimum distance is not equal to its minimum weight.
6. Produce a non-linear code over $\mathbb{F}_2$ that can correct two errors.
7. Produce a linear code over $\mathbb{F}_5$ that can correct two errors.

11.3   Orthogonal Codes

In this section, we shall investigate the dual of code with respect to the Euclidean inner product. This dual code has numerous important properties and works in tandem with the code in terms of correcting errors.

The space $\mathbb{F}_q^n$ has an inner product attached to it. It is

$$[\mathbf{v}, \mathbf{w}] = \sum v_i w_i, \tag{11.1}$$

where the multiplication and addition are done in $\mathbb{F}_q$.

Theorem 11.5 *For* $\mathbf{u}, \mathbf{v}, \mathbf{w} \in \mathbb{F}_q^n$ *and* $\alpha, \beta \in \mathbb{F}_q$ *we have*

1. $[\mathbf{u}, \mathbf{v}] = [\mathbf{v}, \mathbf{u}]$;
2. $[\alpha\mathbf{u}, \beta\mathbf{v}] = \alpha\beta[\mathbf{u}, \mathbf{v}]$;
3. $[\mathbf{u} + \mathbf{v}, \mathbf{w}] = [\mathbf{u}, \mathbf{w}] + [\mathbf{v}, \mathbf{w}]$.

Proof For the first statement, we have $[\mathbf{u}, \mathbf{v}] = \sum u_i v_i = \sum v_i u_i = [\mathbf{v}, \mathbf{u}]$.
For the second statement, $[\alpha\mathbf{u}, \beta\mathbf{v}] = \sum \alpha u_i \beta v_i = \alpha\beta \sum u_i v_i = \alpha\beta[\mathbf{u}, \mathbf{v}]$.
For the third statement, $[\mathbf{u} + \mathbf{v}, \mathbf{w}] = \sum (u_i + v_i)w_i = \sum (u_i w_i + v_i w_i) = [\mathbf{u}, \mathbf{w}] + [\mathbf{v}, \mathbf{w}]$. $\qquad\square$

For any code C over $\mathbb{F}_q$ we can define the orthogonal of the code by

$$C^\perp = \{\mathbf{w} \mid [\mathbf{w}, \mathbf{v}] = 0 \text{ for all } \mathbf{v} \in C\}. \tag{11.2}$$

Theorem 11.6 *If* C *is a code over* $\mathbb{F}_q$, *then* $C^\perp$ *is a linear code.*

Proof If $\mathbf{u}$ is a vector in C, $\mathbf{v}$ and $\mathbf{w}$ are vectors in $C^\perp$ and $\alpha \in \mathbb{F}_q$, then by Theorem 11.5 we have

$$[\mathbf{u}, \mathbf{v} + \mathbf{w}] = [\mathbf{u}, \mathbf{v}] + [\mathbf{u}, \mathbf{w}] = 0 + 0 = 0$$

and

$$[\mathbf{u}, \alpha\mathbf{v}] = \alpha[\mathbf{u}, \mathbf{v}] = \alpha 0 = 0.$$

So both $\mathbf{v} + \mathbf{w}$ and $\alpha\mathbf{v}$ are in $C^\perp$, and it is a linear vector space. $\qquad\square$

Note that we have not assumed that the code C is linear. What we have proven is that the orthogonal code is linear even if C is not. We can also say the following about the orthogonal of a non-linear code.

Theorem 11.7 *If* C *is a non-linear code, then* $C^\perp = \langle C \rangle^\perp$. *That is, the orthogonal of a non-linear code is equal to the orthogonal of the linear code that it generates.*

Proof Let $\mathbf{v}, \mathbf{u} \in C$. Then $\mathbf{w} \in C^\perp$ if and only if $[\mathbf{w}, \mathbf{v}] = [\mathbf{w}, \mathbf{u}] = 0$. This gives that $[\mathbf{w}, \alpha\mathbf{v} + \beta\mathbf{u}] = \alpha[\mathbf{w}, \mathbf{v}] + \beta[\mathbf{w}, \mathbf{u}] = 0 + 0 = 0$. This gives $\mathbf{w} \in \langle C \rangle^\perp$. Therefore, $C^\perp \subseteq \langle C \rangle^\perp$.
We have $C \subseteq \langle C \rangle$ which gives $\langle C \rangle^\perp \subseteq C^\perp$. Then we have that $\langle C \rangle^\perp = C^\perp$. $\qquad\square$

Theorem 11.8 *If* C *is a linear* $[n, k]$ *code over* $\mathbb{F}_q$, *then* $C^\perp$ *is a linear* $[n, n - k]$ *code.*

Proof Consider the generator matrix of an $[n, k]$ code of the form $(I_k \mid A)$. To construct a vector in $C^\perp$, make it anything on the $n - k$ coordinate corresponding to the matrix A. Call this length $n - k$ vector $\mathbf{v}$. Then to make it orthogonal, place in the i-th coordinate, for $1 \leq i \leq k$, the value $-[\mathbf{v}, A_i]$, where A_i is the i-th row of A. Then this vector is orthogonal to every row of the generator matrix and therefore in $C^\perp$. Since there are $n - k$ degrees of freedom in making the vector, the dimension of $C^\perp$ is $n - k$. □

Example 11.5 Let C be the binary code of length n, generated by $(1, 1, \ldots, 1)$. The code C has dimension 1 and minimum weight n. The dual code $C^\perp$ has dimension $n - 1$ and consists of all vectors of length n with even weight. This code is denoted by E_n.

Theorem 11.9 *If C is a code generated by $(I_k \mid A)$, then $C^\perp$ is generated by* $(-A^T \mid I_{n-k})$.

Proof By Theorem 11.8, we know that $C^\perp$ must have dimension $n - k$. A code generated by $(-A^T \mid I_{n-k})$ has dimension $n - k$. The only thing that must be shown then is that the vectors in the code generated by $(-A^T \mid I_{n-k})$ are orthogonal to the vectors in C. The inner product of the i-th row of $(I_k \mid A)$ with the j-th row of $(-A^T \mid I_{n-k})$ gives $A_{ij} - A_{ij} = 0$. Therefore, the two rows are orthogonal, and $(-A^T \mid I_{n-k})$ generates $C^\perp$. □

Example 11.6 Consider the code C over $\mathbb{F}_5$ generated by the matrix:

$$\begin{pmatrix} 1 & 0 & 0 & 1 & 2 \\ 0 & 1 & 0 & 3 & 4 \\ 0 & 0 & 1 & 0 & 1 \end{pmatrix}.$$

Then $C^\perp$ has generator matrix:

$$\begin{pmatrix} -1 & -3 & 0 & 1 & 0 \\ -2 & -4 & -1 & 0 & 1 \end{pmatrix} = \begin{pmatrix} 4 & 2 & 0 & 1 & 0 \\ 3 & 1 & 4 & 0 & 1 \end{pmatrix}.$$

The inner product of the first row of the first matrix with the second row of the second matrix gives $2 - 2 = 0$. Notice that $A_{1,2} = 2$.

Definition 11.2 We say that a code C is self-orthogonal if $C \subseteq C^\perp$ and we say that it is self-dual if $C = C^\perp$.

Example 11.7 The code $C = \{(0, 0), (1, 1)\}$ satisfies $C = C^\perp$ and is a self-dual code. The code $D = \{(0, 0, 0), (1, 1, 0)\}$ satisfies $D \subseteq D^\perp$ and is a self-orthogonal code.

Theorem 11.10 *If C is a self-dual code over $\mathbb{F}_q$ of length n, then n must be even.*

Proof If C is a self-dual code, then $\dim(C) + \dim(C^\perp) = n$ and $\dim(C) = \dim(C^\perp)$ since $C = C^\perp$. This gives $2\dim(C) = n$ and so n must be even. $\square$

We define $C \times D = \{(\mathbf{v}, \mathbf{w}) \mid \mathbf{v} \in C, \mathbf{w} \in D\}$.

Lemma 11.2 *If C is a self-dual code of length n and D is a self-dual code of length m, then $C \times D$ is a self-dual code of length $n + m$.*

Proof We have $C = C^\perp$, $D = D^\perp$ and $|C| = q^{\frac{n}{2}}$, $|D| = q^{\frac{m}{2}}$. Then $|C \times D| = q^{\frac{n+m}{2}}$. Therefore, the code has the right cardinality. We only need to show self-orthogonality. Given $(v, w), (v', w') \in C \times D$, then $[(v, w), (v', w')] = [v, v'] + [w, w'] = 0 + 0 = 0$. Therefore, it is self-orthogonal and of the right cardinality so the code is self-dual. $\square$

Theorem 11.11 *If $p \equiv 1 \pmod 4$ is a prime, then there exist self-dual codes of length n for all even n over $\mathbb{F}_p$.*

Proof If $p \equiv 1 \pmod 4$, then we know -1 is a square. That is, there exists α such that $\alpha^2 = -1$. We have $[(1, \alpha), (1, \alpha)] = 1 + \alpha^2 = 0$. Therefore, the code C generated by $(1, \alpha)$ is self-orthogonal. The code C has p elements, and so $|C|^2 = p^2 = |\mathbb{F}_p^2|$ and so the code is self-dual.

Then, by Lemma 11.2, we can take the cross product of the code with itself t times to get a self-dual code for all lengths of the form $2t$. $\square$

Theorem 11.12 *If $p \equiv 3 \pmod 4$ is a prime, then there exist self-dual codes of length n for all $n \equiv 0 \pmod 4$ over $\mathbb{F}_p$.*

Proof If $p \equiv 3 \pmod 4$, then we know that there exists $\alpha, \beta \in \mathbb{F}_p$ with $\alpha^2 + \beta^2 = -1$. The code generated by the following matrix

$$\begin{pmatrix} 1 & 0 & \alpha & \beta \\ 0 & 1 & -\beta & \alpha \end{pmatrix}$$

is a self-orthogonal $[4, 2]$ code and is therefore self-dual.

Then by Lemma 11.2, we can take the cross product of the code with itself t times to get a self-dual code for all lengths of the form $4t$. $\square$

Exercises

1. Use the inner product to prove that the number of $[n, k]$ codes over $\mathbb{F}_q$ is equal to the number of $[n, n - k]$ codes over $\mathbb{F}_q$.
2. Take the code over $\mathbb{F}_3$ generated by the vectors $(1, 2, 1, 2), (1, 1, 1, 1)$, and $(0, 0, 1, 2)$. Give the generator matrix of this code in standard form. Give the generator matrix of its orthogonal.

3. Prove that if $k > \frac{n}{2}$ then there is no self-orthogonal $[n, k]$ code.
4. Let $\mathbf{v}$ be a vector in $\mathbb{F}_q^n$. Determine the number of vectors $\mathbf{w}$ that are orthogonal to $\mathbf{v}$. Prove your answer.
5. Assume $\mathbf{v}$ and $\mathbf{w}$ are binary vectors where $wt(\mathbf{v}) \equiv wt(\mathbf{w}) \equiv 0 \pmod 4$ with $[\mathbf{v}, \mathbf{w}] = 0$. Prove that $wt(\mathbf{v} + \mathbf{w}) \equiv 0 \pmod 4$.
6. Prove that there are no self-orthogonal vectors of length 1 over $\mathbb{F}_q$.

11.4   Syndrome Decoding

In this section, we shall show how many of the ideas of finite geometry and linear algebra are used to actually correct errors. Instead of starting with a generator matrix for a code, we shall start with a generator matrix for its orthogonal. Assume $H = (I_{n-k} \mid B)$ generates the $[n, n-k]$ code $C^\perp$. It follows immediately that $\mathbf{v} \in C$ if and only if $H\mathbf{v}^\mathsf{T} = \mathbf{0}$. This matrix H is known as the parity check matrix of C.

If you think of the linear transformation $T : \mathbb{F}_q^n \to \mathbb{F}_q^n$, given by $T(\mathbf{v}) = H\mathbf{v}^\mathsf{T}$, then the code C is the kernel of the linear transformation. The vectors in the code C are precisely the vectors that carry information that we want to send across an electronic channel. We define a coset of C to be

$$(C + \mathbf{w}) = \{\mathbf{v} + \mathbf{w} \mid \mathbf{v} \in C\}.$$

Lemma 11.3 *Let C be a linear code in $\mathbb{F}_q^n$. Let $\mathbf{w}$ be any vector in $\mathbb{F}_q^n$. Then $|C| = |(C + \mathbf{w})|$.*

Proof Define the function $\psi : C \to (C + \mathbf{w})$ by $\psi(\mathbf{v}) = \mathbf{v} + \mathbf{w}$. We note that if $\psi(\mathbf{v}) = \psi(\mathbf{u})$ then $\mathbf{v} + \mathbf{w} = \mathbf{u} + \mathbf{w}$ which gives $\mathbf{v} = \mathbf{u}$. Therefore, the map is injective. The map is surjective by construction since every element in $(C + \mathbf{w})$ is of the form $\mathbf{v} + \mathbf{w}$ for some $\mathbf{v} \in C$. Then ψ is a bijection and $|C| = |(C + \mathbf{w})|$. □

Lemma 11.4 *Let C be a linear code in $\mathbb{F}_q^n$. Let $\mathbf{w}, \mathbf{u}$ be vectors in $\mathbb{F}_q^n$. Then either $(C + \mathbf{w}) = (C + \mathbf{u})$ or $(C + \mathbf{w}) \cap (C + \mathbf{u}) = \emptyset$.*

Proof Assume there exists a vector $\mathbf{z} \in (C + \mathbf{w}) \cap (C + \mathbf{u})$. Then $\mathbf{z} = \mathbf{v} + \mathbf{w}$ and $\mathbf{z} = \mathbf{v}' + \mathbf{u}$ for some $\mathbf{v}, \mathbf{v}' \in C$. This gives $\mathbf{v} + \mathbf{w} = \mathbf{v}' + \mathbf{u}$ and $\mathbf{v} - \mathbf{v}' = \mathbf{u} - \mathbf{w}$. Therefore, $\mathbf{u} - \mathbf{w} = \mathbf{c} \in C$.

If $\mathbf{x} \in (C + \mathbf{w})$, then $\mathbf{x} = \mathbf{v}'' + \mathbf{w}$ for some $\mathbf{v}'' \in C$. Then $\mathbf{x} = \mathbf{v}'' + \mathbf{u} - \mathbf{c} = (\mathbf{v}'' - \mathbf{c}) + \mathbf{u}$ and therefore $(C + \mathbf{w}) \subseteq (C + \mathbf{u})$. Since by Lemma 11.3, we have that their cardinalities are identical, this gives $(C + \mathbf{w}) = (C + \mathbf{u})$. Therefore, if the intersection is not empty then they are identical. □

Theorem 11.13 *Let C be a linear* $[n, k]$ *code in* $\mathbb{F}_q^n$. *Then*

$$\mathbb{F}_q^n = \bigcup_{i=0}^{q^{n-k}-1} (C + \mathbf{w}_i),$$

where $\mathbf{w}_0 = \mathbf{0}$ *and the* $\mathbf{w}_i$ *are distinct vectors in* $\mathbb{F}_q^n$.

Proof By Lemma 11.4, the cosets are disjoint. Moreover, for any $\mathbf{w} \in \mathbb{F}_q^n$, we have $\mathbf{w} \in (C + \mathbf{w})$. Therefore, the cosets form a partition of the space.   □

Lemmas 11.3, 11.4, and Theorem 11.13 are known collectively as LaGrange's theorem.

In this scenario, the $\mathbf{w}_i$ are known as the coset leaders. We can choose any vectors at all; however, in terms of the application of coding theory, we generally choose the $\mathbf{w}_i$ that have the smallest Hamming weights.

Example 11.8 Consider the binary code C with parity check matrix:

$$\begin{pmatrix} 1 & 0 & 1 & 0 & 1 \\ 0 & 1 & 0 & 1 & 0 \end{pmatrix}.$$

The code C is then a $[5, 3]$ code. We shall write the ambient space as the code in the first column, followed by the non-trivial cosets of the code in the next three columns using the top vector in the column as the coset leader.

$$
\begin{array}{llll}
(0,0,0,0,0) & (1,0,0,0,0) & (0,1,0,0,0) & (0,0,0,1,1) \\
(1,0,1,0,0) & (0,0,1,0,0) & (1,1,1,0,0) & (1,0,1,1,1) \\
(0,1,0,1,0) & (1,1,0,1,0) & (0,0,0,1,0) & (0,1,0,0,1) \\
(1,0,0,0,1) & (0,0,0,0,1) & (1,1,0,0,1) & (1,0,0,1,0) \\
(0,0,1,0,1) & (1,0,1,0,1) & (0,1,1,0,1) & (0,0,1,1,0) \\
(1,1,0,1,1) & (0,1,0,1,1) & (1,0,0,1,1) & (1,1,0,0,0) \\
(0,1,1,1,1) & (1,1,1,1,1) & (0,0,1,1,1) & (0,1,1,0,0) \\
(1,1,1,1,0) & (0,1,1,1,0) & (1,0,1,1,0) & (1,1,1,0,1)
\end{array}
$$

Notice for the first two cosets, we chose as coset leaders vectors with weight 1, the smallest weight possible for a non-trivial coset leader. However, by the fourth coset we had no more weight 1 vectors at our disposal, so we had to choose a vector with higher weight.

Definition 11.3 Let C be a code in $\mathbb{F}_q^n$ with parity check matrix H. Then the syndrome of a vector $\mathbf{v} \in \mathbb{F}_q^n$ is $S(\mathbf{v}) = H\mathbf{v}^T$.

Theorem 11.14 *Let C be a code in* $\mathbb{F}_q^n$ *with parity check matrix* H *and let* $\mathbf{v}, \mathbf{w} \in \mathbb{F}_q^n$. *We have* $S(\mathbf{v}) = S(\mathbf{w})$ *if and only if* $\mathbf{v}$ *and* $\mathbf{w}$ *are in the same coset of* C.

Proof If $S(\mathbf{v}) = S(\mathbf{w})$, then $H\mathbf{v}^T = H\mathbf{w}^T$ which implies $H(\mathbf{v}^T - \mathbf{w}^T) = \mathbf{0}$. This gives that $\mathbf{v} - \mathbf{w} \in C$. Therefore, $\mathbf{v} \in (C + \mathbf{w})$ and $\mathbf{w} \in (C + \mathbf{w})$; therefore, they are in the same coset of C.

If $\mathbf{u}$ and $\mathbf{x}$ are both in $(C + \mathbf{w})$, then $\mathbf{u} = \mathbf{c} + \mathbf{w}$ and $\mathbf{x} = \mathbf{c}' + \mathbf{w}$, where $\mathbf{c}, \mathbf{c}' \in C$. Then

$$S(\mathbf{u}) = S(\mathbf{c} + \mathbf{w}) = H(\mathbf{c}^T + \mathbf{w}^T) = H\mathbf{c}^T + H\mathbf{w}^T = \mathbf{0} + H\mathbf{w}^T = S(\mathbf{w})$$

and

$$S(\mathbf{x}) = S(\mathbf{c}' + \mathbf{w}) = H((\mathbf{c}')^T + \mathbf{w}^T) = H(\mathbf{c}')^T + H\mathbf{w}^T = \mathbf{0} + H\mathbf{w}^T = S(\mathbf{w}).$$

This gives the result.   □

Assume a vector $\mathbf{v}$ is transmitted from person A to person B. Along the channel errors are introduced into the vector, so that $\mathbf{w} = \mathbf{v} + \mathbf{e}$ is received, where $\mathbf{e}$ is the error vector. Then person B computes $S(\mathbf{w})$ which is equal to $S(\mathbf{e})$. Then to decode the vector, we take $\mathbf{w} - \mathbf{e}$ to get $\mathbf{v}$ as was originally sent. The difficulty is that we may not be sure that $\mathbf{e}$ represents the actual errors that occurred. In essence, we look at the array of elements as was given in Example 11.8. We assume that the error vectors are the coset leaders which appear at the top of each column. For example, assume $(0, 1, 1, 1, 0)$ is received, then it is located at the bottom of column 2. We assume that the error vector is $(1, 0, 0, 0, 0)$ and we decode the vector to $(1, 1, 1, 1, 0)$.

In general, there is no reason to list all of the elements in the array. Instead, as long as we can determine what the coset leaders are (noticing there is not a unique choice), then we only need to compute the syndrome. In the next section, we shall give a specific example of how this can be done.

Exercises

1. Find all vectors in the binary code with parity check matrix

$$\begin{pmatrix} 1 & 0 & 0 & 1 & 0 \\ 0 & 1 & 0 & 1 & 1 \\ 0 & 0 & 0 & 0 & 1 \end{pmatrix}.$$

   Write the space $\mathbb{F}_2^5$ as the union of cosets of this code. Use these cosets to decode the vector $(1, 1, 1, 1, 1)$.
2. Using the array in Example 11.8, decode the following vectors: $(1, 1, 1, 0, 0)$, $(0, 0, 1, 1, 1)$, and $(1, 1, 1, 0, 1)$.
3. Prove that each possible syndrome occurs.
4. Let C be the binary code generated by (10110) and (01011). Produce the standard array for C with error vectors: $00000, 10000, 01000, 00100, 00010, 00001, 11000,$ and 10001. Use the array to decode the following vectors: $11110, 01111, 01110,$ and 11010.

11.5   The Binary Hamming Code and the Projective Plane of Order 2

In this section, we shall show a connection between a finite projective plane and algebraic coding theory. That is, we shall examine the space $\mathbb{F}_2^7$. There are $2^7 = 128$ such vectors, since there are two choices for each coordinate.

The following matrix is the parity matrix for the code. The only acceptable codewords are those that have an inner product of 0 with each of the three rows. Recall the numbers 1 to 7 in base 2: $1, 10, 11, 100, 101, 110, 111$. We notice that the columns are formed by simply writing these numbers as vectors in $\mathbb{F}_2^3$.

Therefore, we let

$$H = \begin{pmatrix} 0\,0\,0\,1\,1\,1\,1 \\ 0\,1\,1\,0\,0\,1\,1 \\ 1\,0\,1\,0\,1\,0\,1 \end{pmatrix}. \tag{11.3}$$

Consider a vector $(1, 1, 1, 1, 1, 1, 0)$. The inner products are

$$[(1,1,1,1,1,1,0), (0,0,0,1,1,1,1)] = 1$$

$$[(1,1,1,1,1,1,0), (0,1,1,0,0,1,1)] = 1$$

$$[(1,1,1,1,1,1,0), (1,0,1,0,1,0,1)] = 1$$

so the vector is not an acceptable codeword.

Consider the vector $(1, 1, 1, 0, 0, 0, 0)$. The inner products are

$$[(1,1,1,0,0,0,0), (0,0,0,1,1,1,1)] = 0$$

$$[(1,1,1,0,0,0,0), (0,1,1,0,0,1,1)] = 0$$

$$[(1,1,1,0,0,0,0), (1,0,1,0,1,0,1)] = 0$$

so the vector is an acceptable codeword.

We see that since the parity check matrix has dimension 3, this gives that the code has dimension $7 - 3 = 4$, and hence there are $2^4 = 16$ vectors in the code. They are the following:

$$(0,0,0,0,0,0,0) \quad (1,1,1,1,1,1,1)$$

$$(1,1,1,0,0,0,0) \quad (0,0,0,1,1,1,1)$$

$$(0,0,1,1,0,0,1) \quad (1,1,0,0,1,1,0)$$

$$(1,0,0,0,0,1,1) \quad (0,1,1,1,1,0,0)$$

$$(0,1,0,1,0,1,0) \quad (1,0,1,0,1,0,1)$$

$$(0,1,0,0,1,0,1) \quad (1,0,1,1,0,1,0)$$

$$(0,0,1,0,1,1,0) \quad (1,1,0,1,0,0,1)$$

$$(1,0,0,1,1,0,0) \quad (0,1,1,0,0,1,1).$$

Suppose you want to send the message 1110000. If it is received as 1110000 then it is assumed to be correct. This is not because the recipient sees it is in the code but rather because the message has inner product 0 with each of the rows of the parity check matrix. In general, you do not want to search through an entire code, because usually the codes contain an extremely large amount of vectors and searching takes a large amount of computation. Checking the parity check matrix takes very little computation. Assume that one coordinate was changed and it became 1110010. Let us check the parity matrix for this vector.

$$[1110010, 0001111] = 1$$

$$[1110010, 0110011] = 1$$

$$[1110010, 1010101] = 0.$$

The vector 110 formed from these three operations is known as the syndrome. Now we read this as a number in base 2. It is 6. So we know that the sixth place is where the error was made and we change 1110010 to 1110000 which is read correctly. The mathematical technique saw that an error was made, determined where it was made and corrected it. To continue with the example, assume 1110000 was sent but 0110000 was read. The parity vector is

$$[0110000, 0001111] = 0$$

$$[0110000, 0110011] = 0$$

$$[0110000, 1010101] = 1.$$

The 001 is 1 so it is the first coordinate that was changed.

The reason it works is essentially geometry. Consider a floor that is made up of square tiles. Put a mark in the center of each tile. Now pick a mark and stand on it. If you move a distance that is less than the distance of half the diagonal of the square, then you are still in that square. An observer can easily determine your original mark.

The reason the code works so nicely is for exactly the same reason. On the floor no two squares overlap and a very similar thing happens for the code.

We have our seven-dimensional space which we already know has $2^7 = 128$ elements (points). We have 16 points in our code. For each vector in the Hamming code, there are seven vectors that are distance 1 from it (formed by changing each of the seven coordinates) and 1 vector that is distance 0 from it (itself). So in each tile there are eight vectors in it and the vector from the Hamming code is in the center. There are 16 vectors in the code each with 8 in their tile so there are $16 \cdot 8 = 128$ vectors represented.

We have shown that each point in the space is either in the code or it is distance from a unique point in the code. Such a code is known as a perfect code and is quite rare. They will be discussed in the next section.

The Hamming code we have described has 16 possible messages and if zero mistakes or one mistake is made the code will not only detect it but correct it. If two mistakes are made we will not be so lucky. We will still be able to detect that there is an error since it will not have all zeros when applying the inner product to the parity check matrix but we will not be able to correct it. So we say that it can correct one error and detect two errors. If you try to correct something that has two errors, then you will correct it to the wrong message.

For example, say you want to send

$$1111111$$

and two errors are made sending

$$1111100.$$

$$[1111100, 0001111] = 0$$

$$[1111100, 0110011] = 0$$

$$[1111100, 1010101] = 1.$$

It *corrects* the message to 0111100, which is of course the wrong one. The code will work remarkably well if we have a high probability that either no or at most one mistake is made. If the probability is too great, then we need to build a bigger code with a higher minimum distance between vectors.

It may be surprising that the Hamming code was partially discovered by gamblers long before it was of use to electronic communication. Often gamblers play the following kind of game. Pick seven Sunday football games that interest you and pick the winner in each of the seven. Bet one dollar and if you get all seven right then you will win 50 dollars and if you get only one wrong then you win 20 dollars. To the person setting up this game, it seems like a fair bet from his standpoint. The gamblers realized that they could pick their bets according to the vectors in the Hamming code and be sure of either winning all 7 or at least 6 of 7. Betting 16 dollars would insure at least a win of 20 dollars. For example, if there are seven games, with a_i playing b_i in the i-th game, then if there is a 1 in the i-th coordinate then you say that a_i will and if there is a 0 you say that b_i will win. Each vector represents a different 1 dollar bet. Of course, soon enough the bookmakers understood what they were doing and changed the rules.

We shall show how to visualize the Hamming code geometrically in a very different way than was described above. Look at the following representation of the projective plane of order 2.

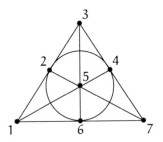

The lines of the plane are

$$\{1,2,3\}$$

$$\{3,4,7\}$$

$$\{1,6,7\}$$

$$\{2,4,6\}$$

$$\{2,5,7\}$$

$$\{3,5,6\}$$

$$\{1,4,5\}$$

The numbers on the points will also indicate a position in a coordinate. If you take the first line and put a 1 for each position on that line you get 1110000 and for the second you get 0011001. This is known as the characteristic function of a line, that is, label the coordinates $v_1, v_2, \ldots, v_7$ and then the define the characteristic function as

$$\chi_L = \begin{cases} 1 & \text{if } (v_i, L) \in I, \\ 0 & \text{if } (v_i, L) \notin I. \end{cases}$$

Notice that if you do it for each of the lines you will get the 7 vectors in the Hamming code that have three ones in them. If you put a 0 in each position on the line and a 1 elsewhere you get the 7 vectors that have 4 ones. Assume you want to send the message 1110000 and a mistake is made so that you receive 1100000. All you need do is find the closest line or complement to a line (meaning the four points that are not on a specific line). That is, by changing one element you can always get to a line or a complement to a line or to the all 0 vector or the all 1 vector.

Recall that the complements of the lines in the projective plane of order 2 are the hyperovals of the plane. This means that the vectors in the code are the zero vector, the all one vector, the characteristic functions of lines, and the characteristic functions of hyperovals.

The geometry of the plane tells you exactly how to decode this code. Assume you want to send the message 1110000 and a mistake is made so that you receive 1100000. All you need do is find the closest line or hyperoval. That is, by changing one element you can always get to a line or a complement to a line. The geometry of the plane and the algebra in the code are actually doing the exact same thing.

Another interesting connection is that if you take the subspace of $\mathbb{F}_2^n$ that is generated by the characteristic functions of the lines of the planes then you get the space that consists of the 16 vectors of the Hamming code.

From a broad perspective, these two seemingly different entities are in reality the same thing.

We shall describe why this works. The matrix H given in 11.3 generates a dimension 3 subspace of $\mathbb{F}_2^7$, since the three rows are linearly independent. Its orthogonal, the code C, then has dimension 4. We know then that C has $2^4 = 16$ vectors and we have displayed these vectors. Hence, H generates $C^\perp$. Consider the vectors ψ_i defined by

$$(\psi_i)_j = \begin{cases} 1 & \text{if } i = j, \\ 0 & \text{if } i \neq j. \end{cases} \tag{11.4}$$

We know that none of the seven vectors ψ_i is in the code C.

Let $\mathbf{w}$ be a vector not in C. Then if $\mathbf{x}$ is in $\mathbf{w} + C$ we have $\mathbf{x} = \mathbf{w} + \mathbf{c}_1$ where $\mathbf{c}_1 \in C$. Then $H(\mathbf{w} + \mathbf{c}_1)^T = H\mathbf{w}^T + H\mathbf{c}_1^T = H\mathbf{w}^T$. Hence, for any vector in $\mathbf{w} + C$ its product with H is the same.

Each coset of C has cardinality 16, and the cosets $\psi_i + C$ are distinct for distinct i. So the entire space $\mathbb{F}_2^7 = \cup(\psi_i + C)$.

Now a vector $\mathbf{c}$ in the code is altered by changing the i-th coordinate then the new vector is $\mathbf{c} + \psi_i$. Then $H(\mathbf{c} + \psi_i)$ is the i-th column of H which we have made the binary representation of i. Hence, we know that if we get the i-th column of H by taking the inner product with the three rows of H then we know that it was the i-th coordinate of c that was changed.

Exercises

1. Produce the parity check matrix for the Hamming code of length 15. As a hint, try writing out the numbers base 2 from 1 to 15. It might take a while to write out all the vectors in this code, there are 2048 of them! But it is easy to find a few vectors in the code and change a coordinate and find which one it is.
2. Use the parity check matrix from the previous exercise to decode the vectors: 111001101101101 and 001100110011010.

11.6   Projective Geometry and Coding Theory

There is a natural connection between codes and projective geometry. It can be used to describe the generalized Hamming codes and several other important results. We shall begin with a description of the generalized Hamming codes.

Recall that the points in $PG_n(\mathbb{F}_q)$ are the vectors in $\mathbb{F}_q^{n+1} - \{(0, 0, \ldots, 0)\}$ moded out by the equivalence, where two vectors are equivalent if one is the scalar multiple of the other. We saw that there were precisely $\frac{q^{n+1}-1}{q-1}$ points in $PG_n(\mathbb{F}_q)$.

Lemma 11.5 *Let* H *be the parity check matrix of a linear code* C *over* $\mathbb{F}_q$. *A weight* s *vector in* C *exists if and only if there are* s *linearly dependent columns in* H.

Proof Assume $v = (v_i)$ is a vector in the code C and $H_1, H_2, \ldots, H_n$ are the columns of H.

Let v be a weight s vector. Since v is orthogonal to the rows of H then $\sum v_i H_i = \mathbf{0}$. Since exactly s of the v_i are non-zero then the corresponding H_i are linearly dependent.

If there are s linearly dependent columns of H, say H_j, $j \in A$ where A is a subset of $\{1, 2, \ldots, n\}$ of cardinality s, then there exists α_j with $\sum \alpha_j v_j = \mathbf{0}$. Let v be a vector where $v_j = \alpha_j$ if j is in A and 0 otherwise. Then $\sum v_i H_i = \mathbf{0}$ and therefore v is orthogonal to the rows of H which gives that $v \in C$.   □

Lemma 11.6 *Let* $v_1, v_2, \ldots, v_s$ *be distinct points of* $PG_n(\mathbb{F}_q)$. *The vectors*

$$v_1, v_2, \ldots, v_s$$

are linearly independent as vectors in $\mathbb{F}_q^{n+1}$ *if and only if the* s *points do not lie on projective* $s - 2$*-dimensional space in* $PG_n(\mathbb{F}_q)$.

Proof Assume $v_1, v_2, \ldots, v_s$ are linearly independent vectors, then they form an s-dimensional vector space in $\mathbb{F}_q^{n+1}$ which is an $s - 1$ projective dimensional space in $PG_n(\mathbb{F}_q)$ which cannot lie on an $s - 2$ projective dimensional space.

If the points do not lie on an $s - 2$ projective space, then they must lie on a space of projective dimension $s - 1$. Hence, as vectors they must generate at least an s-dimensional subspace of $\mathbb{F}_q^{n+1}$. Since there are s vectors generating an s-dimensional space, they must be linearly independent.   □

As an example, three vectors are linearly independent if they do not lie on a projective line which is a one-dimensional projective space.

Theorem 11.15 *An* $[n, k, d]$ *code over* $\mathbb{F}_q$ *exists if and only if there exist* n *points in* $PG_{n-k-1}(\mathbb{F}_q)$ *such that no* $d - 1$ *points lie on a* $d - 3$ *projective space but some* d *points lie on a projective* $d - 2$ *space.*

Proof Assume an $[n, k, d]$ code C exists. Then there exists a generator matrix H for $C^\perp$ with $n - k$ rows. View the columns of H as points in $PG_{n-k-1}(\mathbb{F}_q)$. By Lemma 11.6, no $d - 1$ points lie on a $d - 3$ projective space but some d points lie on a projective $d - 2$ space.

Assume that there exists n points in $PG_{n-k-1}(\mathbb{F}_q)$ such that no $d - 1$ points lie on a projective $d - 3$ projective space but some d points lie on a projective $d - 2$ space. Construct the matrix H with columns as these n points. Then, again by Lemma 11.6 we have the result. $\square$

We can now construct the generalized Hamming codes. Let $v_1, v_2, \ldots, v_n$, $n = \frac{q^r - 1}{q - 1}$, be the distinct points in $PG_{r-1}(\mathbb{F}_q)$, that is, they are vectors of length r in $\mathbb{F}_q^r$. Let H be the r by n matrix where the columns of H are precisely the vectors $v_1, v_2, \ldots, v_n$. Let $Ham(r, q)$ be the code that has H as a parity check matrix, that is, if H generates a code C then $Ham(r, q) = C^\perp$.

Theorem 11.16 *The code $Ham(r, q)$ is a $[\frac{q^r - 1}{q - 1}, \frac{q^r - 1}{q - 1} - r, 3]$ code.*

Proof The length is $\frac{q^r - 1}{q - 1}$ because this is the number of points in the projective geometry. The matrix H has r linearly independent rows so the dimension of $Ham(r, q)$ is $\frac{q^r - 1}{q - 1}$ since the sum of the dimension of the code and its orthogonal equals the length. Finally, since the points are distinct but there are three points on a line by Lemma 11.6 the minimum distance is 3. $\square$

Exercises

1. Modify the decoding algorithm in the previous section to show how to correct a single error using $Ham(r, q)$. Hint: Hv^T may not be a column of H but perhaps a multiple of a column. Determine which error produces that vector and the decoding algorithm follows.
2. Determine the weights of the vectors in $Ham(r, 2)^\perp$.
3. Construct the parity check matrix for $Ham(2, 5)$.
4. Construct the parity check matrix for $Ham(2, 11)$.

11.7 Sphere Packing Bound and Perfect Codes

We shall describe an interesting bound on the size of the code. The bound depends simply on counting principles. We use a few lemmas to prove it and to describe some interesting aspects of the ambient space.

Lemma 11.7 *If v is a vector in $\mathbb{F}_p^n$ then there are $C(n, s)(p - 1)^s$ vectors in $\mathbb{F}_p^n$ that have Hamming distance s from v.*

Proof There are $C(n, s)$ ways of choosing s coordinates from the n coordinates of the ambient space in which to change the entry of **v** and there are $p - 1$ choices for each coordinate to which the entry can be changed.   □

A sphere around a vector **c** of radius t is the set of all vectors whose Hamming distance from **v** is less than or equal to t.

Lemma 11.8 *If* **v** *is a vector in* $\mathbb{F}_p^n$, *then there are* $\sum_{s=0}^{t} C(n, s)(p - 1)^s$ *in a sphere of radius* t *around* **v**.

Proof The possible Hamming distances are from 0 to t and we use Lemma 11.7 to count them.   □

These lemmas can now produce the sphere packing bound.

Theorem 11.17 (Sphere Packing Bound) *Let* C *be a code with minimum weight* $2t + 1$ *then*

$$|C|(\sum_{s=0}^{t} C(n, s)(p - 1)^s) \leq p^n. \qquad (11.5)$$

Proof We know that all of the spheres are distinct since the spheres are radius t and the minimum distance is $2t + 1$. Hence, the number of vectors in all of the spheres must be less than or equal to the number of vectors in the ambient space which is p^n.   □

Definition 11.4 A code with equality in the sphere packing bound is said to be a perfect code.

It is clear why such codes would be thought to be *perfect*. In these codes, every vector in the ambient space is of distance less than or equal to t with a unique vector in the code. In this situation, the spheres of radius t around the codewords partition the space.

We claimed in the previous section that the $[7, 4, 3]$ Hamming code was perfect, we can now prove it.

Corollary 11.1 *The binary* $[7, 4, 3]$ *Hamming code is a perfect code.*

Proof The left side of the inequality 11.5 is $|C| \sum_{s=0}^{1} C(7, s)(2 - 1)^2 = 16(1 + 7) = 128$. The right side is $2^7 = 128$. Hence, the code is perfect.   □

More generally we have the following.

Theorem 11.18 *The generalized Hamming codes over* $\mathbb{F}_p$ *have length* $\frac{p^s - 1}{p - 1}$ *and have dimension* $\frac{p^s - 1}{p - 1} - s$ *and minimum weight 3. These codes are perfect codes.*

Proof We have $n = \frac{p^s - 1}{p - 1}$ and $M = p^{\frac{p^s - 1}{p - 1} - s}$. Then

$$p^{\frac{p^s - 1}{p - 1} - s}\left(1 + \frac{p^s - 1}{p - 1}(p - 1)\right) = p^{\frac{p^s - 1}{p - 1} - s}(1 + p^s - 1)$$

$$= p^{\frac{p^s - 1}{p - 1} - s} p^s = p^{\frac{p^s - 1}{p - 1}}.$$

Therefore, the code is perfect.   □

Remarkably Golay came up with these codes in a single one-page paper early in the history of coding theory [40]. The binary Golay code is closely related to the Witt designs and the Mathieu groups.

Exercises

1. The ternary Golay code is a $[11, 6, 5]$ code. Prove that it is perfect.
2. Find all vectors in $\mathbb{F}_2^4$ in a sphere of radius 2 around the vector $(1, 0, 1, 0)$. Find all vectors in $\mathbb{F}_3^3$ in a sphere of radius 1 around the vector $(1, 2, 1)$.
3. The binary Golay code is a $[23, 12, 7]$ code. Prove that it is perfect.
4. Prove that if the minimum distance of a code is $2t + 1$ then two spheres around codewords of radius less than or equal to t are distinct.

11.8 MDS Codes

One of the most important and interesting classes of codes is MDS codes. They also have very interesting connections to combinatorics. In fact, one of the open problems introduced in an earlier chapter can be phrased in terms of these codes.

We begin with a theorem that gives a bound on how big the minimum distance can be for a code. We use this theorem to define MDS codes. Notice that the bound depends only on combinatorial properties and does not use the underlying algebra of the code.

Theorem 11.19 (Singleton) *Let C be a code of length* n *over an alphabet A of size* q *with minimum Hamming distance* d *and* q^k *elements then*

$$d \leq n - k + 1. \tag{11.6}$$

Proof Delete the first $d - 1$ coordinates of each code word. The vectors formed by the remaining $n - d + 1$ coordinates must be distinct, since otherwise the distance of their corresponding codewords would be less than d. Hence, $q^{n-d+1} \geq q^k$ which gives that $d \leq n - k + 1$ and the result follows.   □

Notice that we did not need to assume that the alphabet A had any algebraic structure. For example, we did not have to assume that it was a field. For the first few connections, we need not assume anything about the algebraic structure; later, however, we shall assume that $A = \mathbb{F}_q$.

A code meeting the bound in Equation 11.6, that is, $d = n - k + 1$, is called a Maximum Distance Separable (MDS) code. They are very important objects in the study of algebraic coding theory.

There are three examples usually referred to as the trivial examples of MDS codes.

- The code $\mathbb{F}_q^n$ has length n, dimension n, and minimum weight 1 and so is an $[n, n, 1]$ MDS code.
- The code generated by $(1, 1, \ldots, 1)$ has length n, dimension 1, and minimum weight n and so is an $[n, 1, n]$ MDS code.
- The code that is orthogonal to the second code, i.e., $C = \{v \mid [v, (1, 1, \ldots, 1)] = 0\}$ has length n, dimension $n - 1$, minimum weight 2, and so is an $[n, n - 1, 2]$ MDS code.

It is not easy to know when MDS codes exist but the following is a well-known conjecture.

Conjecture 11.1 If C is an $[n, k, n - k + 1]$ MDS code over $\mathbb{F}_p$ then $n \leq p + 1$.

We shall now give some interesting connections to some combinatorial objects that we have already studied.

Theorem 11.20 A set of s MOLS of order q is equivalent to an $[s + 2, q^2, s + 1]$ MDS code.

Proof We shall prove one direction and leave the other direction which is basically the reverse as an exercise. Let $L^1, L^2, \ldots, L^s$ be a set of MOLS of order q. Let a vector beginning with (i, j) have in its $(h + 2)$-nd coordinate $L_{i,j}^h$ for all $i, j \in \mathbb{Z}_q$. This gives q^2 vectors of length $s + 2$.

If two vectors agree in the first or second coordinate, then they must disagree in the $(h + 2)$-nd coordinate since no row or column of a Latin square can have the same element twice. If two vectors disagree in the first and second coordinates, then if they agree in the $(h + 2)$-nd coordinate and the $(h' + 2)$-nd coordinate then $(L_{i,j}^h, L_{i,j}^{h'}) = (L_{i',j'}^h, L_{i',j'}^{h'})$ with $i \neq i'$ and $j \neq j'$ contradicting that L^h and $L^{h'}$ are orthogonal. Hence, any two vectors must disagree in at least $s + 1$ places. This is also the largest the minimum weight that can be by the Singleton bound. Therefore, the code is an $[s + 2, 2, s + 1]$ code, and hence an MDS code.   □

The next connection to MDS codes will only relate to codes over finite fields. So from this point we shall assume that the alphabet is a finite field $\mathbb{F}_q$ of order q.

We shall first describe how we can define a code in terms of a parity check matrix. Let H be a matrix with n columns and r linearly independent rows. Let C be a code formed from all of the vectors orthogonal to each row. That is, let $T(\mathbf{v}) = H\mathbf{v}^T$ be a linear transformation. Then the code C is defined by

$$C = \{\mathbf{v} \mid T(\mathbf{v}) = \mathbf{0}\} = \mathrm{Ker}(T). \tag{11.7}$$

Then, C is a linear $[n, n - r, d]$ code.

It is easy to see that the parity check matrix generates $C^\perp$. This gives the following.

Theorem 11.21 *Let C be a linear $[n, k, d]$ code, then $C^\perp$ has length n and dimension $n - k$.*

We can now relate MDS codes and algebraic properties of their parity check matrix.

Lemma 11.9 *Let C be an $[n, k, n - k + 1]$ MDS code with H its parity check matrix. Then any $n - k$ columns of H must be linearly independent.*

Proof Any vector v in the code C represents a linear combination of $wt(v)$ columns of H summing to $\mathbf{0}$. Hence, the minimum weight of the code relates to the largest number of linearly independent columns of H. □

Lemma 11.10 *If C is a linear MDS code over $\mathbb{F}_q$, then $C^\perp$ is a linear MDS code over $\mathbb{F}_q$.*

Proof Let C be an $[n, k, n - k + 1]$ MDS code, and let H be its parity check matrix. Then H is a $n - k$ by n matrix that also generates $C^\perp$ with the property that any $n - k$ columns are linearly independent. Let $\mathbf{v}$ be a vector in $C^\perp$, with $\mathbf{v} \neq \mathbf{0}$. This gives that the maximum number of coordinates with a 0 in $\mathbf{v}$ is $n - k - 1$. This gives that the minimum weight is $k + 1$ and $C^\perp$ is an $[n, n - k, k + 1]$ MDS code. □

These lemmas lead naturally to the following theorem.

Theorem 11.22 *Let C be an $[n, k, d]$ code over $\mathbb{F}_q$ with parity check matrix H. Then the following are equivalent:*

- *C is an MDS code;*
- *every $n - k$ columns of the H are linearly independent;*
- *every k columns of the generator matrix are linearly independent.*

Proof Follows from Lemma 11.9 and Lemma 11.10. □

We can view the columns of H as points in projective space. Consider n points in $PG_{k-1}(\mathbb{F}_q)$ with the property such that no k points lie in a hyperplane $PG_{k-2}(\mathbb{F}_q)$, that is, any k are linearly independent as vectors. These n points are known as an n-arc.

Theorem 11.23 *An* $[n, k, n-k+1]$ *MDS code is equivalent to an* n *arc in* $PG_{n-k-1}(\mathbb{F}_q)$.

Proof By Theorem 11.22, we know that an $[n, k, n-k+1]$ MDS code is equivalent to a parity check matrix with any $n-k$ linearly independent columns, so simply use the n points as the columns of H.   □

Exercises

1. Construct the MDS code from two MOLS of order 3.
2. Prove the remaining direction of Theorem 11.20.

11.9 Weight Enumerators and the Assmus-Mattson Theorem

One of the most interesting and important connections between codes and designs comes from the Assmus-Mattson theorem, which first appeared in [5]. This theorem has been used both in constructing designs and in understanding the structure of codes. In order to state the theorem, we shall first need a discussion of the weight enumerator of a code. There are numerous weight enumerators that can be defined for a code. For example, the complete weight enumerator, the symmetric weight enumerator, the joint weight enumerator, and the higher weight enumerator are all examples of weight enumerators. We shall define and use the simplest which is the Hamming weight enumerator.

Let C be a code over the finite field $\mathbb{F}_q$. We define the Hamming weight enumerator of C as follows:

$$W_C(x, y) = \sum_{c \in C} x^{n-wt(c)} y^{wt(c)}, \tag{11.8}$$

where $wt(c)$ is the Hamming weight of the vector c. The reader should be aware that in some texts the roles of x and y are reversed, for example, it is this way in [4]. It follows immediately that the weight enumerator can also be seen as the following:

$$W_C(x, y) = \sum_{i=0}^{n} A_i x^{n-i} y^i, \tag{11.9}$$

where there are A_i vectors in C of weight i.

As an example, the weight enumerator of the $[7, 4, 3]$ Hamming code is

$$W_C(x, y) = x^7 + 7x^4 y^3 + 7x^3 y^4 + y^7. \tag{11.10}$$

The weight enumerator of a code is related to the weight enumerator of the dual code by the MacWilliams relations. These relations are some of the most interesting and useful theorems in coding theory. These were first proved in the Ph.D. thesis of Jesse MacWilliams [61].

We shall give a proof of the theorem in the binary case and state it for the more general case. The proof for the general case is not that much more difficult but requires a bit more algebra. We need a bit of machinery and a lemma.

Let f be a function with $\mathbb{F}_2^n$ as its domain. The Hadamard transform $\widehat{f}$ of f is defined by

$$\widehat{f}(\mathbf{u}) = \sum_{\mathbf{w} \in \mathbb{F}_2^n} (-1)^{\mathbf{u} \cdot \mathbf{w}} f(\mathbf{w}), \tag{11.11}$$

where $\mathbf{u} \cdot \mathbf{w}$ is the usual dot product.

Lemma 11.11 *If C is an* $[n, k]$ *binary linear code then*

$$\sum_{\mathbf{w} \in C^\perp} f(\mathbf{w}) = \frac{1}{|C|} \sum_{\mathbf{u} \in C} \widehat{f}(\mathbf{u}). \tag{11.12}$$

Proof We have

$$\sum_{\mathbf{u} \in C} \widehat{f}(\mathbf{u}) = \sum_{\mathbf{u} \in C} \sum_{\mathbf{w} \in \mathbb{F}_2^n} (-1)^{\mathbf{u} \cdot \mathbf{w}} f(\mathbf{w})$$

$$= \sum_{\mathbf{w} \in \mathbb{F}_2^n} f(\mathbf{w}) \sum_{\mathbf{u} \in C} (-1)^{\mathbf{u} \cdot \mathbf{w}}.$$

If $\mathbf{w} \in C^\perp$ then $(-1)^{\mathbf{v} \cdot \mathbf{w}} = 1$ for all $\mathbf{u} \in C$. If $\mathbf{w} \notin C^\perp$ then $(-1)^{\mathbf{v} \cdot \mathbf{w}}$ is 1 and -1 equally often and hence cancels to 0 over the summation. This gives

$$\sum_{\mathbf{u} \in C} \widehat{f}(\mathbf{u}) = |C| \sum_{\mathbf{w} \in C^\perp} f(\mathbf{w}). \tag{11.13}$$

Thus we have the result.   □

This brings us to one of the most important theorems in all of coding theory.

Theorem 11.24 (MacWilliams Relations) *Let C be a linear code over* $\mathbb{F}_2$. *Then*

$$W_{C^\perp}(x, y) = \frac{1}{|C|} W_C(x + y, x - y). \tag{11.14}$$

Proof Apply Lemma 11.11 to the function $f(\mathbf{v}) = x^{n-wt(\mathbf{v})}y^{wt(\mathbf{v})}$. Then we have

$$\widehat{f}(\mathbf{v}) = \sum_{\mathbf{w}\in\mathbb{F}_2^n} (-1)^{\mathbf{v}\cdot\mathbf{w}} x^{n-wt(\mathbf{v})} y^{wt(\mathbf{v})}$$

$$= \sum_{\mathbf{w}\in\mathbb{F}_2^n} (-1)^{v_1 w_1 + v_2 w_2 + \cdots + v_n w_n} \prod_{i=1}^n x^{1-w_i} y^{w_i}$$

$$= \sum_{w_1=0}^{1} \sum_{w_2=0}^{1} \cdots \sum_{w_n=0}^{1} \prod_{i=1}^{n} (-1)^{v_i w_i} x^{1-w_i} y^{w_i}$$

$$= \prod_{i=1}^{n} \sum_{\alpha=0}^{1} (-1)^{v_i \alpha} x^{1-\alpha} y^{\alpha}.$$

If $v_i = 0$ then inner sum is $x + y$. If $v_i = 1$ then the inner sum is $x - y$.
This gives that $\widehat{f}(\mathbf{v}) = (x + y)^{n-wt(\mathbf{v})} (x - y)^{wt(\mathbf{v})}$.   □

More generally, the theorem can be stated as follows.

Theorem 11.25 (MacWilliams Relations) *Let C be a linear code over $\mathbb{F}_q$. Then*

$$W_{C^{\perp}}(x, y) = \frac{1}{|C|} W_C(x + (q-1)y, x - y). \tag{11.15}$$

We often let $x = 1$ and simply write the weight enumerator as $W_C(y) = \sum A_i y^i$. This allows for the weight enumerator to be written in a simpler manner.

Given a vector $\mathbf{v}$, the support of $\mathbf{v}$ consists of those coordinates where the vector is non-zero. For example, if $p_1, p_2, \ldots, p_8$ are the coordinates of the space and $(3, 1, 0, 2, 0, 0, 1, 1)$ is a vector then its support is $\{p_1, p_2, p_4, p_7, p_8\}$. Viewing the coordinates of $\mathbb{F}_q^n$ as points we can construct designs using the supports as blocks. We can now state the Assmus-Mattson theorem, without proof, as it appears in [4].

Theorem 11.26 (Assmus-Mattson Theorem) *Let C be an $[n, k, d]$ code over $\mathbb{F}_q$ and $C^{\perp}$ an $[n, n - k, d^{\perp}]$ code. Let $w = n$ when $q = 2$ and if q is not 2 let w be the largest integers satisfying*

$$w - \left(\frac{w + q - 2}{q - 1}\right) < d.$$

Let $w^{\perp} = n$ when $q = 2$ and if q is not 2 let $w^{\perp}$ be the largest integers satisfying

$$w^{\perp} - \left(\frac{w^{\perp} + q - 2}{q - 1}\right) < d^{\perp}.$$

Suppose there is an integer t with $0 < t < d$ that satisfies the following: If $W_{C^{\perp}} = \sum B_i y^i$ at most $d - t$ of $B_1, B_2, \ldots, B_{n-t}$ are non-zero. Then for each i with

$d \le i \le w$ *the supports of the vectors of weight* i *in* C, *provided there are any, give a* t-*design. Similarly, for each* j *with* $d^\perp < j \le \min\{w^\perp, n - t\}$ *the supports of the vectors of weight* j *in* $C^\perp$, *provided there are any, give a* t-*design.*

We shall apply the previous theorem to the binary self-dual $[24, 12, 8]$ Golay code. A self-dual code is a code that satisfies $C = C^\perp$. Here $w = w^\perp = n$ and $d = d^\perp = 8$. If t is 5 then $d - t = 3$. We see that of $B_1, B_2, \ldots, B_{19}$ only B_8, B_{12} and B_{16} are non-zero. Hence, only $d - t$ of $B_1, B_2, \ldots, B_{n-t}$ are non-zero. This proves that the supports of the vectors of weight 8, weight 12 and weight 16, each form 5-designs. The λ_i^j table for the $5 - (24, 8, 1)$ design formed by the weight eight vectors is given in 11.16.

We shall consider the design formed by the weight 12 vectors. There are 24 points and 2576 blocks. We have that $\lambda_0 = 2576$. Using the relation that $\lambda_s = \frac{v-s}{k-s}\lambda_{s+1}$ we get that $\lambda_1 = 1288$, $\lambda_2 = 616$, $\lambda_3 = 280$, $\lambda_4 = 120$, and $\lambda_5 = 48$.

Then we can construct the table for λ_i^j:

$$
\begin{array}{ccccccccccc}
 & & & & & 2576 & & & & & \\
 & & & & 1288 & & 1288 & & & & \\
 & & & 616 & & 672 & & 616 & & & \\
 & & 280 & & 336 & & 336 & & 280 & & \\
 & 120 & & 160 & & 176 & & 160 & & 120 & \\
48 & & 72 & & 88 & & 88 & & 72 & & 48
\end{array}
\qquad (11.16)
$$

We shall now examine a non-binary case. Consider the ternary self-dual $[12, 6, 6]$ code. It has weight enumerator:

$$W_C(y) = 1 + 264y^6 + 440y^9 + 24y^{12}. \qquad (11.17)$$

We have $d = d^\perp = 6$. We must find the largest w with

$$w - \left(\frac{w + q - 2}{q - 1}\right) < d$$

$$w - \left(\frac{w + 1}{2}\right) < 6$$

$$\frac{w}{2} - \frac{1}{2} < 6.$$

This gives that $w = w^\perp = 12$. If t is 5 then $d - t = 1$. We see that of $B_1, B_2, \ldots, B_7$ only B_6 is non-zero. Hence, only $d - t$ of $B_1, B_2, \ldots, B_{n-t}$ are non-zero. This proves that the supports of the vectors of weight 6, weight 9, and weight 12 each form 5-designs.

1. Find the weight enumerator of the orthogonal of the $[7, 4, 3]$ Hamming code with weight enumerator given in Equation 11.10.
2. Determine the parameters for the designs formed by the Assmus-Mattson theorem for the ternary self-dual Golay code.
3. Find the weight enumerator for the code of length n generated by the all one vector. Use the MacWilliams relations to find the weight enumerator of its orthogonal. Could this have been computed directly?
4. The self-dual $[48, 24, 12]$ binary code has weight enumerator:

$$W_C(y) = 1 + 17296\, y^{12} + 535095\, y^{16} + 3995376\, y^{20} + 7681680\, y^{24}$$
$$+ 3995376\, y^{28} + 535095\, y^{32} + 17296\, y^{36} + y^{48}.$$

- Use the Assmus-Mattson theorem to prove that the supports of any non-trivial weight form a 5-design.
- Prove that the vectors of weight 12 form a $5 - (48, 12, 8)$ design.
- Prove that the vectors of weight 16 form a $5 - (48, 16, 1365)$ design.
- Prove that the vectors of weight 20 form a $5 - (48, 20, 3176)$ design.
- Prove that the vectors of weight 24 form a $5 - (48, 24, 190680)$ design.
- Produce the λ_i^j table for the $5 - (48, 12, 8)$ design.

11.10 Codes of Planes

Codes have often been very useful in understanding finite designs. In this section, we shall describe some codes that come from designs, and we shall show how codes can be used to give a short proof of part of the Bruck-Ryser theorem.

Recall that a finite projective plane Π of order n is a set of points $\mathcal{P}$, a set of lines $\mathcal{L}$, and an incidence relation I between them, where $|\mathcal{P}| = |\mathcal{L}| = n^2 + n + 1$, and any two points are incident with a unique line, and any two lines are incident with a unique point.

Let Π be a projective plane of order n, and let p be a prime dividing n.

The characteristic function of a line L is

$$v_L(p) = \begin{cases} 1 \text{ if } p \text{ is incident with } L, \\ 0 \text{ if } p \text{ is not incident with } L. \end{cases} \tag{11.18}$$

The code generated by the characteristic functions of lines over F_p is denoted by $C_p(\Pi)$. That is

$$C_p(\Pi) = \langle v_L \mid L \in \mathcal{L} \rangle. \tag{11.19}$$

We define the Hull to be

$$\text{Hull}_p(\Pi) = C_p(\Pi) \cap C_p(\Pi)^{\perp}. \tag{11.20}$$

It is immediate that the Hull is a self-orthogonal code that is contained in $C_p(\Pi)$. In general, we always take the prime p to be a divisor of n. The interesting codes are all over $\mathbb{F}_p$ where p divides n.

Lemma 11.12 *Let Π be a projective plane of order n and let p be a prime dividing n. Then*

$$\sum_{L \in \mathcal{L}} v_L = \mathbf{j},$$

where $\mathbf{j}$ is the all one vector.

Proof Through any point there are $n + 1$ lines and $n + 1 \equiv 0 \pmod{p}$ when p divides n which gives that in each coordinate of $\sum_{L \in \mathcal{L}} v_L$ is a 1. □

The following result can be found in [4].

Theorem 11.27 *Let Π be a projective plane of order n, and let p be a prime dividing n. We have that*

$$\text{Hull}_p(\Pi) = \langle v_L - v_M | \ L \text{ and } M \text{ lines of } \Pi \rangle \tag{11.21}$$

and $C_p(\Pi) = \langle \text{Hull}_p(\Pi), \mathbf{j} \rangle$, where $\mathbf{j}$ is the all one vector.

Proof For any three lines L, M, and T of Π, we have $[v_T, v_L - v_M] = 0$, and therefore $v_L - v_M$ is in $C_p(\Pi)^\perp$ for any two lines v_L, v_M of Π. It is clear that these vectors are in $C_p(\Pi)$ and hence in $\text{Hull}_p(\Pi)$.

This gives that $\text{Hull}_p(\Pi)$ is at most of codimension 1 in $C_p(\Pi)$ since $C_p(\Pi) = \langle \text{Hull}_p(\Pi), v_L \rangle$ for any line L. We note that $\mathbf{j}$, the all one vector, is in $C_p(\Pi)$ since $\sum_{L \in \mathcal{L}} v_L = \mathbf{j}$, and $[\mathbf{j}, v_L] = 1$ for any line L, and therefore $\mathbf{j}$ is not in $\text{Hull}_p(\Pi)$ giving that $C_p(\Pi) = \langle \text{Hull}_p(\Pi), \mathbf{j} \rangle$. □

The following lemma is well known, see [4].

Lemma 11.13 *Let Π be a projective plane of order n. If p sharply divides n, that is, p divides n but p^2 does not, then $\dim(C_p(\Pi)) = \frac{n^2+n+2}{2}$.*

It follows immediately from this lemma and by the previous theorem that
$\dim(C_p(\Pi)^\perp) = n^2 + n + 1 - \frac{n^2+n+2}{2} = \frac{n^2+n+2}{2} - 1 = \dim(\text{Hull}_p(\Pi))$
which gives that $C_p(\Pi) = \text{Hull}_p(\Pi)^\perp$.

The following is a standard theorem in coding theory, see [62] for a proof.
Recall that a self-dual code is a code C with $C = C^\perp$.

Lemma 11.14 *Let p be a prime with $p \equiv 3 \pmod{4}$; if C is a self-dual code of length m over $\mathbb{F}_p$, then 4 divides m.*

For the remainder we assume that $p \equiv 3 \pmod 4$ and that p sharply divides n, the order of the plane Π. We have that

$$C_p(\Pi) = Hull_p(\Pi) \cup (Hull_p(\Pi) + \mathbf{j}) \cup (Hull_p(\Pi) + 2\mathbf{j})$$
$$\cup \cdots \cup (Hull_p(\Pi) + (p-1)\mathbf{j}),$$

since $C_p(\Pi) = \langle Hull_p(\Pi), \mathbf{j} \rangle$, where $(Hull_p(\Pi) + i\mathbf{j})$ denotes the coset of $Hull_p(\Pi)$ in $C_p(\Pi)$ formed by adding $i\mathbf{j}$ to every vector in $Hull_p(\Pi)$. Let $H_i = (Hull_p(\Pi) + i\mathbf{j})$, giving that $H_i \cap H_k = \emptyset$ for $i \neq k$ and

$$\bigcup_{i=0}^{p-1} H_i = C_p(\Pi).$$

Let $v \in H_i$, $w \in H_k$ then $v = h + i\mathbf{j}$ and $w = h' + k\mathbf{j}$ where $h, h' \in Hull_p(\Pi)$. Then

$$[v, w] = [h + i\mathbf{j}, h' + k\mathbf{j}] = [h, h'] + [h, k\mathbf{j}] + [i\mathbf{j}, h'] + [i\mathbf{j}, k\mathbf{j}] = ik. \quad (11.22)$$

To the vectors in H_i adjoin some vector v_i of length 3; and set $C' = \bigcup(H_i, v_i)$. To insure the linearity of this new code once we have chosen v_1 and then v_i is forced to be iv_1. To insure that C' is self-orthogonal we must have $[v_1, v_1] = -[H_1, H_1] = -1$, where $[H_i, H_k]$ is the inner product of any vector in H_i with any vector in H_k.

Let $v_1 = (x, y, 0)$ where $x^2 + y^2 = -1$. It is well known that this has a solution in F_p when $p \equiv 3 \pmod 4$, with neither x nor y equal to 0. Then

$$[v_i, v_k] = ikx^2 + iky^2 = ik(x^2 + y^2) = -ik = -[H_i, H_k].$$

Let w be the vector of length $n^2 + n + 4$ of the form $(\mathbf{0}, a, b, c)$. We want $ax + by = 0$; hence, choose a non-zero a, and then we have $b = -a(\frac{x}{y})$. Then, we want $a^2 + b^2 + c^2 = 0$ giving that

$$a^2 + a^2 \frac{x^2}{y^2} + c^2 = 0$$

$$c^2 = -(1 + \frac{x^2}{y^2})a^2$$

$$c^2 = \frac{a^2}{y^2}.$$

Then $c = \frac{a}{y}$. Note w is self-orthogonal, orthogonal to every vector in C' and is linearly independent over C'. Let $D = \langle C', w \rangle$, then D has length $n^2 + n + 4$ and has dimension $\frac{n^2+n+2}{2} + 1 = \frac{n^2+n+4}{2}$. This gives that D is a self-dual code. This implies that $n^2 + n + 4$ is divisible by 4, which implies $n^2 + n + 1 \equiv 1 \pmod 4$. If $n \equiv 1$ or $2 \pmod 4$ then $n^2 + n + 1$ is not $1 \pmod 4$ giving a special case of the Bruck-Ryser theorem.

Theorem 11.28 *If* $n \equiv 1$ *or* $2 \pmod 4$ *and* p *is a prime sharply dividing* n *with* p *a prime and* $p \equiv 3 \pmod 4$, *then there does not exist a projective plane of order* n.

Exercises

1. Determine which order less than or equal to 30 are eliminated by Theorem 11.28.

Cryptology

12

One of the most interesting and important modern applications of discrete combinatorial mathematics is cryptology. Cryptology is the science of sending and receiving information that is meant to be secret. It has been in use since the time of Julius Caesar and has been played an important role in historical events including both world wars. More recently, beginning in the late twentieth century, it has become foundational in Internet security and commerce. Cryptology has never been so widely used as it has been in the twenty-first century. Every time someone purchases an item on the Internet, cryptology is used. It is often used in numerous electronic communications every day.

Several terms are used to describe the various parts of this discipline. The term cryptography refers to the science of creating a system to keep information secret, whereas cryptanalysis refers to the science of breaking a system designed to keep information secret. However, the terms cryptography and cryptology are often used interchangeably without much confusion.

Cryptology has had a very long and interesting history. It is one branch of mathematics that has had an immediate impact on world events. For example, during World War I, the German Zimmerman made an overture to Mexico, via a secret message, to attack the United States with German supplies and logistical help. They believed this would keep America busy in the western hemisphere and keep them out of the European conflict. The Americans intercepted and decoded the message which helped push America toward involvement in the war. The system used to encrypt the message was a straightforward combinatorial technique. During World War II, every major nation involved developed their own cryptographic system, but perhaps the most important was the German Enigma machine. This machine looked like a typewriter and used a combinatorial system to encode the messages. Famously, the British set up an organization at Bletchley park to decode the Enigma machine, and many historians attribute to this decoding a shortening of the war by years. During

S. T. Dougherty, *Combinatorics and Finite Geometry*,
Springer Undergraduate Mathematics Series,

the sixteenth century, Mary Queen of Scots used a simple cryptographic system in the Babington plot, which was decoded by Queen Elizabeth's codebreaker. The results of this codebreaking were used in her trial which ended with her execution in 1587.

For this text, we shall be primarily interested in the application of combinatorial techniques to this study. From the very beginning, combinatorial techniques were at the very core of the construction of cryptographic systems. In many ways, it is the combinatorial explosion, namely, the rapidly increasing number of ways that an event can happen that allows for secrecy.

We shall begin by describing various substitution ciphers and then move on to more modern and sophisticated techniques.

12.1   Substitution Ciphers

Caesar Cipher

The most basic and possibly least effective cryptographic technique is a simple substitution cipher. This has its origins in the Caesar cipher used by Julius Caesar 2000 years ago. Caesar took a message written in Latin and simply cycled the letters of the message three spaces (assuming the standard ordering of letters a, b, c, d, …, z). (We are using the ordering of the letters as is generally done in English, ignoring the fact that the Romans did not use the letter j.) For example, the message

all of gaul is divided into three parts

becomes

DOO RI JDXO LV GLYLGHG LQWR WKUHH SDUWV.

Notice that the original message is written in small letters and the encoded message, called the ciphertext, is written in capital letters. We shall continue using this convention throughout this chapter.

As usual, for early cryptographic systems, this was used largely for military purposes. This technique was very effective for Caesar because few people were even literate in Latin among his enemies. So even this very simple technique was likely to work. At that time, messages were carried by hand by an individual. This means that intercepting the message meant intercepting the person carrying the message. It was much more likely that the person carrying the message could destroy it if they thought they were about to be captured, so even having the message intercepted was less likely. However, anyone who knew that Caesar used this technique to encode messages would have been able to instantly decode it. Given this fact, it was not a very secure system. In this case, decoding this system is easy. Namely, if you know that the writer is simply cycling the letters by some amount, then once you establish one letter you can break the entire system. In a language like English, where there are two prominent one letter words, namely, a and I, it is a simple matter to determine one letter if the writer has made the mistake of indicating ends of words. Even if the words are all written together without spaces, finding the letter corresponding to the most common letter in English, namely, e, is again an easy matter. In fact, even an ancient codebreaker would know the most common letter in any written language they are reading.

Generalizing this technique is to simply make an arbitrary permutation of the 26 letters, rather than simply cycling them. We know there are $26! \approx 4 \times 10^{26}$ possible permutations of the letters. It would seem at first glance that with so many possibilities this would be an effective system. However, it is an extremely insecure system.

If you were sending random collections of letters, then the system would be much more secure. However, generally a message is written in a human language like English. In English, as in every language written with letters, certain letters are used far more often than other languages. Frequency tables in most languages are very well known and easily determined. This makes decrypting a secret message using a simple cipher quite easy. As long as the message is long enough, all that is necessary is to make a frequency table of the symbols in the encoded message and match them with the frequency table for the language. Then using some simple reasoning in terms of the language, decryption becomes quite simple.

For example, in modern terms, it is a simple matter to find an electronic text written in the language of your choice, perhaps news stories or a book in electronic form. Then, have the computer simply count the number of occurrences of each letter to establish their frequency. Given a document that has used a substitution cipher, you perform the same computation. Then, the probability is very high, if the message is long enough, that the symbol used most often is e. The second is most likely t, and the third is most likely a. It is possible that this is not true, but the probability that the top three are e, t, and a in some order is very high. This technique is much less likely to work to establish which symbol corresponds to x, z, and q, which occur rather infrequently. However, given a correct decoding of the more common letters, these letters will come from context.

It is a fairly simple exercise to write a computer program using these techniques to decrypt messages encrypted with a simple cipher. Given the existence of a computer, the simple cipher can be broken almost immediately by competent cryptographers.

We can make a more systematic technique for producing a substitution using modular arithmetic. We can attach numerical values 0 to 25 based on the letters' position in the standard ordering of the English alphabet. Then, the Caesar cipher can be seen as simply $\phi(x) = x + 3 \pmod{26}$. Then, an affine substitution cipher can be made by defining any function of the form

$$\psi(x) = \alpha x + \beta.$$

We need $\gcd(\alpha, 26)$ to be 1; otherwise, the function is not injective and has no inverse, which would be necessary to decode the message. To decode the message you use the inverse function

$$\psi^{-1}(x) = \alpha^{-1}(x - \beta).$$

Since this produces a simple substitution cipher, it is still quite an easy matter to break the system.

By the nineteenth century, a simple substitution was no longer be used in any serious situation. Next, we shall examine several most sophisticated techniques that were used, which are based on substitution ciphers.

Vigenère Cipher

The Vigenère cipher goes back to the sixteenth century, and many thought it was secure even up to the twentieth century. However, Babbage and Kasiski had already shown how to attack it during the nineteenth century and Friedman developed further attacks in the 1920s. This cipher uses the same sort of modular arithmetic as was done in the affine substitution cipher but with a twist. Take a given message and turn the text into numbers modulo 26. Then take a keyword, which will form the key for the system. We shall take the word geometry. This word corresponds to the vector $(6, 4, 14, 12, 4, 19, 17, 24)$. We note that the length of this vector is 8.

Assume we want to send the following message:

$$donotgogentleintothatgoodnight.$$

Then we change the message into its corresponding numbers and make a sequence of length 30. To each number with index i in the sequence modulo 8, we add the number corresponding to the ith coordinate in the vector. For our example, we take the following:

```
d    o    n    o    t    g    o    g    e    n    t    l    e    i    n
3   14   13   14   19    6   14    6    4   13   19   11    4    8   13
6    4   14   12    4   19   17   24    6    4   14   12    4   19   17
9   18    1    0   23   25    5    4   10   17    7   23    8    1    4
J    S    B    A    X    Z    F    E    K    R    H    X    I    B    E
```

```
t    o    t    h    a    t    g    o    o    d    n    i    g    h    t
19   14   19    7    0   19    6   14   14    3   13    8    6    7   19
24    6    4   14   12    4   19   17   24    6    4   14   12    4   19
17   20   23   21   12   23   25    5   12    9   17   22   18   11   12
R    U    X    V    M    X    Z    F    M    J    R    W    S    L    M
```

Then the ciphertext is

$$JSBAXZFEKRHXIBERUXVMXZFMJRWSLM.$$

While this is based on the substitution cipher which is highly insecure, there are two more cryptographic techniques at play. First, it is a different substitution cipher for each of the coordinates modulo the length of the key. Second, the length of the key is a secret.

We can use probability and a geometric function to determine the length of the key. Take the vector of the letters' frequencies in English and call it $\mathbf{v}_0$. That is,

$(\mathbf{v}_0)_i = p_i$ where p_i is the probability that the letter i is used in a standard English text. Let $\mathbf{v}_j$ be the vector formed by shifting $\mathbf{v}_0$ cyclicly j times to the right.

It is a simple, but time-consuming exercise, to determine that the maximum value of $[\mathbf{v}_i, \mathbf{v}_j]$, where this indicates the standard inner product, occurs when $i = j$. Moreover, this is intuitive, since the values are probabilities between 0 and 1. Therefore, when taking the inner product of $\mathbf{v}_i$ with $\mathbf{v}_i$, the larger values are multiplied by larger numbers maximizing their effect of being large and the smaller numbers are multiplied by smaller numbers minimizing their effect of being small.

First, we need to determine the length of the keyword. To do this write the ciphertext on two sheets of paper and begin cycling the second one under the first and count the number of occurrences where the same letter appears. Whichever shift has the most such coincidences is probably the length of the key. This is a direct consequence of the fact that the maximum value of the above inner product is $[\mathbf{v}_i, \mathbf{v}_j]$.

Once one knows the length of the key, one can use frequency analysis on the coordinates that are the same modulo the length of the keyword. Alternatively, you can take the frequency of letters in the jth coordinate modulo the length of the keyword and divide each number by the total numbers counted. Call this vector $\mathbf{w}$. Then take $[\mathbf{v}_i, \mathbf{w}]$ for all i. Whichever i gives the largest inner product is most likely the shift used for that substitution cipher which gives the value for each letter in the keyword.

Playfair Cipher

The Playfair system was constructed by Charles Wheatstone in 1854 and named it for Baron Playfair of St. Andrews, since he convinced the British government to use it. It was used by the British military in both the Boer War and World War I. The system begins by building a matrix which is used as a key.

Take a word, then write the letters used in the word to begin a 5 by 5 square, discarding any letters that have already been used. Then complete the square using the remaining letters in the alphabet (equating i and j). Using the word erudite, we obtain the following matrix:

$$\begin{array}{ccccc} e & r & u & d & i \\ t & a & b & c & f \\ g & h & k & l & m \\ n & o & p & q & s \\ v & w & x & y & z \end{array}$$

Then split the message into groups of two letters and encode the message using the following rules:

1. If a double letter appears, place an x between them. If there is an odd number of letters, then place an x at the end.
2. When two letters are not in the same row nor the same column, replace each letter by the letter that is in its row and in the other letter's column.
3. When two letters are in the same row, replace each letter with the letter that is to its right in the row (wrapping around the row if necessary).
4. When two letters are in the same column, replace each letter with the letter that is below it in the column (wrapping around the column if necessary).

We shall use this to encode the following message: the eagle lands at midnight. We begin by splitting the message into groups of two letters.

th ex ea gl el an ds at mi dn ig ht

Then apply the cipher to get the following.

AG UV RT HM DG TO IQ BA SF EQ EM GA

To decode a message, simply reverse the previous instructions.
For example, assume you received the message:

GD PO OH UI RD GD UW

Finding G and D in the table, you find the letters in their own rows and in the others column and you get l and e. Finding P and O, we see that they are in the same row, so we go backward in the row to get o and n. Continuing with this process, we see that the original message is

leonhard euler.

In essence, this system is a glorified substitution cipher. Instead of substituting letters one at a time, they are substituted two at a time. This type of system is easily defeated since the frequencies of pairs of letters in English are well established as well. Moreover, even though there are $26^2 = 676$ pairs of letters in English, many of them never occur or occur so infrequently that they can be discarded. For example, the pair q followed by any letter other than u either never occurs or occurs in only a few words which are unlikely to be used and of foreign origin so that one can reasonably abandon them. It becomes a fairly simple matter to decode a message using this system, given enough time. Moreover, if you can determine the keyword, then decoding the messages becomes a trivial matter. The Germans during World War I had a better system which we now describe.

ADFGX Cipher
To begin this system, you create a 5 by 5 matrix and fill it with the alphabet (equating i and j as usual) and indexing the rows and columns by the letters A, D, F, G, and X. These letters are chosen since the Morse codes of these letters were distinct and not easily confused. Of course, at the time, one of the most used methods of communications was the telegraph, and so this would have been of great importance. We shall assume we have the following matrix:

	A	D	F	G	X
A	m	u	d	c	n
D	f	v	w	r	p
F	q	b	s	e	y
G	t	k	g	i	x
X	p	l	a	z	h

For each letter in the message, you replace it with the row and column index of that letter. For example, the letter e would become FG. At this point, it is only a very simple substitution cipher. However, the complexity is introduced in the next step.

Take a keyword, say Munich, and write the encoded message as a matrix with the letters of the keyword indexing the columns. Then rearrange the columns by putting the letters of the keyword in alphabetical order. The last step is to read the message down the columns.

Assume our original message is: "strike up the band". We first change it into its representation in terms of A, D, F, G, and X. We get

$$FFGADGGGGDFGADDXGAXXFGFDXFAXAF.$$

Then, we write the message in the following matrix:

M	U	N	I	C	H
F	F	G	A	D	G
G	G	G	D	F	G
A	D	D	X	G	A
X	X	F	G	F	D
X	F	A	X	A	F

Next, we rearrange the matrix by putting the columns in alphabetical order:

C	H	I	M	N	U
D	G	A	F	G	F
F	G	D	G	G	G
G	A	X	A	D	D
F	D	G	X	F	X
A	F	X	X	A	F

Then the ciphertext is

$$DFGFAGGADFADXGXFGAXXGGDFAFGDXF.$$

To decode, the process is simply done in reverse.

We note that the security of this system lies in the fact that a frequency count is not useful on the cipher text, since the pairs of letters corresponding to the original message may have been scrabbled in the final ciphertext. Later, this technique was replaced by the ADFGVX system, where all 26 letters and 10 digits could be used in a message.

This technique is certainly more effective than the Playfair system, but it was broken by Georges Painvin and the French Bureau du Chiffre using various techniques. For example, if you can intercept two different ciphertexts sent around the same time that agree in the first few characters. This is not as unlikely as it may seem, since often military messages have similar beginnings. This was especially true for the German military of the time. Hopefully, this means that the messages agree for the first few words. This would mean that the top several entries of the columns are identical.

Next search for other instances of identical occurrences. Hopefully, this indicates the beginning of columns. If we are fortunate enough to be correct, this will indicate the length of columns. If the columns are different lengths, the longer ones should be in the beginning and the shorter ones at the end. Then try various arrangements of the columns, which will give a straightforward substitution cipher. Applying frequency analysis to each should decode one message and indicate the proper ordering of the columns. While a computer could do this quite simply and quickly, by hand it is no easy matter. However, the French were able to decode a significant number of German transmissions.

Block Ciphers

The final substitution cipher that we shall discuss is the block cipher. We have seen that substituting one letter at a time falls easily to frequency analysis. Even substituting two at a time, generally falls to frequency analysis as well, noting that the frequency of pairs of letters is equally well known. One might then think to increase the number of letters taken at a time to make frequency analysis less effective. This can be done with a block cipher.

Take a message and break it into blocks of size n. Then write each letter as a number modulo 26. This assumes that we are using the standard English alphabet. Of course, one can use as large a modulus as you like, using as many symbols as you like. To construct a substitution cipher for n letters at a time, construct an n by n invertible matrix M. Then take a vector $\mathbf{v}$ of n symbols constructed from n letters. Form the vector $M\mathbf{v}^T = \mathbf{w}$, which is then the ciphertext. To decode the message, we take $M^{-1}\mathbf{w} = M^{-1}M\mathbf{v}^T = \mathbf{v}^T$.

Example 12.1 Take the message "send money" and split it into groups of size 3 and take the numerical value modulo 26. This gives $(18, 4, 13)(3, 12, 14)(13, 4, 24)$. Take the matrix

$$M = \begin{pmatrix} 3 & 1 & 2 \\ 2 & 1 & 4 \\ 5 & 3 & 3 \end{pmatrix}.$$

Then $M(18, 4, 13)^T = (6, 14, 11)^T$, $M(3, 12, 14)^T = (23, 22, 15)^T$, and $M(13, 4, 24)^T = (13, 22, 19)^T$. Then the ciphertext is $(6, 14, 11)(23, 22, 15)(13, 22, 19)$ which gives

GOLXWPNWT.

The inverse matrix is

$$M = \begin{pmatrix} 15 & 21 & 14 \\ 20 & 19 & 22 \\ 7 & 24 & 7 \end{pmatrix}.$$

Then $M^{-1}(6, 14, 11)^T = (18, 4, 13)^T$, $M^{-1}(23, 22, 15)^T = (3, 12, 14)$, and M^{-1} $(13, 22, 19) = (13, 4, 24)$, which gives

$$sendmoney.$$

Notice that the strength of the system is in the diffusion. For example, "sen" was sent to "GOL". If we take the word "ten", then $M(19, 4, 13)^T = (9, 16, 16)^T$ which corresponds to "JQQ". We note that "sen" and "ten" are sent to two very distant elements.

Even with just a 3 by 3 matrix, this technique is far superior to a simple substitution cipher. However, the computations can be quite lengthy even for a small message without a computer. For example, it would be hard for someone during World War I to compute these correctly for a long message even for a small matrix. With a computer, you can use a very large matrix which will give a fair amount of security using very simple linear algebra.

Other Techniques

We shall describe some other techniques that make substitution methods more effective.

The first technique is used to combat the effectiveness of frequency analysis. In this system, instead of replacing the most common element "e" by a single element, you replace it with a sequence of elements, $\epsilon_1, \epsilon_2, \epsilon_3, \ldots, \epsilon_{\kappa_e}$, where ϵ_i refers to symbol in a set. That is, on the ith occurrence modulo κ_e, you replace "e" with ϵ_i. Replace "t" with $\tau_1, \tau_2, \tau_3, \ldots, \tau_{\kappa_t}$ and "a" with $\alpha_1, \alpha_2, \alpha_3, \ldots, \alpha_{\kappa_a}$. Then continue, by replacing letters with a number of symbols that corresponds to its frequency in the language. In other words, κ is chosen for each letter so that each symbol has nearly the same frequency in a standard document.

While this technique is far superior to a substitution cipher, there are attacks against it. However, they are much more sophisticated than standard frequency analysis.

Another technique is to construct a one time pad. This is a generalization of the technique used in the Vigenère cipher. In this system, a prearranged sequence a_i of ones and zeros is made. The sequence is of length n. Then take a message and turn it into a sequence of ones and zeros, m_i. Then take $s_i = a_i + m_i$ (mod 2) and send it. The message is decoded by the same technique, namely, $s_i + a_i = m_i$ (mod 2). Essentially, there is a substitution cipher for every coordinate. It is as if you have a Vigenère cipher keyword of length n.

The difficulty with this system is that it can only be used once. Each time the system is used, it becomes less secure. Moreover, you need a very long sequence a_i to apply it to a meaningful message and you must exchange it ahead of time. It has been rumored that this type of system was used between the president of the United States and the premier of the Soviet Union during the cold war.

It is also possible to try and hide a secret message, which is called steganography. This technique is often used to send secret photos. Think of a photo as a matrix of zeros and ones, which are read to produce an image. In general, all computer files can be thought of as being of this form. Two people agree before communicating on which coordinates will be where the message is hidden. Then a much larger matrix is also shared. For example, two people take a series of very high-definition pictures which are viewed as n by n binary matrices $M_1, M_2, \ldots, M_s$. They agree on which k^2 coordinates are where the secret photo will be hidden. Then they separate. The spy can then take a photo with a resolution that fits into the k by k matrix. This matrix S_i is written on the pre-approved coordinates. Then the matrix $P_i = M_i + S_i \pmod 2$. The picture can then be sent to headquarters or simply uploaded to the Internet. At headquarters, they receive P_i and add M_i to it to obtain $P_i + M_i = S_i$. Then they have the secret photo taken by the spy.

One might believe that by adding the photo to M_i, this original photo will look altered. However, if k is small enough compared to n it is almost impossible to tell that the photo has been altered. Given also that millions of photos are uploaded to the internet every day, it would be difficult, if not impossible, to test every photo if there are some indications of this. To human sight, it is usually impossible to see any difference between M_i and P_i.

Example 12.2 We shall show a very simple example to illustrate this steganographic technique.

Assume the spy and headquarters agree on the following 2^2 coordinates in a 4 by 4 matrix:

$$\begin{pmatrix} & & & X \\ X & & & \\ & X & & \\ & & X & \end{pmatrix}.$$

Then assume the matrix M is given by the following:

$$\begin{pmatrix} 1 & 1 & 1 & 1 \\ 0 & 1 & 1 & 0 \\ 0 & 0 & 1 & 1 \\ 1 & 0 & 1 & 0 \end{pmatrix}.$$

Headquarters receive the following matrix:

$$\begin{pmatrix} 1 & 1 & 1 & 0 \\ 1 & 1 & 1 & 0 \\ 0 & 0 & 1 & 1 \\ 1 & 0 & 0 & 0 \end{pmatrix}.$$

and adds it to M. The result is the following matrix:

$$\begin{pmatrix} & & & 1 \\ 1 & & & \\ & 0 & & \\ & & 1 & \end{pmatrix}.$$

The information is read as $(1, 1, 0, 1)$ which is interpreted as

$$\begin{pmatrix} 1 & 1 \\ 0 & 1 \end{pmatrix}.$$

Exercises

1. Use the word Britain to construct a Playfair matrix and encode the message: "we shall meet under the apple tree" using this matrix.
2. Use the given ADFGX matrix and the word Berlin to encode the message: "we shall meet under the apple tree".
3. Determine the inverse of the affine cipher $\phi(x) = 3x + 9$.
4. Use the affine cipher $\psi(x) = 5x + 7$ to encode the message: "this food is delicious". Find the inverse function and use it to decode the message.
5. Count the number of possible affine ciphers using (mod 26). Prove your answer.
6. Try to encode the message: "this food is delicious" using the map $\psi(x) = 13x + 2$ to produce a convincing example as to why this does not work as a cipher.
7. Prove that the matrix M in the block cipher must satisfy $\gcd(\det(M), 26) = 1$.
8. Use frequency analysis to decode the following possibly treacherous message written in English:

MGXS HS LGX ZQKICX QY GKAES XJXSLC, HL WXZQAXC SXZX-CCEIP YQI QSX BXQBRX LQ FHCCQRJX LGX BQRHLHZER WESFC MGHZG GEJX ZQSSXZLXF LGXA MHLG ESQLGXI, ESF LQ ECCKAX EAQSV LGX BQMXIC QY LGX XEILG, LGX CXBEIELX ESF XOKER CLELHQS LQ MGHZG LGX REMC QY SELKIX ESF QY SELKIXÕC VQF XSLHLRX LGXA, E FXZXSL IXCBXZL LQ LGX QBHSHQSC QY AESTHSF IXOKHIXC LGEL LGXP CGQKRF FXZREIX LGX ZEKCXC MGHZG HABXR LGXA LQ LGX CXBEIELHQS.

MX GQRF LGXCX LIKLGC LQ WX CXRY-XJHFXSL, LGEL ERR AXS EIX ZIXELXF XOKER, LGEL LGXP EIX XSFQMXF WP LGXHI ZIX-ELQI MHLG ZXILEHS KSERHXSEWRX IHVGLC, LGEL EAQSV LGXCX EIX RHYX, RHWXILP ESF LGX BKICKHL QY GEBBHSXCC.Ð LGEL LQ CXZKIX LGXCX IHVGLC, VQJXISAXSLC EIX HSCLHLK-LXF EAQSV AXS, FXIHJHSV LGXHI UKCL BQMXIC YIQA LGX ZQSCXSL QY LGX VQJXISXF, LGEL MGXSXJXI ESP YQIA QY VQJX-ISAXSL WXZQAXC FXCLIKZLHJX QY LGXCX XSFC, HL HC LGX IHVGL QY LGX BXQBRX LQ ERLXI QI LQ EWQRHCG HL, ESF LQ HSCLHLKLX SXM VQJXISAXSL, REPHSV HLC YQKSFELHQS QS CKZG BIHSZHBRXC ESF QIVESHDHSV HLC BQMXIC HS CKZG YQIA, EC LQ LGXA CGERR CXXA AQCL RHTXRP LQ XYYXZL LGXHI CEYXLP ESF GEBBHSXCC.

12.2   German Enigma Machine

We now move to one of the most famous cases of breaking a cryptographic system, namely, the German Enigma machine that was used in the Second World War. It is one of the most important applications of combinatorial ideas (in terms of world events) that has ever occurred. Successful breaking of the Enigma code is often said to have reduced the Second World War by years. It was used in various forms by the different branches in the German military throughout the duration of World War II.

We begin by describing the main components of the machine and determine the number of possible initial configurations.

- Keyboard
  The operator types on the keyboard exactly as one would do for a usual typewriter. When one letter is typed, a different letter lights up. The one that the operator types is the message and the one that lights up is the ciphertext. The operation is reversed by the receiver, namely, they press the button that was sent as the ciphertext and the original message lights up. This gives the reader an indication of the ease of using this machine. The operators needed no special skills or intelligence, just the ability to type and write down which letters were lit up. This simplicity allowed it to be used throughout the German military.
- Plugboard
  The plugboard uses six pairs of plugs that can be used to interchange six pairs of letters. In other words, 6 pairs of letters are chosen from 26 letters and they are exchanged. There are

$$C(26,2)C(24,2)C(22,2)C(20,2)C(18,2)C(16,2)/6! = 100,391,791,500$$

  ways of interchanging six pairs of letters.
  The plugboard was not part of the original design but was added to significantly increase the number of initial possible configurations.
- Static Rotor
  The static rotor does nothing cryptographic. It performs a functional role in that it simply turns mechanical action into electrical contacts.
- Rotors
  The rotors are circular with letters written around the outside of the wheel. For the Enigma used by the army, there are three possible rotors that can be used in any order for the three rotor positions: right, middle, and left. Each has an inner and outer contact. Hence, there $C(5,3) = \frac{5!}{2!3!} = 10$ ways to pick three from five. The German navy used a machine with more rotors which increased the number of possibilities. We shall focus on the army version for our discussion. Each time a letter is typed the rotors move one position. These act the way the hands of a clock move. In other words, after a certain number of turns a rotor will cause the next rotor to move one space over. So three rotors can act in a manner similar to a second hand, minute hand, and hour hand. Each rotor can be placed with any of

the 26 letters at the top. Hence, there are $26^3 = 17576$ possible first positions for the three rotors and $3! = 6$ possible orderings of the three. Therefore, there are

$$C(5,3)(6)(26^3) = 1,054,560$$

possible initial positions for the rotors.

- Reflector

  There are two possible reflectors. The reflector must scramble the letters; otherwise, it would send the letter back on the same path un-encrypted.

- Lampboard

  The plugboard outputs the message to the lampboard. The operator records the letter and sends it in the message.

Essentially, what the Enigma machine does is use a different substitution cipher for each letter of the message. It takes advantage of the vast number of possibilities to make it extremely difficult to decode the message. Moreover, the substitution cipher used for each letter is different every day, and since the first thing operators did when sending a message was to choose a setting using the standard setting of the day, each cipher was different for each letter for each message. Therefore, all classical techniques of decoding were useless against this machine.

Next, we shall count the number of possible initial configurations of the Enigma machine.

Theorem 12.1 *The number of possible initial configurations is*

$$(1,054,560)(100,391,791,500) = 105,869,167,644,240,000.$$

Proof We saw that there were

$$C(26,2)C(24,2)C(22,2)C(20,2)C(18,2)C(16,2)/6! = 100,391,791,500$$

possible ways or arranging the plugboard and

$$C(5,3)(6)(26^3) = 1,054,560$$

possible ways of arranging the rotors.

This number is greater than 10^{18}, so it is clear that any type of straightforward attempt of running through the possibilities would not be feasible, not even with a high-speed computer. Even if we assume that a computer could try 10^9 possibilities, it would still take $10^9 = 1,000,000,000$ seconds to go through them. However, there are only $86,400$ in each day. It is precisely this combinatorial explosion that gave this cryptosystem its high level of security.

In actual practice, each operator had a book for the month with the codes that were going to be used for each day in that month. Then he put it on this initial setting and chose three letters (say r, f, v) and then sent $rfvrfv$ as his first message to a receiver. This would indicate to the receiver that r, f, and v were the settings for the three rotors. Of course, to do this the receiver must have the same codebook that the sender has.

The reason that rfv is repeated is to ensure that if a mistake is made they will realize it and ask for a retransmission. This prevents an entire message of garbled information from being sent. However, this duplication of the key was the biggest help to the cryptanalysts.

Let us assume the first six permutations that are performed by the Engima machine are A, B, C, D, E, F. Then if $xyzxyz$ gets sent to $dmqvbn$, it means that A interchanges x and d and D interchanges x and v. Therefore, the product DA (doing A first) interchanges d and v. We have eliminated the unknown x. The same can be done for permutations EB and FC.

Then, with enough data, we can write these permutations as products of cycles like

$$(dvpfkxgzyo)(eijmunqlht)(bc)(rw)(a)(s). \qquad (12.1)$$

These rely only on the daily settings of the plugboard and the rotors, not the established key that was found in the codebook. If the plugboard settings are changed while the initial positions of the rotors remain the same, then the cycle lengths of these permutations in Eq. 12.1 remain unchanged.

Polish mathematicians, Rejewski, Zygalski and Różycki, compiled a catalog of all 105, 456 initial settings (this was when there were only three rotors possible) along with the cycle lengths of the permutations formed from these before World War II began. They smuggled this information to Britain.

Essentially, the codebreakers had to determine what the possible permutations were based on the messages that were intercepted and the list of possible settings with lengths of permutations. Each day this process had to be repeated to find the daily keys. This attack was mechanized by building a "bombe", which was an early computer. The design is attributed to Alan Turing and is often considered to be one of the first electronic computers. Given a modern computer, this search would be quite simple, but for the technology of the day, this computation was quite daunting.

This codebreaking was done at Bletchley park in England. The British government gathered together mathematicians, problem-solvers, and others to find help the initial setting for each day, decode the messages, and translate them into English. The activities at Bletchley Park were kept secret for decades after the end of the war. Little credit was given to those who worked there nor to the Poles who had shown them how to attack the Enigma years earlier.

Exercises

1. How many possible plugboard settings would there be if 10 pairs of letters could be switched?

2. How many possible rotor settings would there be if there were k rotors chosen from n possible rotors, and each one could be placed in 26 different ways (meaning a different letter at the top)?
3. Assume that the switchboard could interchange 13 pairs of letters, and that there are five rotors chosen from eight possible rotors. How many initial configurations would there be?

12.3 Public-Key Cryptography

The notion of public-key cryptography was first brought out by Diffie and Hellman in [26]. The main goal of public-key cryptography is to construct a system which is so secure that the whole world can know how the information is encoded and yet the information can only be decoded by the person who constructs the system. This type of system is necessary for things like Internet commerce. Consider a company with a website that allows a customer to purchase items. In order to purchase the item, the customer must enter information like their card number, their address, and their name. This information must be kept secret. The company's website must send information to the computer of the user as to how this information is to be encrypted. Anyone can purchase items on this website so anyone can figure out how the information is encrypted. In other words, the company must tell the entire world how to send them information. The system must be designed so that even though you know how the information is encrypted, it is still impossible to decode the information. It is this type of system which allows for such business to occur over the Internet. We begin with some number-theoretic results which allow us to construct such systems.

We shall now prove the well-known Chinese remainder theorem, which we shall use to prove important results needed for the RSA cryptosystem.

Theorem 12.2 *Let n and m be natural numbers with $\gcd(n, m) = 1$. Then*

$$x \equiv a \pmod{m}$$
$$x \equiv b \pmod{n}$$

has a unique solution $\mod mn$.

Proof If $x \equiv a \pmod{m}$ then $x = a + mk$ for some $k \in \mathbb{Z}$. Then $x \equiv b \pmod{n}$ gives $a + mk \equiv b \pmod{n}$ and then $mk \equiv b - a \pmod{n}$. Then since m is relatively prime to n there is a unique solution for k, call it s. So $k = s + jn$ for some $j \in \mathbb{Z}$. Then $x = a + m(s + jn) = a + ms + jmn$ giving that $x \equiv a + ms \pmod{mn}$ and we have the result.

Example 12.3 We shall solve the following system:

$$x \equiv 5 \pmod 7$$
$$x \equiv 6 \pmod{11}.$$

If $x \equiv 5 \pmod 7$ we have that $x = 5 + 7k$ for some $k \in \mathbb{Z}$. Then we get $5 + 7k \equiv 6 \pmod{11}$ which simplifies to $7k \equiv 1 \pmod{11}$. Then multiplying both sides of the equation by 8 we get $k \equiv 8 \pmod{11}$. This gives that $k = 8 + 11s$ for some $s \in \mathbb{Z}$. Then

$$x = 5 + 7k = 5 + 7(8 + 11s) = 5 + 56 + 77s = 61 + 77s.$$

This gives the answer that $x \equiv 61 \pmod{77}$.

It is imperative that the moduli are relatively prime to ensure that there is a unique solution. For example, if one were to try to solve

$$x \equiv 1 \pmod 4$$
$$x \equiv 0 \pmod 8,$$

we see immediately that there are no solutions to this system. Moreover, if one were to try to solve

$$x \equiv 0 \pmod 4$$
$$x \equiv 0 \pmod 6,$$

we would have as answers both 0 and 12 $\pmod{24}$.

Applying the Chinese remainder theorem inductively, we get the following corollary.

Corollary 12.1 *Let* $m_1, m_2, \ldots, m_t$ *be natural numbers with* $\gcd(m_i, m_j) = 1$ *if* $i \neq j$. *Then*

$$x \equiv a_1 \pmod{m_1}$$
$$x \equiv a_2 \pmod{m_2}$$
$$\vdots$$
$$x \equiv a_t \pmod{m_t}$$

has a unique solution $\pmod{m_1 m_2 \cdots m_t}$.

Example 12.4 We shall solve the following system:

$$x \equiv 2 \pmod 5$$
$$x \equiv 5 \pmod 9$$
$$x \equiv 3 \pmod{11}.$$

We begin by solving the first two equations simultaneously. Starting with $x = 2 + 5k$, we write $2 + 5k \equiv 5 \pmod 9$. This gives $5k \equiv 3 \pmod 9$. Multiplying both sides by 2, we obtain $k \equiv 6 \pmod 9$. This gives $k = 6 + 9s$ and $x = 2 + 5(6 + 9s) = 32 + 45 s$. Therefore, the first two equations give that $x \equiv 32 \pmod{45}$. Then we set $32 + 45 s \equiv 3 \pmod{11}$, which gives $s \equiv 4 \pmod{11}$. Thus, $s = 4 + 11t$, and $x = 32 + 45(4 + 11t) = 212 + 495t$. Finally, we have $x \equiv 212 \pmod{495}$.

For a much more general version of the Chinese remainder theorem and its application to algebraic coding theory, see [29].

The following is a generalization of Theorem 2.10.

Theorem 12.3 *If* m *and* n *are relatively prime, then* $\phi(mn) = \phi(m)\phi(n)$.

Proof Let $\mathcal{U}_r$ be the set of units in $\mathbb{Z}_r$. Then if $a \in \mathcal{U}_{mn}$ then a is relatively prime to m and relatively prime to n. Since m and n are relatively prime, there is a unique solution to

$$x \equiv a \pmod{m}$$
$$x \equiv b \pmod{n}$$

for each $a \in \mathcal{U}_m, b \in \mathcal{U}_n$. Therefore, $\phi(mn) = \phi(m)\phi(n)$.

Example 12.5 Consider $n = 4(3) = 12$. Then numbers that are relatively prime to 4 are 1 and 3. By the previous theorem, we have $\phi(12) = \phi(4)\phi(3) = 2(2) = 4$. The numbers that are relatively prime to 3 are 1 and 2. Hence, writing numbers as pairs modulo 4 and 3 and applying the Chinese remainder theorem, we have $CRT(1, 1) = 1, CRT(3, 1) = 7, CRT(1, 2) = 5,$ and $CRT(3, 2) = 11$ which are the 4 numbers relatively prime to 12.

Theorem 12.4 *Let* p *be a prime. Then* $\phi(p^e) = (p - 1)p^{e-1}$.

Proof The only numbers that are not relatively prime to p^e between 1 and $p^e - 1$ are the $p^{e-1} - 1$ multiples of p. Thus, the number of elements relatively prime to p^e in this range is

$$p^e - 1 - (p^{e-1} - 1) = p^e - p^{e-1} = p^{e-1}(p - 1).$$

This gives the result.

Example 12.6 Consider $27 = 3^3$, the numbers that are not relatively prime to 27 between 1 and 26 are $3, 6, 9, 12, 15, 18, 21, 24$. Hence, the number that are relatively prime is $26 - 8 = 18 = (3 - 1)3^2$.

Using the two previous theorems, we can now find the Euler ϕ function for any positive integer.

Theorem 12.5 *Let* $n = \prod p_i^{e_i}$, *where* p_i *is a prime with* $p_i \neq p_j$ *if* $i \neq j$. *Then*

$$\phi(n) = \prod (p_i - 1) p_i^{e_i - 1}.$$

Proof From Theorem 12.3, we have $\phi(n) = \prod \phi(p_i^{e_i})$. Then by Theorem 12.4, we have $\phi(p_i^{e_i}) = p_i^{e_i - 1}$ which gives the result.

We know by Fermat's Little Theorem that if $a \not\equiv 0 \pmod{p}$ then $a^{p-1} \equiv 1 \pmod{p}$. But this does not mean that $a^i \not\equiv 1$ for $0 < i < p - 1$. For example, consider any prime $p > 3$, then $(p - 1)^2 \equiv 1 \pmod{p}$ with $2 < p - 1$. We are interested in those a for which this does not happen and we make the following definition.

Definition 12.1 Let p be a prime. An element $a \pmod{p}$ is a *primitive root* $\pmod{p}$ if $a^{p-1} \equiv 1 \pmod{p}$ but $a^i \not\equiv 1 \pmod{p}$ for $0 < i < p - 1$.

Example 12.7 Consider the prime p. We shall give the smallest value for which $a^i \equiv 1 \pmod{7}$ for each of the non-zero values of a. We have $2^3 \equiv 1 \pmod{7}$, $3^6 \equiv 1 \pmod{7}$, $4^3 \equiv 1 \pmod{7}$, $5^6 \equiv 1 \pmod{7}$, and $6^2 \equiv 1 \pmod{7}$. Therefore, 3 and 5 are primitive elements $\pmod{7}$.

Notice that in the previous example, there are two primitive roots and $\phi(7 - 1) = 2$. We shall prove this result in general but first we need a lemma.

Lemma 12.1 *Let* a *be a primitive root* $\pmod{p}$, *then* a^i *is a primitive root if and only if* $\gcd(i, p - 1) = 1$.

Proof Assume $\gcd(i, p - 1) = 1$ and $(a^i)^j \equiv 1 \pmod{p}$. Then $a^{ij} \equiv 1 \pmod{p}$ giving that $ij \equiv 0 \pmod{p - 1}$. Since $\gcd(i, p - 1) = 1$ this implies that $j \equiv 0 \pmod{p - 1}$. Then a^i is a primitive root $\pmod{p}$.

Assume $\gcd(i, p - 1) \neq 1$ and $(a^i)^j \equiv 1 \pmod{p}$. Then there exists $k < p - 1$ with $ik \equiv 0 \pmod{p - 1}$. Then $(a^i)^k \equiv a^{ik} \equiv 1$. Then a^i is not a primitive root $\pmod{p}$.

Theorem 12.6 *The number of primitive elements* $\pmod{p}$ *is* $\phi(p - 1)$.

Proof First, we know that the multiplicative group of $\mathbb{F}_p$ is a cyclic group. This means that the generator of the cyclic group a has the property that $a^{p-1} \equiv 1 \pmod{p}$ and $a^h \not\equiv 1 \pmod{p}$ for $0 < h < p - 1$.

By Lemma 12.1, we have that a^i is a primitive root if and only if $\gcd(i, p - 1) = 1$. Then the number of primitive roots is the number of elements less than $p - 1$ that are relatively prime to $p - 1$, that is, $\phi(p - 1)$.

Example 12.8 Let $p = 11$, here 2 is a primitive root. We give the values of 2^i for all i.

i	0	1	2	3	4	5	6	7	8	9	10
2^i	1	2	4	8	5	10	9	7	3	6	1

The numbers less than 10 that are relatively prime to 10 are $1, 3, 7, 9$. Therefore, by Lemma 12.1, we have that $2, 8, 7,$ and 6 are the four primitive roots $\pmod{11}$. We note that $\phi(10) = 4$.

Definition 12.2 Let a be a primitive root $\pmod{p}$. Define $L_a(b) = i$ where i is the unique number with $a^i = b$ and $0 \le i < p - 1$.

This function is defined for each prime but it is read $\pmod{p-1}$. In general, it is very difficult to compute for large primes.

Example 12.9 Continuing Example 12.8, we have that $L_2(3) = 8$, $L_2(5) = 10$, and $L_2(9) = 6$.

Assume a is a primitive root $\pmod{p}$. Then the following rules apply. The proofs are identical to the standard proofs for logarithms. Notice that these are all done $\pmod{p-1}$ since $\phi(p) = p - 1$.

- $L_a(bc) = L_a(b) + L_a(c) \pmod{p-1}$;
- $L_a(bc^{-1}) = L_a(b) - L_a(c) \pmod{p-1}$;
- $L_a(b^e) = eL_a(b) \pmod{p-1}$.

12.3.1 RSA Cryptosystem

We shall now describe one of the most celebrated and profitable results in cryptography. The discovery prompted a large industry using this technique to produce secure Internet commerce. The RSA system was first given by Rivest, Shamir, and Adleman in [75]. Most likely, every time you purchase something online, you are using the RSA system. Virtually every time, very secure information is passed electronically, you are using the RSA system. The strength of the system comes from the fact that it is very difficult to factor large numbers, especially if they are the product of two very large primes. While the theoretical foundations for these systems are, in fact, very old, it would not have been possible to use this system before the advent of computers. This is because the computations that are used involve numbers with over 100 digits, so performing these computations by hand would have been impossible.

We begin by assuming we have a message that we wish to transmit. Any message that one may want to send can be changed into a numeric message. There are numerous techniques for doing this and they need not be cryptographically secure to apply public-key encryption. Therefore, we shall assume that all of our messages are numeric and we shall now show how to use this technique to send a numeric message M.

Let p and q be primes. In practice, to make the system secure, these primes will often be over 100 digits long. For added security, the primes can always be made to be of longer length. It may seem that it is difficult to find primes that are over 100 digits long, but in reality, it is quite easy. One can use a test like the Miller-Rabin primality test, to find very large primes in a short period of time with very limited computer time. It is not difficult to write a short program on a personal computer to find this size of prime quite quickly. Then, we let $n = pq$. In practice, the number n will have more than 200 digits. Again, this is no problem for any computer. It is immediate from Theorems 12.3 and 12.4 that $\phi(n) = (p-1)(q-1)$. If we know p and q, this is trivial to compute, but if you are handed a 200-digit number that is the product of two large primes, then without knowing this factorization, it is computationally infeasible to find $\phi(n)$. Next choose an integer e that is relatively prime to $\phi(n)$, of course this is easy to find. There are practical considerations for choosing e but the system will work for an e that is relatively prime to $\phi(n)$; however, certain values of e make the system less secure.

Since e is relatively prime to $\phi(n)$, there exists a d with $ed \equiv 1 \pmod{\phi(n)}$ by Lemma 2.2. That is, there exists some integer k with $ed = \phi(n)k + 1$. In general, this d is easy to find. One could use the Euclidean algorithm or one could simply use Euler's theorem, noting that $e^{\phi(\phi(n))} \equiv 1 \pmod{\phi(n)}$ so that $e^{\phi(\phi(n))-1}e \equiv 1 \pmod{\phi(n)}$. Then $d = e^{\phi(\phi(n))-1}$.

To send the message M, we calculate $M^e \pmod{n}$ and send this value. Note that the value of e is told to anyone who wants it. If you want to send a message, all you need to know is e and n and, of course, your message M.

The receiver, who knows d, takes

$$(M^e)^d \equiv M^{ed} \pmod{n}$$
$$\equiv M^{\phi(n)k+1}$$
$$\equiv (M^{\phi}(n))^k M^1$$
$$\equiv 1^k M \equiv M.$$

For the implementation of this technique, we allow e and n to be known by anyone. Then, if you want to send M to the receiver, only the receiver knows p and q, and therefore only they know $\phi(n)$ and d. Here, the key is public so that anyone can send the information but only the receiver can decode it.

Public for RSA	Private for RSA
n, e	$\phi(n), d$

The key to this system is that given the large number $n = pq$, in practice it is at least $(10^{100})^2 = 10^{200}$, it is very hard to factor the number. If you were able to factor n, then it is easy to compute $\phi(n) = (p-1)(q-1)$. It remains an extremely difficult problem to factor large numbers even with a high-powered computer. One possible way would be if a quantum computer were constructed, then Shor's algorithm would factor the numbers, see [79] for a complete description.

Example 12.10 We shall take a small example to illustrate the cryptosystem. We shall use very small primes so that the reader can more clearly see the computations. Let $p = 131$ and $q = 157$. It is easy to verify that both of these numbers are prime. Compute $n = 131(157) = 20567$. Then $\phi(n) = (p-1)(q-1) = 130(156) = 20280$. We choose $e = 37$. Then computing $37^{\phi(\phi(n))-1} = 4933$, we obtain $37(4933) \equiv 1 \pmod{20280}$. Then $d = 4933$.

Let assume that the message we want to send is 22. Therefore, we send $M^e \pmod{n} = 22^{37} \pmod{20567} = 13524$. When 13524 is received, the receiver computes $(M^e)^d \pmod{n} = 13524^{4933} \pmod{20567} = 22$ and the correct message is received.

12.3.2 El Gamal Cryptosystem

The next public-key cryptosystem that we shall discuss is the El Gamal Cryptosystem. It is equally as powerful as RSA but is not used as much in actual practice, see [31]. While the security of RSA comes from the difficulty in factoring large numbers, the security of El Gamal comes from the difficulty of taking a discrete logarithm.

We begin by letting p be a prime number. In practice, the prime p generally has over 200 digits. As we stated before when discussing RSA it is not difficult to find such primes. We recall that by Euler's theorem or Fermat's Little Theorem that $\alpha^{p-1} \equiv 1 \pmod{p}$ for all $\alpha \not\equiv 0 \pmod{p}$. An element α is a primitive root $\pmod{p}$ if $\alpha^k \not\equiv 1 \pmod{p}$ for all k with $1 \leq k \leq p-2$. We let α be a primitive root $\pmod{p}$.

As before, we can assume that any message can be seen as a numeric message; therefore, let M be an integer that represents the message $0 \leq M < p$.

Let a be an integer which the receiver keeps secret. This integer must be between 1 and $p-1$. Compute $\beta = \alpha^a$.

The sender chooses a secret integer k and computes $r \equiv \alpha^k \pmod{p}$ and $t \equiv \beta^k M \pmod{p}$. Then (r, t) is sent to the receiver.

The receiver computes

$$tr^{-a} \equiv \beta^k M(\alpha^k)^{-1} \equiv (\alpha^a)^k M\alpha^{-ak} \equiv M \pmod{M}.$$

Then the message has been correctly decoded with perfect secrecy, since knowing α and β, it is computationally impractical to compute the value of a.

Public for El Gamal	Private for El Gamal
p, α, β	a

Example 12.11 We shall illustrate this cryptosystem with a small example. Let $p = 101$. We choose $\alpha = 3$ as our primitive root. We take $a = 71$ as our secret parameter.

Then $\beta = 3^{71} \equiv 51 \pmod{101}$. We let $p = 101, \alpha = 3, \beta = 51$ be made public. Assume the message we want to send is 43, we pick our secret $k = 13$. Then $r = \alpha^k$ $(\mod p) = 3^{13} \pmod{101} = 38$. Then $t = \beta^k M \pmod{p} = 51^{13} \pmod{101} = 59$.

The pair $(38, 59)$ is sent. The receiver computes $tr^{-a} \pmod{p} = 59(38^{-71})$ $(\mod 101) = 59(38^{29}) \pmod{101} = 424$, and the correct message is received.

12.3.3 Diffie-Hellman Key Exchange

Another technique which uses these ideas is the Diffie-Hellman key exchange. The point of the Diffie-Hellman key exchange is not to exchange a large amount of information but rather to establish a key that both parties can use for a system to exchange information secretly. Assume Alice and Bob would like to establish a key. They choose a prime p and an integer a which is a primitive root $\pmod{p}$.

Alice chooses an exponent α and computes $a^{\alpha} \pmod{p}$ and sends it to Bob. Bob chooses an exponent β and computes a^{β} and sends it to Alice.

When Alice receives a^{β} she computes $(a^{\beta})^{\alpha} \pmod{p} \equiv a^{\alpha\beta} \pmod{p}$. When Bob receives a^{α} he computes $(a^{\alpha})^{\beta} \pmod{p} \equiv a^{\alpha\beta} \pmod{p}$. Then both Alice and Bob both have the number $a^{\alpha\beta} \pmod{p}$, and no one else could have this number without knowing α and β and being able to compute a discrete logarithm. Notice that Alice never knows what β is and Bob never knows what α is; however, they both know $a^{\alpha\beta}$ which they can then use as a key.

Example 12.12 We shall show an example using small numbers. Let $p = 13$ and let $a = 2$ be a primitive root in $\mathbb{Z}_{13}$. Alice can take $\alpha = 5$ and Bob can take $\beta = 7$. Then Alice sends $2^5 = 6$ and Bob sends $2^7 = 11$. Alice receives 11 and computes $11^5 = 7$. Bob receives 6 and computes $6^7 = 7$.

Exercises

1. Solve the following systems of equations using the Chinese remainder theorem:
   (a)

   $$x \equiv 4 \pmod{7}$$
   $$x \equiv 5 \pmod{11}$$

   (b)

   $$x \equiv 3 \pmod{13}$$
   $$x \equiv 7 \pmod{17}$$

(c)

$$x \equiv 12 \quad (\text{mod } 19)$$
$$x \equiv 15 \quad (\text{mod } 22)$$

(d)

$$x \equiv 3 \quad (\text{mod } 8)$$
$$x \equiv 2 \quad (\text{mod } 13)$$
$$x \equiv 11 \quad (\text{mod } 15)$$

2. Determine all primitive roots (mod 11) and (mod 13).
3. Determine $L_2(a)$ for all values of a modulo 13.
4. Determine the number of primitive roots modulo 1601.
5. Determine $\phi(n)$ for all n with $2 \le n \le 25$.
6. Using the prime 17, with $a = 11$, encode the message 9 using the El Gamal cryptosystem. Determine r and t using $k = 7$, then decode the received message.
7. Using primes $p = 11$ and $q = 17$, with $e = 7$, encode the message $M = 23$ using the RSA cryptosystem. Determine the parameter d and use it to decode the message back to M.
8. Find a primitive root (mod 23) and use it in the Diffie-Hellman key exchange to exchange the key 19 showing all steps.

12.4   McEliece Cryptographic System

In this section, we shall describe a cryptographic system that withstands quantum attacks and is based on the theory of error-correcting codes. It was first described by R. J. McEliece in [63]. This system uses the decoding of vectors using a given code to uncover the message. It is certainly not used as much as the RSA or ElGamal systems, but if a quantum computer were built then these two systems would become immediately useless, as a quantum computer would defeat both of them. However, at present, no quantum algorithm defeats this system so it would be a reliable system should this occur.

We shall assume that the receiver chooses an $[n, k, d]$ binary code, with $d = 2t + 1$, and finds a generator matrix for the code denoted by G. Of course, G has k rows and n columns. We know that if $d = 2t + 1$ then the code can correct t errors correctly using nearest neighbor decoding. This is a key step in implementing the system. Then choose an invertible k by k matrix S and an n by n permutation matrix P. Then take $G' = SGP$ which is again a k by n matrix. The matrix G' is made public.

The sender wants to send a message $\mathbf{x}$, which is a vector of length k. The sender picks a vector $\mathbf{e}$ with Hamming weight no more than t which acts as the error vector. The sender forms $\mathbf{y} = \mathbf{x}G' + \mathbf{e}$ and sends this as the message. If someone intercepts this message, they must be able to determine what this error message is in order to determine the sent message. It is obvious then that the system must be set up with fairly large n and d to make the system effective.

The receiver performs the following computations:

1. Compute $\mathbf{y}' = \mathbf{y}P^{-1} = \mathbf{x}SG + \mathbf{e}'$, where $\mathbf{e}' = \mathbf{e}P$. Note that $\mathbf{e}'$ is also a weight t vector since P is a permutation matrix.
2. Apply the decoding algorithm to $\mathbf{y}'$ and correct it to $\mathbf{x}'$.
3. Compute the vector $\mathbf{u}$ such that $\mathbf{u}G = \mathbf{x}'$.
4. Compute $\mathbf{x} = \mathbf{u}S^{-1}$.

It is immediate that the code we choose must have an efficient decoding algorithm to make the system effectively computable. The security comes from the difficulty of decoding $\mathbf{y}'$ to $\mathbf{x}'$. In order to make this difficult, it is best to have a very large d. As an example, the Goppa codes are often used, as their parameters are $n = 2^m$, $k = n - mt$, and $d = 2t + 1$.

Example 12.13 The only codes that we have specified an easily computable decoding algorithm in this text are the Hamming codes. Therefore, we shall use the $[7, 4, 3]$ Hamming code to illustrate the McEliece cryptographic system. Recall that the code is given by the parity check matrix:

$$H = \begin{pmatrix} 0\,0\,0\,1\,1\,1\,1 \\ 0\,1\,1\,0\,0\,1\,1 \\ 1\,0\,1\,0\,1\,0\,1 \end{pmatrix}. \tag{12.2}$$

We shall use the matrix H in the decoding algorithm. We shall take as the generator matrix the following matrix:

$$G = \begin{pmatrix} 1\,0\,0\,0\,0\,1\,1 \\ 0\,1\,0\,0\,1\,0\,1 \\ 0\,0\,1\,0\,1\,1\,0 \\ 0\,0\,0\,1\,1\,1\,1 \end{pmatrix}. \tag{12.3}$$

We note the first three rows correspond to lines in $PG_2(\mathbb{F}_2)$ and the last row corresponds to a hyperoval. Next, we choose a 4 by 4 invertible matrix S:

$$S = \begin{pmatrix} 1\,0\,0\,1 \\ 0\,1\,1\,1 \\ 1\,1\,0\,0 \\ 1\,1\,1\,1 \end{pmatrix} \tag{12.4}$$

whose inverse is

$$S^{-1} = \begin{pmatrix} 0\,1\,0\,1 \\ 0\,1\,1\,1 \\ 1\,1\,1\,0 \\ 1\,1\,0\,1 \end{pmatrix}. \tag{12.5}$$

Next, we need a 7 by 7 permutation matrix:

$$P = \begin{pmatrix} 0\,0\,0\,0\,0\,1\,0 \\ 0\,0\,0\,1\,0\,0\,0 \\ 0\,1\,0\,0\,0\,0\,0 \\ 0\,0\,0\,0\,0\,0\,1 \\ 1\,0\,0\,0\,0\,0\,0 \\ 0\,0\,1\,0\,0\,0\,0 \\ 0\,0\,0\,0\,1\,0\,0 \end{pmatrix}. \tag{12.6}$$

We note that there is precisely one 1 in each row and column. Next we compute $G' = SGP$, which is

$$G' = \begin{pmatrix} 1\,0\,0\,0\,0\,1\,1 \\ 1\,1\,0\,1\,0\,0\,1 \\ 1\,0\,1\,1\,0\,1\,0 \\ 1\,1\,1\,1\,1\,1\,1 \end{pmatrix}. \tag{12.7}$$

This matrix is made public.

Let us suppose we wish to send the message $\mathbf{x} = (1,1,1,0)$. We compute $\mathbf{x}G'$ which is $(1,1,1,0,0,0,0)$. Next we pick an error message with $t = 1$ error. We choose $\mathbf{e} = (0,0,0,0,0,1,0)$. This gives $\mathbf{y} = (1,1,1,0,0,1,0)$ which is the sent message.

The receiver takes this message and begins by computing $\mathbf{y}' = \mathbf{y}P^{-1} = (1,0,1,0,1,1,0)$. Apply the decoding algorithm to thiss vector, that is, compute $H(\mathbf{y}'^{T})$, which gives $\begin{pmatrix} 0 \\ 0 \\ 1 \end{pmatrix}$. The error $\mathbf{e}'$ is $(1,0,0,0,0,0,0)$ since it occurs in the first coordinate. Hence, the corrected vector is $\mathbf{x}' = (0,0,1,0,1,1,0)$. Then $\mathbf{u} = (0,0,1,0)$ since $\mathbf{u}G = \mathbf{x}'$. Finally, we compute $\mathbf{u}S^{-1} = (1,1,1,0)$ which was the sent message. Hence, we have secretly sent this message via this public-key encryption method.

Exercises

1. Prove that a binary k by k matrix is invertible if and only if the determinant is 1.
2. Construct an example of the McEliece cryptographic system using any binary code.
3. Compute the determinant of the matrix S given in Eq. 12.4.
4. Verify that S and S^{-1} given in Eqs. 12.4 and 12.5 are inverse matrices.
5. Find the inverse of the matrix P given in Eq. 12.6.

Games and Designs

Games have often given rise to interesting combinatorial questions. For example, determining whether there is a winning strategy in chess has proven to be an extremely difficult mathematical question. Part of the reason for the difficulty is the combinatorial explosion that we have noticed before. Chess is played on an 8 by 8 board with 32 pieces. These seemingly small numbers are large enough to make it so that the number of possible games is far larger than any person or computer can handle. It is estimated that the number of possible chess games is larger than 10^{120} which is far larger than any computer could evaluate even if running for billions of years. Moreover, the game theory behind programming a computer to play chess well requires a great deal of combinatorics as well as some very clever programming.

Games can also be useful in developing geometric and combinatorial intuition. In this chapter, we shall describe a game using designs and a technique for determining if the game has a winning strategy. The game we shall describe is a generalization of a game played on finite planes that was introduced in the paper "Tic-Tac-Toe on a Finite Plane" [19]. This game was further developed in [20,21].

We shall begin by describing some basics of combinatorial game theory. Then we shall describe the specifics of the game. We introduce weight functions and use them to examine the strategies of the game.

Some games that people play often have a probabilistic aspect to them. For example, the play in poker or bridge is dependent upon the cards that each player receives in the deal. These games require a different approach than a game that is purely combinatorial. As an example, in a game of poker a player can attempt to bluff his opponent by making him think that his hand is much better than it actually is. However, the same sort of bluffing is not possible in a game of checkers where all of the information of the game is available to each player. In these games of perfect information, there is no way to convince your opponent that your position is any different than it actually is because nothing is hidden. There is a large literature devoted to

S. T. Dougherty, *Combinatorics and Finite Geometry*, Springer Undergraduate Mathematics Series

probabilistic games and combinatorial games. There are also numerous applications of these games in economics and political science. For a reference on game theory, we suggest Rapoport's book "Two-Person Game Theory" [70] or Straffin's book "Game Theory and Strategy" [88] and for a description of combinatorial games we suggest "Winning Ways for your Combinatorial Plays" by Elwyn R. Berlekamp, John Horton Conway, and Richard K. Guy [8].

We shall only consider games like chess or checkers that have no probabilistic aspect and where there is perfect information. In other words, each player knows the complete situation of the game, there is nothing hidden. Games can also be played by various numbers of players. We consider only games played by two people.

We shall describe some of the basic terms of game theory. A strategy is an algorithm for play. Specifically, given a specific state of the game a strategy will determine what play the player should make at that point. It is called a winning strategy if it assures that its application will give a win for the player, and it is called a drawing strategy if it assures that a player can force a draw. For the positional games that we will describe, there are two possibilities. The first is that the first player has a winning strategy. The second is that the second player can force a draw. The reason for this is what is called strategy stealing. If the second player could have a winning strategy, then the first player could simply assume that a play has been made and then use the strategy that the second player would have used as their own. While this does not work for all games, it does apply for the positional games that we describe.

13.1 The Game

Most children have played the game Tic-Tac-Toe. The game is played on a 3 by 3 grid where the first player marks with an X and the second with an O, and they continue until someone completes a horizontal or vertical line or a diagonal or until each space is filled. Anyone who has played the game realizes after about 5 min. that the second player can force a draw. We shall generalize this game to be played over a design.

Let $D = (\mathcal{P}, \mathcal{B}, \mathcal{I})$ be a $t - (v, k, \lambda)$ design. The first player plays by marking a point in the design with an X. Player 2 plays by marking a point with an O. A player wins the game if they have marked all of the points on a block of the design. The game is a draw if all points have been marked and no block has all points with the same mark.

As an example consider the game played on the projective plane of order 2. Throughout this section, we shall denote the i-th play of the first player by X_i and the i-th play of the second player by O_i.

This is a win for the first player since X_1, X_3 and X_4 complete a line in the plane (Fig. 13.1).

We shall now show that X has a winning strategy on the projective plane of order 2. The first player places an X on any of the seven points. Then O places a point on any of the six remaining points. At this point the first player places an X on the remaining point of the line through the point chosen by the first and second player.

Fig. 13.1 A win for X

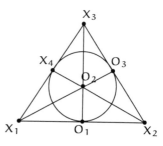

Fig. 13.2 A win for X

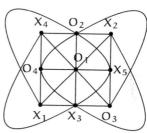

This line now has two points with an X and one with an O. The second player now places an O anywhere. The first player places an X on the line between the two points containing an O. This prevents O from completing a line. Now there is a line between X_1 and X_3 and a line between X_2 and X_3. The second player can block only one of these lines, and hence X_4 will be a win for X. This is exactly what happened in the previous example.

We shall now show that the second player has a drawing strategy on the biplane of order 2. The second player simply places their first two points anywhere. After the first players' third move, there are two remaining points on the biplane. At least one of them will not be on the line through O_1 and O_2 when viewing the biplane as a projective plane of order 2. The second player places an O on that point. Then the first player cannot complete a block since each block is the complement of a line. Thus, the second player has a drawing strategy on the biplane of order 2. For the biplane of order 1, it is easy to see that there is a drawing strategy since a block has three points and the first player only has two plays.

Consider the game in Fig. 13.2 on the affine plane of order 3. Even though it has the same number of points as the usual game of Tic-Tac-Toe, it is quite different. The usual strategy of the second player fails to force a draw.

A draw is a configuration on the plane such that each point is marked X or O in which no block has each point marked the same.

The fundamental question involved in this game is for which designs is there a winning strategy and for which are there drawing strategies.

1. Prove that the second player can force a draw for the standard game of Tic-Tac-Toe.
2. Use the strategy exhibited in the previous to prove that the first player has a winning strategy on the affine plane of order 3.
3. Prove that there are no draws in the projective plane of order 2 and the affine planes of orders 2 and 3.
4. Prove that first player has a winning strategy on the affine plane of order 2.

13.2 Weight Functions

In this section, we shall describe a technique which will enable us to determine in most cases whether there is a winning strategy or not. The technique uses weight functions defined on a stage of the game. This is a standard technique used in the theory of games. For example, in chess, a weight function can be developed to decide what the next move should be. Namely, it gives a weight to the strengths and weaknesses of a given move in terms of which pieces you can now attack and from which you are now open to attack. Therefore, when teaching a computer to play chess, the weight of the board for all possible moves can be examined and then the move that gives the most favorable weight is chosen. We shall make a similar type of weight function, which we shall use to determine where the next move should be. We will be able to show that in many instances, playing by the results of this weight function, we will be able to show that either the first player has a winning strategy or the second player will have a drawing strategy.

The following result of Erdös and Selfridge [32] gives conditions for when the second player can force a draw in many positional games, of which our version of Tic-Tac-Toe is an example.

Theorem 13.1 *Let* P *be a finite set of points. Let* $L = \{A_1, A_2, \ldots, A_m\} \subseteq 2^P$ *be a collection of winning sets, where* $|A_i| = k$ *for all* $1 \leq i \leq m$. *Two players take turns selecting points from* P *until either one player has all of the points of a winning set or there is a draw. If* $m < 2^{k-1}$, *then the second player can force a draw.*

The proof, given later, employs the following sets and weight functions. We shall adopt the techniques to our own setting. Namely, we let $D = (\mathcal{P}, \mathcal{B}, \mathcal{I})$ be a design. Assume the current state of the game is $[(X_1, \ldots, X_i), (O_1, \ldots, O_{i-1})]$, that is, X_i is the point marked by the first player in the i-th move and O_i is the point marked by the second player in the i-th move. Let $\mathcal{B}$ represent the set of blocks and let $\mathcal{B}_i = \{\ell - \{X_1, \ldots, X_i\} \mid \ell \in L, \ \ell \cap \{O_1, \ldots, O_{i-1}\} = \emptyset\}$. In other words, the set $\mathcal{B}_i$ consists of all partial blocks that are not blocked by the second player as of the i-th move of the second player. Recall that if there is an O placed on even one point of the block, then it is blocked since the first player can not win by completing this block.

Therefore, $\mathcal{B}_i$ contains sets of points, s, where $s \subseteq b$ for some block $b \in \mathcal{B}$. Additionally, we define $\mathcal{B}_\infty$ to be the set when no more moves can be made in the game. It is clear that if $\emptyset \in \mathcal{B}_i$, then the first player has won since there is a block that has an X on each of its points.

We define $\mathcal{P}_i = \mathcal{P} - \{X_1, \ldots, X_i, O_1, \ldots, O_{i-1}\}$. Namely, it is the set of unmarked points at the i-th move of the second player.

For points $p, q \in \mathcal{P}_i$, we define weight functions. What they are essentially giving is the weight given that the position of play is $\mathcal{B}_i$ and then a player marks the point q. They are defined by

$$w(\mathcal{B}_i) = \sum_{s \in \mathcal{B}_i} 2^{-|s|}, \tag{13.1}$$

$$w(q|\mathcal{B}_i) = \sum_{s \in \mathcal{B}_i, q \in s} 2^{-|s|}, \tag{13.2}$$

and

$$w(p, q|\mathcal{B}_i) = \sum_{s \in \mathcal{B}_i, \{p, q\} \subseteq s} 2^{-|s|}. \tag{13.3}$$

Theorem 13.2 *We have that in any design,*

$$w(\mathcal{B}_i) - w(\mathcal{B}_{i+1}) = w(O_i|\mathcal{B}_i) - w(X_{i+1}|\mathcal{B}_i) + w(X_{i+1}, O_i|\mathcal{B}_i). \tag{13.4}$$

Proof The left side of the equation gives the change in the weight from $\mathcal{B}_i$ to $\mathcal{B}_{i+1}$. The right side represents a standard inclusion-exclusion counting technique. It adds the weights of all partial blocks that contain O_i since those blocks are now blocked. It subtracts the weights of the blocks through X_{i+1} since these weights are now $2^{-|s+1|}$ in the new configuration. Then it adds the weight of the block that contained both points (if such a block had not yet been blocked). □

Notice that the closer the first player is to winning on a certain block, the higher its corresponding weight is.

The relevant point is that if $\emptyset \in \mathcal{B}_i$, then $w(\mathcal{B}_i) \geq 1$ since $2^0 = 1$ and the first player has won the game. If $w(\mathcal{B}_i) \leq 1$, then $\emptyset \notin \mathcal{B}_i$ and so the first player has not won. This leads to a natural way for the second player to play the game. Namely, the second player plays by minimizing the value of $w(\mathcal{B}_i)$ after the second player makes their i-th move. If it is possible for the second player to keep this weight less than 1, then they have an algorithm to force a draw.

Consider Eq. 13.4. The second player can make their choice so that

$$w(O_i|\mathcal{B}_i) \geq w(X_{i+1}|\mathcal{B}_i).$$

In fact, this is how the second player chooses their play by maximizing this weight. This shows that the second player can play so that the weights $w(\mathcal{B}_i)$ are a non-increasing sequence. Therefore, if the second player can ever get the weight of the game below 1, they can be assured of forcing a draw.

We can now give a proof of Theorem 13.1. Although we have defined everything in terms of designs, it is easy to see that it extends to the more general notation of the theorem.

Proof of Theorem 13.1

Proof Consider the weight of the game after the first player has made a single move. We notice that some blocks have an X on a single point and some do not. Hence, the size of the sets in $\mathcal{B}_1$ are either $k-1$ or k. But we know that $2^{-(k-1)} > 2^{-k}$. This gives the following:

$$
\begin{aligned}
w(\mathcal{B}_1) &\leq \sum_{b \in \mathcal{B}} 2^{-(k-1)} \\
&= |B| 2^{-(k-1)} \\
&= \frac{m}{2^{k-1}}.
\end{aligned}
$$

Since we have assumed that $m < 2^{k-1}$, we know that $\frac{m}{2^{k-1}} < 1$ and thus $w(\mathcal{B}_1) < 1$. The above discussion now gives that the second player has a drawing strategy. $\square$

It is an immediate corollary that every design where the second player has a drawing strategy can be marked so that each point has either an X or an O and no block has all the same marking. In other words, we are able to show that a draw exists. We are able to conclude this combinatorial information from a game-theoretic argument.

We shall examine the situation for various designs that we have encountered in the text.

13.2.1 The Game on Planes

The situation for planes was described in the paper [19]. We have already seen that there is a winning strategy for the first player for the projective plane of order 2 and the affine planes of orders 2 and 3.

Projective planes of order n have $n^2 + n + 1$ winning sets and $n + 1$ points in a winning set. For projective planes of order $n > 4$, we have $n^2 + n + 1 < 2^n$ and so there is a drawing strategy for the second player in these cases.

Affine planes of order n have $n^2 + n$ winning sets and n points in a winning set. For projective planes of order $n > 6$, we have $n^2 + n + 1 < 2^n$ and so there is a drawing strategy for the second player in these cases. Of course, there is no plane of order 6, but this tells us that there is a drawing strategy for all affine planes of order $n \geq 7$.

This leaves us with the cases of the affine planes of orders 4 and 5 and the projective planes of orders 3 and 4.

We can examine the weight function more closely to examine the projective plane of order 4. After the first player places an X, there are n^2 lines with no markings and $n + 1$ with a single X. Then

$$
\begin{aligned}
w(\mathcal{B}_i) &= (n^2)2^{-(n+1)} + (n+1)2^{-n} \\
&= 16(2^{-5}) + 5(2^{-4}) \\
&= \frac{13}{16} < 1.
\end{aligned}
$$

This gives that there is a drawing strategy on this plane.

The affine plane of order 4 is perhaps the most interesting case. It has draws yet the first player has a winning strategy. The fact that the first player has such a strategy was shown by computer, independently by three different people (see [19], for details). As far as actual play, this is probably the most interesting plane on which to play.

We summarize the results in the following theorem.

Theorem 13.3 *The second player can force a draw in all projective planes of order $n > 2$, and the first player has a winning strategy on the projective plane of order 2. The second player can force a draw in all affine planes of order $n > 4$, and the first player has a winning strategy on all affine planes of order $n \leq 4$.*

13.2.2 The Game on Biplanes

Recall that a biplane of order n is a symmetric $2 - (\frac{n^2+3n+4}{2}, n+2, 2)$ design.

Theorem 13.4 *There is a drawing strategy for all biplanes.*

Proof We have shown that there is a drawing strategy for the biplane of order 1 and for the biplane of order 2 (which will follow from the next argument as well). The size of the blocks is $n + 2$, and the number of blocks (winning sets) is $\frac{n^2+3n+4}{2}$. If $n \geq 2$ we have that

$$
\frac{n^2 + 3n + 4}{2} < 2^{n+1}
$$

and so by Theorem 13.1 there is a drawing strategy.   □

13.2.3 The Game on Symmetric Designs

In this subsection, we shall only consider $\lambda > 2$ since we have already considered the other cases.

In a symmetric design, there are $\frac{(n+\lambda)(n+\lambda-1)}{\lambda}$ blocks and each block has $n + \lambda$ points.

It is easy to see that for $\lambda > 2$ we have $\frac{(n+\lambda)(n+\lambda-1)}{\lambda} < 2^{n+\lambda-1}$ for all n. Then applying Theorem 13.1 gives the following result.

Theorem 13.5 *The second player has a drawing strategy on all symmetric designs with* $\lambda > 2$.

13.2.4 The Game on Hadamard Designs

We shall now consider the case of Hadamard 3-designs and 2-designs. The Hadamard 3-design of order 1 consists of four points with six blocks each containing two points. In fact, every subset of size 2 of the four points is a block. It is easy to see that the first player has a winning strategy since their second move guarantees a win. In fact, there are no draws on this design. The Hadamard 2-design of order 1 consists of three points and three blocks that consist of one point. Here, the first move gives a victory and again there are no draws. The Hadamard 2-design is a $(7, 3, 1)$ design, and we have already seen that X has a winning strategy there.

We need to examine the Hadamard 3-design of order 2 which is a $3 - (8, 4, 1)$ design with 14 blocks. We see that 14 is not less than $2^3 = 8$, and so the theorem does not apply. The design can be thought of as the completion of the Hadamard 2-design which is a $(7, 3, 1)$ design. That is, we add a single point, and the new blocks are the old blocks adjoined with this new point and the compliment of old blocks. A win for X would consist of four points on a block and X has exactly four plays. Hence, each of the plays must be on a single block. We know that through any three points there is one block. After the first player places his third X, the second player need only marks the remaining point on this block if it has not already been marked with an O. Thus, the second player has a drawing strategy in this case.

A Hadamard 3-design is a design with $8n - 2$ blocks which are winning sets each of size $2n$. For $n > 2$, we have $8n - 2 < 2^{2n-1}$ and so by Theorem 13.1 there is a drawing strategy for the second player.

A Hadamard 2-design is a design with $4n - 1$ blocks which are winning sets each of size $2n - 1$, for $n > 2$ we have $4n - 1 < 2^{2n-2}$ and so by Theorem 13.1 there is a drawing strategy for the second player.

We summarize the results in the following theorem.

Theorem 13.6 *The first player has a winning strategy for the Hadamard 3-designs of order 1, and the first player has a winning strategy for the Hadamard 2-designs of orders 1 and 2. For* $n > 2$, *the second player has a drawing strategy for all Hadamard*

2-designs and for $n > 1$ *the second player has a drawing strategy for Hadamard 3-designs of order* n.

13.2.5 The Game on Nets

We shall examine the situation for k-nets of order n.

Theorem 13.7 *If a k-net of order* n *extends to a k'-net of order* n *where there is a drawing strategy, then there is a drawing strategy on the k-net.*

Proof Since the point sets are the same, you can simply apply the same strategy. Since a winning set in the net is also a winning set in the larger net, then the strategy that forces a draw on the larger net will force a draw on the net as well. □

To apply Theorem 13.1, we need

$$kn < 2^{n-1}.$$

We note that k is bounded above by $n + 1$ and so for $n > 6$ there is a drawing strategy on all k-nets of order n.

We know for $n = 6$ the maximum value of k is 3. In this case $18 < 32$ and so there is a drawing strategy on all nets of order 6 by Theorem 13.1.

For $n = 5$ if $k = 4$ or 5, then we know the net completes to an affine plane. Then by Theorem 13.7, the net must have a drawing strategy. For $k = 3$ we have $15 < 16$ and so by Theorem 13.1 there is a drawing strategy. Theorem 13.7 gives that there is a drawing strategy for $k = 1$ and $k = 2$. Therefore, there is a drawing strategy for all k-nets of order 5.

Consider a 3-net of order 3. The first player places X_1 arbitrarily and the same for O_1. Then the first player places X_2 anywhere that is on a line with X_1 but not on a line with O_1. Then O_2 must be placed to block this line. The first player can then place X_3 on the intersection of the lines through X_1 and X_2 and through X_1 and X_3. The second player can only block one of these lines, and the first player has a winning strategy. Notice that this game has only one more winning set than the usual game of Tic-Tac-Toe, but in reality it is quite different. In the usual game, the center point is special because it has 4 winning sets through it. On a 3-net of order 3 no such point exists. Of course we know that there is a winning strategy on a 4-net of order 3.

Consider a 3-net of order 4. After the first player places X_1 arbitrarily, there are three lines that have three unmarked points and nine that have no marked points. This gives that

$$w(\mathcal{B}_1) = 3(\frac{1}{8}) + 9(\frac{1}{16}) = \frac{15}{16} < 1$$

and so the second player has a drawing strategy.

We have already seen that the first player has a winning strategy for a 5-net of order 4.

We summarize the results in the following theorem.

Theorem 13.8 *The second player can force a draw on all k-nets of order n with n > 4 and for all k-nets with k ≤ 2 except for the 2-net of order 2 in which there is a winning strategy. There is a winning strategy for all k-nets of order 3 if k > 2. There is a drawing strategy for k-nets of order 4 for 1 ≤ k ≤ 3 and a winning strategy for k = 5.*

13.2.6 The Game on Steiner Triple Systems

For the Steiner triple system with three points, there is obviously a drawing strategy. Wherever the second player places, their O blocks the only block in the system.

Steiner triple systems are at least a 2-design. Thus for $v > 3$ the same winning strategy that the first player used on the projective plane of order 2 will work. That is, the first two moves are arbitrary, then X_2 is placed on the block with X_1 and O_1. After any placement of O_2, the first player places X_3 on the block with O_1 and O_2. Then, the second player can only block either the block between X_1 and X_3 or the block between X_2 and X_3 but not both. Then, the first player has a winning strategy.

We summarize this as follows.

Theorem 13.9 *There is a drawing strategy for the second player for the Steiner triple system with $v = 3$ and a winning strategy for the first player for all others.*

Exercises

1. Change the definition of a win from marking all $n + 1$ points on a line on a finite projective plane, to marking at least n points on a line of a finite projective plane. Determine if the first player has a winning strategy or the second player has a drawing strategy in this scenario for the projective planes of order 2, 3, and 4.
2. Change the definition of a win from marking all n points on a line on a finite affine plane, to marking at least $n - 1$ points on a line of a finite projective plane. Determine if the first player has a winning strategy or the second player has a drawing strategy in this scenario for the affine planes of order 2, 3, and 4.
3. Prove that there is a drawing strategy on a 1-net of order 2 and a winning strategy on all other nets of order 2.
4. Prove that there is a drawing strategy of k-nets of order 3 when $0 < k \leq 2$.
5. Prove that there is a drawing strategy on the projective plane of order 3. Proceed as follows. Assume the first three plays are made arbitrarily. Then, the second player makes his second play by placing an O on the line through X_1 and X_2 unless it has already been done in which case O_2 is placed off this line. Then X_3 can be placed arbitrarily giving rise to four possible configurations. Compute $w(\mathcal{B}_3)$ in each of these cases and show that it is less than 1.

6. Construct a draw on the affine plane of order 4.
7. Using the same initial configurations in the previous, show that the second player has a drawing strategy on the affine plane of order 5. The situation becomes much more complicated, and it will probably be necessary to use a computer to evaluate all possible configurations.
8. Describe the drawing strategy on the biplanes of order 3 without using the weight functions, that is, describe it so that a person playing could use the technique to force a draw.
9. Prove that draws exist for all affine planes of order $n \geq 3$.
10. Show how to use a transversal to a k-net of order n, $k \leq n$, to construct a draw on the net.

Epilogue

14

Early in the text we encountered the following diagram:

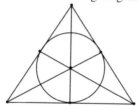

Throughout, we have seen this image as a representation of many of the objects studied in the text. Specifically, it represented the following:

- the projective plane of order 2,
- the biplane of order 2,
- the Hadamard 2-design coming from a Hadamard matrix of size 8,
- a Steiner triple system on seven points,
- the geometric representation of the $[7, 4, 3]$ Hamming code, and
- a design formed from a difference set.

It is precisely this kind of interesting object which fuels the study of finite geometry and combinatorics. We can see many different ways of approaching the object and many different classes of objects for which this is the first example in the class. Often in the study of mathematics in general and combinatorics, in particular, different paths of inquiry will lead to the same ideas.

Numerous different questions in geometry, statistics, algebra, and information theory have given rise to many interesting combinatorial structures which we have attempted to illustrate in this text. Students wishing to further their study have a variety of paths to follow.

S. T. Dougherty, *Combinatorics and Finite Geometry*,
Springer Undergraduate Mathematics Series

1. For a classical text on combinatorics, see Hall's text *Combinatorial theory* [44].
2. To learn more about Latin squares, see the book *Latin squares and their applications* by Keedwell and Dénes [54]. To learn about the connection between Latin squares and many topics of discrete mathematics, see *Discrete mathematics using Latin squares* by Laywine and Mullen [59].
3. To learn more about coding theory, see the text by Huffman and Pless, *Fundamentals of error-correcting codes* [49].
4. To learn about the connection between coding theory, finite geometry and designs, see the text by Assmus and Key, *Designs and their codes* [4].
5. To learn more about cryptography, see the text *Introduction to cryptography with coding theory* by Trappe [92], which is an introduction to many topics in cryptography. To learn more in this area, see the advanced text *Cryptography. Theory and practice* by Stinson [85].
6. To learn more about combinatorial games, see the landmark text by Berlekamp, Conway and Guy, *Winning ways for your mathematical plays* [8].
7. To learn more about combinatorial designs, see Stinson's text *Combinatorial designs. Constructions and analysis* [86]. For an encyclopedic description of designs, see the text edited by Colbourn and Dinitz *The CRC handbook of combinatorial designs* [24].
8. To learn more about matrices in combinatorics, see the text by Brualdi and Ryser *Combinatorial matrix theory* [15].

At present, numerous open questions remain in almost every topic examined in this text, and a large number of new applications are being introduced in information theory, cryptography, computer science, and statistics every year. In general, mathematics often progresses in an attempt to solve major questions. It has often been suggested that what a pure mathematician wants is a ridiculously difficult problem on which incremental advances can be made.

The list of guiding problems that we have encountered have included the following:

1. Determine the maximum number of MOLS of order n for all n.
2. Determine for which n there exists affine and projective planes of order n.
3. Determine the number of non-isomorphic planes for a given order n.
4. Determine when non-Desarguesian planes exist.
5. Determine precisely for which parameters there exists t-(v, k, λ) designs.
6. Determine for which n Hadamard matrices of order n exist.

Solutions to Selected Odd Problems

Section 1.1

(1) If n is even, use the function from the odds to the evens defined by $f(i) = i + 1$. This is a bijection and shows that the cardinalities of the two sets are the same. If n is odd, then use f^{-1}. This function is not surjective, as it leaves n untouched. Therefore, there are more odds than evens in this case.

(3) The image of f must have the same cardinality as the domain since the function is injective. This gives that the cardinality of the image and codomain are the same. Then, since the sets are finite, we have the result.

(5) If it were finite then Exercise 3 would give that the map must be surjective.

(7) If f is a bijection from A to $\{1, 2, \ldots, n\}$ and g is a bijection from A to $\{1, 2, \ldots, m\}$, then $f \circ g^{-1}$ is a bijection from $\{1, 2, \ldots, m\}$ to $\{1, 2, \ldots, n\}$ which gives the result.

(9) We have $C(p, i) = \frac{p!}{(p-i)!i!} = \frac{p(p-1)!}{(p-i)!i!}$. If $0 < i < p$, then the integer p does not appear in the denominator, and hence there is nothing to cancel the p in the numerator. Since we know this value must be an integer we have the result.

(11) We have

$$C(2n, n) = C(2n - 1, n) + C(2n - 1, n - 1)$$
$$= \frac{(2n - 1)!}{(n - 1)!n!} + \frac{(2n - 1)!}{n!(n - 1)!}$$
$$= 2\frac{(2n - 1)!}{(n - 1)!n!}.$$

S. T. Dougherty, *Combinatorics and Finite Geometry*,
Springer Undergraduate Mathematics Series

(13) The number is 2^{k-1}.

(15) The total number is $35^6 = 1,838,265,625$.

(17) $C(17,6) = 12,376$.

(19) $C(16,5) = 4,368$.

(21) The number of games is $9! = 362,880$. The number of final configurations is $C(9,5) = 126$.

(23) Straightforward induction.

(25) $105,869,167,644,240,000$. See Chap. 12 for a complete description.

(27) There are 12 ways where switching works and 6 ways where switching does not work. Therefore one should switch. See Chap. 9 for a complete description of this problem.

Section 1.2

(1) 840.

(3) $10,080, 3,326,400, 34,650$.

Section 1.3

(1) We have $3(8) + 1 = 25$ and $27 > 25$.

(3) $5(6) + 1 = 31$.

(5) The boxes are 4-unit-length line segments forming the line. Then $3(4) + 1 = 13$.

(7) $7(11) + 1 = 78$.

Section 1.4

(1)(a) $(1,2,3,4,5)$.

  (b) $(3,1,5,2,4)$.

  (c) $(1,6)$.

  (d) (1).

  (e) (1).

  (f) $(2,5)(1,3)(4,6)(7,8)$.

(3) We have $(\sigma\tau)(\tau^{-1}\sigma^{-1}) = \sigma(\tau\tau^{-1})\sigma^{-1} = \sigma\sigma^{-1} = (1)$.

(5) Prove it by contradiction. If it could be, then write it in both ways and equate them. Then multiply both sides by the inverse of one of them. Then you would have that the identity is written as an odd permutation which is a contradiction.

(7) To prove that it is injective, $ax = ay$ implies $x = y$ by multiplication on both sides by a^{-1}, which is possible since it is a field. Surjection follows by $a(a^{-1}y)) = y$. The second proof follows from the fact that σ_a and σ_b send 1 to a and b, respectively. Hence, they cannot be equal unless a and b are equal.

(9) The counts are $26!$, $(26^2)!$ and $(26^3)!$.

Section 1.5

(1) 6.

(3) 13.

(5) 8.

(7) Assume the limit is L. Then

$$\lim_{n \to \infty} \frac{\mathcal{F}_{n+1}}{\mathcal{F}_n} = L$$

$$\lim_{n \to \infty} \frac{\mathcal{F}_n + \mathcal{F}_{n-1}}{\mathcal{F}_n} = L$$

$$\lim_{n \to \infty} 1 + \frac{\mathcal{F}_{n-1}}{\mathcal{F}_n} = L$$

$$1 + \frac{1}{L} = L$$

$$L + 1 = L^2.$$

Applying the quadratic formula gives $L = \varphi$.

B.2   Problems of Chap. 2

Section 2.1

(1) (a) 1 (b) 4 (c) 9 (d) 1 (e) 8.

(3) Apply the binomial theorem to get

$$(a + b)^p = \sum_{k=0}^{p} C(p, k) x^{p-k} y^k.$$

If $0 < k < p$, then $C(p, k) = \frac{p!}{k!(p-k)!}$ is divisible by p, and hence is 0 (mod p). If $k = 0$ we get x^p and if $k = p$ we get y^p. This gives the result.

(5) Consider $1 + 1 = 2$ in $\mathbb{Z}_{12}$.

(7) If $ab \equiv 0 \pmod{n}$, n a prime, then if $a \not\equiv 0 \pmod{n}$ there exists a^{-1} satisfying $a^{-1}a = 1$. Multiplying both sides of the equation, we get $a^{-1}ab = 0$ which gives $b = 0$. Therefore, one of a or b must be 0.

(9) In $\mathbb{Z}_4$ the solutions are $1, 3$. In $\mathbb{Z}_8$ the solutions are $1, 3, 5, 7$. In $\mathbb{Z}_{16}$ the solutions are $1, 7, 9, 15$.

(11) (a) 1 (b) 3 (c) 1 (d) 4 (e) 12.

(13) (a) 15 (b) 8, 20 (c) 13 (d) 8 (e) 4, 9, 14.

Section 2.2

(1) Straightforward verification.

(3) Straightforward verification.

(5) It is immediate that $\Phi(ab) = (ab)^p = a^p b^p = \Phi(a)\Phi(b)$. Exercise 3 in Sect. 2.1 gives that $\Phi(a+b) = \Phi(a) + \Phi(b)$.

(7) The element $(0,1)$ is not $(0,0)$ and has no multiplicative inverse. Therefore, it is not a field.

(9) We have: $1^{-1} = 1, 2^{-1} = 6, 3^{-1} = 4, 4^{-1} = 3, 5^{-1} = 9, 6^{-1} = 2, 7^{-1} = 8,$ $8^{-1} = 7, 9^{-1} = 5, 10^{-1} = 10.$

(11) The unique solution is $x = a^{-1}(c - b)$.

(13) If $ba^{-1} = 1$. Then multiplying by a on the right gives $b = a$.

(15) Let $c = a^{-1}b$.

(17) If $x^3 + x + 1$ were reducible then either 1 or 0 would be a root since it would factor into a degree 1 and a degree 2 polynomial. However, neither is a root. The field of order 8 is then $\mathbb{F}_2[x]/\langle x^3 + x + 1\rangle = \{0, 1, x, x+1, x^2, x^2+1, x^2+x, x^2+x+1\}$.

(19) It is easy to see that the case $u^2 = 0$ is not a field since there are zero divisors. For the case $u^2 = u$, it is not a field since $u(1+u) = 0$ and there are zero divisors. The case where $u^2 = 1$ gives $(1+u)^2 = 0$, and we have zero divisors and is therefore not a field. The case where $u^2 = 1 + u$ gives the field of order 4 realized as $\mathbb{F}_2[x]/\langle x^2 + x + 1\rangle$.

Section 2.3.1

(1) The recursive definition is $a_0 = 0, a_{n+1} = a_n + 4n^3 + 6n^2 + 4n + 1$. The proof is straightforward.

(3) The final digit must be 0, 1, or 4.

Section 2.3.2

(1) Straightforward.

Section 2.3.3

(1) If $k = n$, $\begin{bmatrix} n \\ n \end{bmatrix} = 1$ and $\begin{Bmatrix} n \\ k \end{Bmatrix} = 1$. If $k = n - 1$, $\begin{bmatrix} n \\ n-1 \end{bmatrix} = n$ and $\begin{Bmatrix} n \\ k \end{Bmatrix} = n$.

Section 2.3.4

(1) Simply place each disk on a separate pole, placing the largest on the desired pole. Then each disk is moved twice except the largest giving $2n - 1$ moves.

B.3   Problems of Chap. 3

Section 3.1

(1) Let $L_{ij} = i + j$. Then $L_{ij} = L_{ij'} \Rightarrow i + j = i + j' \Rightarrow j = j'$ and $L_{ij} = L_{i'j} \Rightarrow i + j = i' + j \Rightarrow i = i'$. Therefore, it is a Latin square.

(3) Let $L_{ij} = ij$. Then $L_{ij} = L_{ij'} \Rightarrow ij = ij' \Rightarrow j = j'$ and $L_{ij} = L_{i'j} \Rightarrow ij = i'j \Rightarrow i = i'$. Therefore, it is a Latin square.

(5) Begin by showing that a Latin square of order 3 is equivalent to either

$$\begin{pmatrix} 0\ 1\ 2 \\ 1\ 2\ 0 \\ 2\ 0\ 1 \end{pmatrix} \text{ or } \begin{pmatrix} 0\ 1\ 2 \\ 2\ 0\ 1 \\ 1\ 2\ 0 \end{pmatrix}$$

and the result follows.

(7)

$$\begin{pmatrix} 0\ 1\ 2\ 3\ 4 \\ 1\ 2\ 3\ 4\ 0 \\ 2\ 3\ 4\ 0\ 1 \\ 3\ 4\ 0\ 1\ 2 \\ 4\ 0\ 1\ 2\ 3 \end{pmatrix} \quad \begin{pmatrix} 0\ 1\ 2\ 3\ 4 \\ 2\ 3\ 4\ 0\ 1 \\ 4\ 0\ 1\ 2\ 3 \\ 1\ 2\ 3\ 4\ 0 \\ 3\ 4\ 0\ 1\ 2 \end{pmatrix}$$

$$\begin{pmatrix} 0\ 1\ 2\ 3\ 4 \\ 3\ 4\ 0\ 1\ 2 \\ 1\ 2\ 3\ 4\ 0 \\ 4\ 0\ 1\ 2\ 3 \\ 2\ 3\ 4\ 0\ 1 \end{pmatrix} \quad \begin{pmatrix} 0\ 1\ 2\ 3\ 4 \\ 4\ 0\ 1\ 2\ 3 \\ 3\ 4\ 0\ 1\ 2 \\ 2\ 3\ 4\ 0\ 1 \\ 1\ 2\ 3\ 4\ 0 \end{pmatrix} .$$

(9) Let $L_{ij} = ai + j$. Then $L_{ij} = L_{ij'} \Rightarrow ai + j = ai + j' \Rightarrow j = j'$ and $L_{ij} = L_{i'j} \Rightarrow ai + j = ai' + j \Rightarrow i = i'$ since $\gcd(a, n) = 1$. Therefore, it is a Latin square. When $\gcd(a, n) \neq 1$ and $ai - ai' = 0$ gives $a(i - i') = 0$. Let $i - i'$ be b where $ab = 0$, and we have that it is not a Latin square in that case.

(11) Straightforward verification.

(13) There exists $p_1^{e_1} - 1$ MOLS of order $p_1^{e_1}$ and $p_2^{e_2} - 1$ MOLS of order $p_2^{e_2}$. Then the minimum number is the minimum number of these two numbers by the cross-product construction.

Section 3.2

(1) We have $L_{ij} = i + j = j + i = L_{ji}$.

(3) $L_{ij} = ij = ji = L_{ji}$.

(5) Fix any two columns. Each pair in these two columns must have a unique number in the third column. The result follows from this fact.

(7) Let $\perp$ indicate that two squares are orthogonal.

$$L = \begin{pmatrix} 0\ 1\ 2 \\ 2\ 0\ 1 \\ 1\ 2\ 0 \end{pmatrix} \perp \begin{pmatrix} 0\ 1\ 2 \\ 1\ 2\ 0 \\ 2\ 0\ 1 \end{pmatrix}$$

$$L^T = \begin{pmatrix} 0\ 2\ 1 \\ 1\ 0\ 2 \\ 2\ 1\ 0 \end{pmatrix} \perp \begin{pmatrix} 0\ 1\ 2 \\ 1\ 2\ 0 \\ 2\ 0\ 1 \end{pmatrix}$$

$$L^r = \begin{pmatrix} 0\ 1\ 2 \\ 2\ 0\ 1 \\ 1\ 2\ 0 \end{pmatrix} \perp \begin{pmatrix} 0\ 1\ 2 \\ 1\ 2\ 0 \\ 2\ 0\ 1 \end{pmatrix}$$

$$L^c = \begin{pmatrix} 0\ 1\ 2 \\ 1\ 2\ 0 \\ 2\ 0\ 1 \end{pmatrix} \perp \begin{pmatrix} 0\ 1\ 2 \\ 2\ 0\ 1 \\ 1\ 2\ 0 \end{pmatrix}$$

$$(L^c)^T = \begin{pmatrix} 0\ 1\ 2 \\ 1\ 2\ 0 \\ 2\ 0\ 1 \end{pmatrix} \perp \begin{pmatrix} 0\ 2\ 1 \\ 1\ 0\ 2 \\ 2\ 1\ 0 \end{pmatrix}$$

$$(L^r)^T = \begin{pmatrix} 0\ 2\ 1 \\ 1\ 0\ 2 \\ 2\ 1\ 0 \end{pmatrix} \perp \begin{pmatrix} 0\ 1\ 2 \\ 1\ 2\ 0 \\ 2\ 0\ 1 \end{pmatrix}.$$

Section 3.3

(1) The unique completed squares is

$$\begin{pmatrix} 1\ 2\ 3\ 4\ 5 \\ 4\ 3\ 2\ 5\ 1 \\ 3\ 1\ 5\ 2\ 4 \\ 5\ 4\ 1\ 3\ 2 \\ 2\ 5\ 4\ 1\ 3 \end{pmatrix}.$$

(3) Straightforward following the hint.

B.4   Problems of Chap. 4

Section 4.1

(1) The first part follows trivially by replacing every occurrence of the word line with point and vice versa in the proofs. For the second part, an affine plane of order n has n^2 points and $n^2 + n$ lines. Reversing the roles gives the wrong number of points and lines.

(3) Given a point p, there are $n + 1$ lines through p. Each line has n points on it distinct from p. Then the number of points is $n(n + 1) + 1 = n^2 + n + 1$.

(5) Let p be a point and ℓ be a line off p. Through each of the n points on ℓ, there is a line through p and that point. Then there is a unique line m through p parallel to ℓ. This gives $n + 1$ lines through p.

(7) Through A are the lines $\{A, E\}, \{A, C\}$ and $\{A, G\}$. Through G are the lines $\{G, E\}, \{G, C\}$ and $\{A, G\}$. Through E are the lines $\{E, G\}, \{C, E\}$ and $\{A, E\}$. Through C are the lines $\{A, C\}, \{C, E\}$ and $\{G, C\}$.

(9) Through any two points the line between them is the restriction of the line between them from the projective plane. Given two lines, if they meet at a point on L (must be only one) they are now parallel; otherwise, they meet at the same point they met in the projective plane.

(11) The affine plane:

π	p_1	p_2	p_3	p_4	p_5	p_6	p_7	p_8	p_9
ℓ_1	1	1	1	0	0	0	0	0	0
ℓ_2	0	0	0	1	1	1	0	0	0
ℓ_3	0	0	0	0	0	0	1	1	1
ℓ_4	1	0	0	1	0	0	1	0	0
ℓ_5	0	1	0	0	1	0	0	1	0
ℓ_6	0	0	1	0	0	1	0	0	1
ℓ_7	1	0	0	0	1	0	0	0	1
ℓ_8	0	1	0	0	0	1	1	0	0
ℓ_9	0	0	1	1	0	0	0	1	0
ℓ_{10}	0	0	1	0	1	0	1	0	0
ℓ_{11}	1	0	0	0	0	1	0	1	0
ℓ_{12}	0	1	0	1	0	0	0	0	1

The projective plane:

Π	p_1	p_2	p_3	p_4	p_5	p_6	p_7	p_8	p_9	p_{10}	p_{11}	p_{12}	p_{13}
L_1	1	1	1	0	0	0	0	0	0	1	0	0	0
L_2	0	0	0	1	1	1	0	0	0	1	0	0	0
L_3	0	0	0	0	0	0	1	1	1	1	0	0	0
L_4	1	0	0	1	0	0	1	0	0	0	1	0	0
L_5	0	1	0	0	1	0	0	1	0	0	1	0	0
L_6	0	0	1	0	0	1	0	0	1	0	1	0	0
L_7	1	0	0	0	1	0	0	0	1	0	0	1	0
L_8	0	1	0	0	0	1	1	0	0	0	0	1	0
L_9	0	0	1	1	0	0	0	1	0	0	0	1	0
L_{10}	0	0	1	0	1	0	1	0	0	0	0	0	1
L_{11}	1	0	0	0	0	1	0	1	0	0	0	0	1
L_{12}	0	1	0	1	0	0	0	0	1	0	0	0	1
L_{13}	0	0	0	0	0	0	0	0	0	1	1	1	1

Section 4.2

(1) The number of points is $|F \times F| = n^2$. The number of lines is the possibilities of x for $x = c$ which is n and the number of possibilities for m and b in $y = mx + b$ which is n^2. Therefore, the number of lines is $n^2 + n$. The number of points on a line is the number of elements (c, y) where c is fixed which is n or the number of (x, y) satisfying $y = mx + b$. For a given m and b and x, there is a unique y and there are n choices for x; therefore, there are n points on these lines. Through any point (c, d), it is on the line $x = c$ and fixing m there is a unique b satisfying $d = mc + b$. Therefore, there are $n + 1$ lines through a point. There is a parallel class for each m and one for the horizontal lines. Therefore, there are $n + 1$ parallel classes.

(3) We have $(a, b, c) = 1(a, b, c)$ so it is reflexive. If $(a, b, c) = \lambda(d, e, f)$ then $(d, e, f) = \frac{1}{\lambda}(a, b, c)$ so it is symmetric. If $(a, b, c) = \lambda(d, e, f)$ and $(d, e, f) = \mu(g, h, i)$, then $(a, b, c) = \lambda\mu(g, h, i)$ and so it is transitive. Therefore, it is an equivalence relation.

(5) We have $|\mathcal{P}| = \frac{n^3 - 1}{n - 1} = n^2 + n + 1$.

(7) The points are $(0, 0, 1), (0, 1, 0), (0, 1, 1), (0, 1, 2), (1, 0, 0), (1, 0, 1),$ $(1, 0, 2), (1, 1, 0), (1, 1, 1), (1, 1, 2), (1, 2, 0), (1, 2, 1), (1, 2, 2)$.

The lines are

$$L_1 = [0, 0, 1] = \{(0, 1, 0), (1, 0, 0), (1, 1, 0), (1, 2, 0)\};$$
$$L_2 = [0, 1, 0] = \{(1, 0, 0), (0, 0, 1), (1, 0, 2), (1, 0, 1)\};$$
$$L_3 = [0, 1, 1] = \{(0, 1, 2), (1, 0, 0), (1, 1, 2), (1, 2, 1)\};$$
$$L_4 = [0, 1, 2] = \{(1, 0, 0), (0, 1, 1), (1, 1, 1), (1, 2, 2)\};$$
$$L_5 = [1, 0, 0] = \{(0, 1, 1), (0, 1, 0), (0, 0, 1), (0, 1, 2)\};$$
$$L_6 = [1, 0, 1] = \{(1, 0, 2), (0, 1, 0), (1, 1, 2), (1, 2, 2)\};$$
$$L_7 = [1, 0, 2] = \{(1, 0, 1), (0, 1, 0), (1, 2, 1), (1, 1, 1)\};$$
$$L_8 = [1, 1, 0] = \{(0, 0, 1), (1, 2, 0), (1, 2, 1), (1, 2, 2)\};$$
$$L_9 = [1, 1, 1] = \{(1, 1, 1), (1, 2, 0), (1, 0, 2), (0, 1, 2)\};$$
$$L_{10} = [1, 1, 2] = \{(0, 1, 1), (1, 2, 0), (1, 0, 1), (1, 1, 2)\};$$
$$L_{11} = [1, 2, 0] = \{(0, 0, 1), (1, 1, 0), (1, 1, 1), (1, 1, 2)\};$$
$$L_{12} = [1, 2, 1] = \{(1, 1, 0), (1, 0, 2), (1, 2, 1), (0, 1, 1)\};$$
$$L_{13} = [1, 2, 2] = \{(0, 1, 2), (1, 2, 2), (1, 0, 1), (1, 1, 0)\}.$$

Section 4.4

(1) Straightforward.

Section 4.5

(1) Once the grid is formed, there is a unique way to complete a plane.
(3) Follow the hint.

B.5   Problems of Chap. 5

Section 5.1

(1) Start at the lower left and end at the lower right.
(3) The following graph has neither an Euler tour nor an Euler cycle.

(5) For five vertices use C_5. For four vertices use the following:

(7) A tour only exists if $n = 2$. Cycles exist if and only if $n > 1$ is odd.
(9) Prove it by induction by taking a graph that does not have an Euler cycle and adjoin one vertex but only connect it to one vertex of the graph.

Section 5.2

(1) The tour $v_1, v_2, \ldots, v_n$ is an Euler tour if and only if $\Phi(v_1), \Phi(v_2), \ldots, \Phi(v_n)$ is an Euler tour.
(3) If n is odd then every vertex has even degree; therefore, it has an Euler cycle. If n is even then every vertex has odd degree; therefore, it does not have an Euler cycle. Only K_2 has two odd degree vertices with the rest even.
(5) The wheel graph has $2(n - 1)$ edges and the cycle graph has n edges.
(7) The number of edges is ab.
(9) The partition is $\{A\}, \{B, C, D\}, \{E\}, \{F, G, H\}$.
(11) The quotient graph of both is either L_2 or D_2.

(13) Use C_5 and the following:

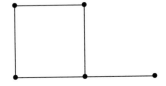

(15) Map the vertices by any bijection ψ. Then $\{v_1, v_2\}$ is mapped to $\{\psi_{v_1}, \psi(v_2)\}$.

Section 5.2

(1) 3.
(3) Use K_4 drawn as follows:

K_4

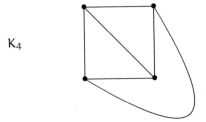

(5) $x(x-1)^2 + x(x-1)(x-2)^2$.
(7) Follow the hint.

Section 5.3

(1) The in and out degrees are as follows:

vertex	in degree	out degree
0	1	2
1	2	1
2	3	2
3	1	1
4	2	3

(3) If $v_1, v_2, \ldots, v_n$ is a tour with $(v_i, v_{i+1}) \in E$. This implies $(v_1, v_3) \in E$ by the transitivity of the relation. Therefore, $v_1, v_3, \ldots, v_n$ is a tour. Then if $v_1, v_i, v_{i+1}, \ldots, v_n$ is a tour then $(v_1, v_{i+1}) \in E$ by the transitivity of the relation. Then by induction $(v_1, v_n) \in E$.

(5) Number the vertices in any manner. From vertex 1 to the next vertex there are n choices for a tour. Then after a tour including i vertices, then there are $n - i$ choices in continuing the tour. Then by induction the number of Hamilton tours is $(n - 1)!$.

B.6   Problems of Chap. 6

Section 6.1

(1) They are not linearly independent: $(1, 2, 1) + 2(2, 0, 1) + 2(2, 2, 0) = (0, 0, 0)$.

(3) The set $(1, 0, 0, \ldots, 0), (0, 1, 0, 0, \ldots, 0), \ldots (0, 0, \ldots, 0, 1)$ is a basis of $\mathbb{F}_q^n$.

(5) If $\mathbf{v}_i = \sum_{j \neq i} \alpha_j \mathbf{v}_j$ then $\sum_{j \neq i} \alpha_j \mathbf{v}_j - \mathbf{v}_i = \mathbf{0}$ is a non-trivial linear combination summing to $\mathbf{0}$. If $\sum \alpha_i \mathbf{v}_i = \mathbf{0}$, then choose an i with $\alpha_i \neq 0$ and $\mathbf{v} = -(\frac{1}{\alpha_i}) \sum_{j \neq i} \alpha_j \mathbf{v}_j$.

(7) If the span of $\mathbf{v}_1, \ldots, \mathbf{v}_s$ is not all of V, then add any vector $\mathbf{w}$ not in the span. Then $\mathbf{v}_1, \ldots, \mathbf{v}_s, \mathbf{w}$ is a linearly independent set.

(9) We have $(x_1, mx_1) + (x_2, mx_2) = (x_1 + x_2, m(x_1 + x_2))$ so it is closed under addition. Then $\alpha(x, mx) = (\alpha x, m(\alpha x))$ and it is closed under scalar multiplication. Therefore, it is vector subspace of F^2. Its dimension is 1.

(11) If $\mathbf{v}, \mathbf{w} \in V \cap W$, then $\alpha \mathbf{v} + \beta \mathbf{w} \in V$ since V is a vector space and $\alpha \mathbf{v} + \beta \mathbf{w} \in W$ since W is a vector space. Therefore, $V \cap W$ is a vector subspace of both V and W.

(13) The set $(1, 0, 0, \ldots, 0), (0, 1, 0, 0, \ldots, 0), \ldots (0, 0, \ldots, 0, 1)$ is a basis of $\mathbb{F}_q^n$ and it has n vectors in it.

(15) We have $T(\mathbf{0}) = T(\mathbf{v} - \mathbf{v}) = T(\mathbf{v}) - T(\mathbf{v}) = \mathbf{0}$.

Section 6.2

(1)

	L_1	L_2	L_3	L_4	L_5	L_6	L_7	L_8	L_9	L_{10}	L_{11}	L_{12}
P_1	1	0	0	1	0	0	1	0	0	0	1	0
P_2	1	0	0	0	1	0	0	1	0	0	0	1
P_3	1	0	0	0	0	1	0	0	1	1	0	0
P_4	0	1	0	1	0	0	0	0	1	0	0	1
P_5	0	1	0	0	1	0	1	0	0	1	0	0
P_6	0	1	0	0	0	1	0	1	0	0	1	0
P_7	0	0	1	1	0	0	0	1	0	1	0	0
P_8	0	0	1	0	1	0	0	0	1	0	1	0
P_9	0	0	1	0	0	1	1	0	0	0	0	1

(3) The number of points is 27, then number of lines is 117, and the number of planes is 39.

(5) The number of points is 625 and the number of hyperplanes is 780.

(7) Let p be a point. Take any k-dimensional subspace V and form the k flat $p + V$.

(9) Take the lines $\{(0,0,0,0),(1,1,1,1)\}$ and $\{(1,0,1,0),(0,1,0,1)\}$. These are parallel.

Section 6.3

(1)

	L_1	L_2	L_3	L_4	L_5	L_6	L_7	L_8	L_9	L_{10}	L_{11}	L_{12}	L_{13}
P_1	1	0	0	1	0	0	1	0	0	0	1	0	0
P_2	1	0	0	0	1	0	0	1	0	0	0	1	0
P_3	1	0	0	0	0	1	0	0	1	1	0	0	0
P_4	0	1	0	1	0	0	0	0	1	0	0	1	0
P_5	0	1	0	0	1	0	1	0	0	1	0	0	0
P_6	0	1	0	0	0	1	0	1	0	0	1	0	0
P_7	0	0	1	1	0	0	0	1	0	1	0	0	0
P_8	0	0	1	0	1	0	0	0	1	0	1	0	0
P_9	0	0	1	0	0	1	1	0	0	0	0	1	0
P_{10}	1	1	1	0	0	0	0	0	0	0	0	0	1
P_{11}	0	0	0	1	1	1	0	0	0	0	0	0	1
P_{12}	0	0	0	0	0	0	1	1	1	0	0	0	1
P_{13}	0	0	0	0	0	0	0	0	0	1	1	1	1

(3) 806.

(5) Let **v** be a vector generating the one-dimensional subspace that is the point. Then extend this to $k + 1$ linearly independent vectors (we can extend to a basis so we can stop at $k + 1$). Then the point is on a k-dimensional object.

(7) 5.

Section 6.4

(1) Straightforward verification.

Section 6.5

(1) $6, 14, 21, 22, 30, 33, 38, 42, 46$.

(3) Straightforward verification.

Section 6.6

(1) Straightforward verification.

(3) Another would be $\{h, b, d, f\}$.

Section 6.7

(1) The image of the subplane of Π is a subplane of Σ.

(3) Use the fact that $\mathbb{F}_q$ is a subfield of $\mathbb{F}_{q^2}$. Construct $PG_2(\mathbb{F}_q)$ and show that this is a Baer subplane of $PG_2(\mathbb{F}_{q^2})$.

Section 6.8

(1) Straightforward verification.

B.7 Problems of Chap. 7

Section 7.1

(1) Straightforward verification.

(3) Straightforward verification.

(5)

$$n^2 + n + 1$$
$$n^2 \qquad n+1$$
$$n^2 - n \qquad n \qquad 1 \tag{B.1}$$

(7) $\lambda = 78$.

(9) Projective: $(n^2 + n + 1, n^2, n^2 - n)$; Affine: $(n^2, n^2 - n, n^2 - n - 1)$.

Section 7.2

(1) We have $\frac{n^2 + 3n + 4}{2} = \frac{n(n+3)}{2} + 2$. If n is even then $\frac{n}{2}$ is an integer. If n is odd, then $n + 3$ is even and $\frac{n+3}{2}$ is an integer. Therefore, the number is always an integer.

Section 7.3

(1) If $\lambda = 1$, then $\frac{(n+\lambda-1)(n+\lambda)}{\lambda} + 1 = \frac{n(n+1)}{1} + 1 = n^2 + n + 1$. If $\lambda = 2$, then $\frac{(n+\lambda-1)(n+\lambda)}{\lambda} + 1 = \frac{(n+1)(n+2)}{2} + 1 = \frac{n^2+3n+4}{2}$.

Then

$$\frac{(n+\lambda-1)(n+\lambda)}{\lambda}+1 = \frac{(q^{m-1}+\frac{q^{m-1}-1}{q-1}-1)(q^{m-1}+\frac{q^{m-1}-1}{q-1})}{\frac{q^{m-1}-1}{q-1}}+1$$

$$= \frac{(\frac{q^m-1}{q-1}-1)(\frac{q^m-1}{q-1})}{\frac{q^{m-1}-1}{q-1}}+1$$

$$= \frac{(q^m-q)(q^m-1)}{(q^{m-1}-1)(q-1)}+1$$

$$= \frac{q^{m+1}-q}{q-1}+\frac{q-1}{q-1}$$

$$= \frac{q^{m+1}-1}{q-1}.$$

Section 7.4

(1) The system with $v = 3$ is the single block $\{\alpha, \beta, \gamma\}$. The system with $v = 7$ is the projective plane of order 2.

Section 7.5

(1) Start with the first horizontal line. On this line there are n choices for a point on a transversal. Given that the first i horizontal lines have been marked, then there are $n - i$ choices to mark on the net horizontal line avoiding each marked vertical line. Then there are $n!$ choices and hence $n!$ transversals.
(3) The parameters are $v = 24$, $b = 36$, $k = 4$, $n = 6$.
(5) The parameters are $v = n^2 + n$, $b = n^2$, $k = n + 1$.
(7) Construct the isomorphism by fixing the parallel class corresponding to vertical lines and by transposing the second and third horizontal lines. Then the isomorphism sends line i in the third parallel class to line i.
(9) Start at each coordinate of the first line and show there are no transversals through that point.

B.8   Problems of Chap. 8

Section 8.1

(1) If $v \cdot w = 0$ then multiplying any coordinate by -1 gives the same dot product and $v \cdot -w = -0 = 0$. Similarly, if $v \cdot v = n$ and $v_i = \pm 1$ then multiplying by -1 does not change this either. Since this applies to columns or rows we have

the first result. Then multiply each coordinate of the first row that has a -1 in it by -1 to get a matrix where the first row is all 1. Then multiply each row that begins with a -1 by -1 and we have a matrix that has first row and column where all entries are 1.

(3) If there were one, then the first row and column would be made all 1 by the first exercise. Then the matrix would be

$$\begin{pmatrix} 1 & 1 & 1 \\ 1 & a & b \\ 1 & c & d \end{pmatrix}.$$

Then we have $a + b = 0$ and $c + d = 0$ giving $b = -a$ and $d = -c$. Then the orthogonality of the last two rows give $ac + ac = 0$ and $2ac = 0$, which is not possible for values of ± 1. Hence, there is no Hadamard matrix with $n = 3$.

(5)

$$\begin{pmatrix} 1 & 1 & 1 & 1 \\ 1 & 1 & -1 & -1 \\ 1 & -1 & 1 & -1 \\ 1 & -1 & -1 & 1 \end{pmatrix}.$$

(7) The squares are $0, 1, 3, 4, 9, 10, 12$. The non-squares are $2, 5, 6, 7, 8, 11$.

(9) Simply follow the hint.

Section 8.2

(1) Straightforward.

Section 8.3

(1) Straightforward verification.

Section 8.4

(1) Each possible k-tuple occurs exactly once. Therefore, reading the k-tuples as numbers in base n give each possible number between 0 and $n^k - 1$ exactly once.

Section 8.5

(1) (a) 24 (b) 2 (c) 2 (d) 4.

(3) (a) 01011 (b) 11111 (c) 11011 (d) 00000.

(5) If they are both Boolean algebras of cardinality 2^n, then they are both isomorphic to $\mathcal{P}(\{1, 2, \dots, n\})$. Then the result follows from the composition of isomorphisms.

(7) In (A, R^{-1}) the role of l.u.b and g.l.b is reversed. Its Hasse diagram is the original one upside down.

(9) If $n > 2$ then the Boolean algebra is isomorphic to $\mathcal{P}(\{1, 2, \ldots, k\})$, where $k > 1$. Then the elements corresponding to $\{1\}$ and $\{2\}$ are not comparable.

Section 8.6

(1)

	00	01	10	11
00	0	1	2	3
01	1	0	3	2
10	2	3	0	1
11	3	2	1	0

It is symmetric since if $a + b = c$ in this group, then $a = c + b$.

(3)

$$A_0 = \begin{pmatrix} 1 & 0 & 0 & 0 & 0 & 0 \\ 0 & 1 & 0 & 0 & 0 & 0 \\ 0 & 0 & 1 & 0 & 0 & 0 \\ 0 & 0 & 0 & 1 & 0 & 0 \\ 0 & 0 & 0 & 0 & 1 & 0 \\ 0 & 0 & 0 & 0 & 0 & 1 \end{pmatrix} \quad A_1 = \begin{pmatrix} 0 & 1 & 0 & 0 & 0 & 0 \\ 1 & 0 & 0 & 0 & 0 & 0 \\ 0 & 0 & 0 & 0 & 0 & 1 \\ 0 & 0 & 0 & 0 & 1 & 0 \\ 0 & 0 & 0 & 1 & 0 & 0 \\ 0 & 0 & 1 & 0 & 0 \end{pmatrix} \quad A_2 = \begin{pmatrix} 0 & 0 & 1 & 0 & 0 & 0 \\ 0 & 0 & 0 & 0 & 1 & 0 \\ 1 & 0 & 0 & 0 & 0 & 0 \\ 0 & 0 & 0 & 0 & 0 & 1 \\ 0 & 1 & 0 & 0 & 0 & 0 \\ 0 & 0 & 0 & 1 & 0 & 0 \end{pmatrix}$$

$$A_3 = \begin{pmatrix} 0 & 0 & 0 & 1 & 0 & 0 \\ 0 & 0 & 0 & 0 & 0 & 1 \\ 0 & 0 & 0 & 0 & 1 & 0 \\ 1 & 0 & 0 & 0 & 0 & 0 \\ 0 & 0 & 1 & 0 & 0 & 0 \\ 0 & 1 & 0 & 0 & 0 & 0 \end{pmatrix} \quad A_4 = \begin{pmatrix} 0 & 0 & 0 & 0 & 1 & 0 \\ 0 & 0 & 0 & 1 & 0 & 0 \\ 0 & 1 & 0 & 0 & 0 & 0 \\ 0 & 0 & 1 & 0 & 0 & 0 \\ 0 & 0 & 0 & 0 & 0 & 1 \\ 1 & 0 & 0 & 0 & 0 & 0 \end{pmatrix} \quad A_5 = \begin{pmatrix} 0 & 0 & 0 & 0 & 0 & 1 \\ 0 & 0 & 1 & 0 & 0 & 0 \\ 0 & 0 & 0 & 1 & 0 & 0 \\ 0 & 1 & 0 & 0 & 0 & 0 \\ 1 & 0 & 0 & 0 & 0 & 0 \\ 0 & 0 & 0 & 0 & 1 & 0 \end{pmatrix}.$$

(5) Use the group that is the n fold product of $(\mathbb{F}_2, +)$.

B.9   Problems of Chap. 9

Section 9.1

(1) $\frac{5}{18}$.

(3) $\frac{7}{128}$.

(5) $\frac{169}{324}$.

(7) $\frac{5}{6}$.

(9) $\frac{63}{256}$.

(11) $\frac{11}{1024}$.

(13) $\frac{1}{11}$.

(15) $\approx \frac{1}{8(10^{67})}$.

(17) $\frac{(p-1)(q-1)(r-1)}{pqr}$.

(19) $\frac{n^2+n+1}{C(n^2+n+1,3)}$.

Section 9.2

(1) $\frac{1}{13}$.

(3) $\frac{3}{13}$.

B.10   Problems of Chap. 10

Section 10.1

(1) Closure is immediate, associativity follows from the associativity of addition in the integers, the identity is 1, and the inverse of a is $n - a$.

(3) Closure follows from the fact that $\frac{a}{b} + \frac{c}{d} = \frac{da+bc}{bd}$ and $bd \neq 0$ if neither b nor d are 0. Associativity is straightforward. The identity is $\frac{0}{1}$ and the inverse of $\frac{a}{b}$ is $\frac{-a}{b}$.

(5) If n is a composite, then $n = ab$, then $ab = 0 \notin \mathbb{Z}_n - \{0\}$. This means that it is not closed and so not a group. If n is prime then, for any $a \in \mathbb{Z}_n - \{0\}$, there exists b, c with $ab + cn = 1$ since $\gcd(n, a) = 1$. Then reading $\pmod{n}$ we have $ab \equiv 1 \pmod{n}$. The remainder of the proof is straightforward.

(7) Assume e and e' are both identities. Then we have $e = ee' = e'$.

(9) If σ and τ both fix a point p, the $\sigma(\tau(p)) = \sigma(p) = p$ so it is closed. If $\sigma(p) = p$ then taking σ^{-1} of both sides gives $p = \sigma^{-1}(\sigma(p)) = \sigma^{-1}(p)$. Therefore, it is a subgroup.

(11) The isomorphism sends a^i to $i \pmod{n}$.

(13) Any subgroup would have to have its order divide p but only 1 and p divide p so the only subgroups are trivial.

Section 10.2

(1, 3, 5, 7) Straightforward verifications.

Section 10.3

(1) A loop must have an identity e. Then the map $f(x) = xa$ must be injective by the right cancelation property. This gives that there must be an element b with $ba = e$.

(3) Similar to Theorem 10.2.

Section 10.4

(1) Straightforward verification.

(3) Straightforward.

B.11   Problems of Chap. 11

Section 11.2

(1) Take any code with a generator matrix of the from $(I_k|A)$, where I_k is the k by k identity matrix and A is any k by $n - k$ matrix.

(3) It reduces to $\begin{pmatrix} 1\,0\,0\,0 \\ 0\,1\,0\,0 \\ 0\,0\,1\,0 \\ 0\,0\,0\,1 \end{pmatrix}$.

(5) The code $\{(1111),(1110)\}$ has minimum distance 1 but minimum weight 3.

(7) The following code can correct two errors: $\{(00000),(11111),(22222),(33333),(44444)\}$.

Section 11.3

1. Define a function from the set of k dimension codes to the set of $n - k$ dimension codes by $\Phi(C) = C^{\perp}$. Then $\Phi(C) = \Phi(D)$ implies $C^{\perp} = D^{\perp}$ which gives $C = D$ and the map is injective. Then $\Phi(D^{\perp}) = D$ for any code D. Therefore, Φ is a bijection and the cardinality of the two sets is the same.

3. We have $\dim(C) \le \dim(C^{\perp})$ and $\dim(C) + \dim(C^{\perp}) = n$. Then if $k > \frac{n}{2}$ this gives a contradiction since the sum of the two dimensions must be greater than n.

5. We have $wt(\mathbf{v} + \mathbf{w}) = wt(\mathbf{v}) + wt(\mathbf{w}) - 2|\mathbf{v} \wedge \mathbf{w}|$. Since $[\mathbf{v}, \mathbf{w}] = 0$ we have $|\mathbf{v} \wedge \mathbf{w}| \equiv 0 \pmod 2$. Then we have $wt(\mathbf{v} + \mathbf{w}) \equiv 0 \pmod 4$.

Section 11.4

1. The code is $\{00000, 11110, 00100, 11010\}$.
   The space is partitioned as follows:

$$
\begin{array}{c|c|c|c|c|c|c|c}
00000 & 10000 & 01000 & 00010 & 00001 & 10001 & 00011 & 01001 \\
11110 & 01110 & 10110 & 11100 & 11111 & 01111 & 11101 & 10111 \\
00100 & 10100 & 01100 & 00110 & 00101 & 10101 & 00111 & 01101 \\
11010 & 01010 & 10010 & 11000 & 11011 & 01011 & 11001 & 10011
\end{array}
$$

   The vector 11111 is decoded to 11110.

3. The number of cosets of a k-dimensional code is $\frac{q^n}{q^k} = q^{n-k}$. The number of possible syndromes is q^{n-k}. Then since every coset has a unique syndrome we have the result.

Section 11.5

(1) The matrix is

$$
\begin{pmatrix}
0 & 0 & 0 & 0 & 0 & 0 & 1 & 1 & 1 & 1 & 1 & 1 & 1 & 1 \\
0 & 0 & 1 & 1 & 1 & 1 & 0 & 0 & 0 & 0 & 1 & 1 & 1 & 1 \\
0 & 1 & 1 & 0 & 0 & 1 & 1 & 0 & 0 & 1 & 1 & 0 & 0 & 1 & 1 \\
1 & 0 & 1 & 0 & 1 & 0 & 1 & 0 & 1 & 0 & 1 & 0 & 1 & 0 & 1
\end{pmatrix}.
$$

Section 11.6

(1) The syndrome is a scalar multiple of a column of the parity check matrix. Then subtract that scalar from the position of that column to decode the vector.

(3) The matrix is

$$
\begin{pmatrix}
0 & 1 & 1 & 1 & 1 & 1 \\
1 & 0 & 1 & 2 & 3 & 4
\end{pmatrix}.
$$

Section 11.7

(1) We have $3^6(1 + 11(2) + 55(2)^2) = 3^6(3^5) = 3^{11}$.

(3) We have $2^{12}(1 + 23 + 253 + 1771) = 2^{12}(2048) = 2^{12}2^{11} = 2^{23}$.

Section 11.8

(1) We list the codewords as an array:

$$
\begin{array}{cccc}
0 & 0 & 0 & 0 \\
0 & 1 & 1 & 1 \\
0 & 2 & 2 & 2 \\
1 & 0 & 2 & 1 \\
1 & 1 & 0 & 2 \\
1 & 2 & 1 & 0 \\
2 & 0 & 1 & 2 \\
2 & 1 & 2 & 0 \\
2 & 2 & 0 & 1
\end{array}
$$

Section 11.9

(1) The weight enumerator is $x^7 + 4x^3y^4$.

(3) We have $W_C(x, y) = x^n + y^n$. Then $W_{C^{\perp}}(x, y) = \frac{1}{2}((x + y)^n + (x - y)^n)$
$= \sum_{i=1}^{\lfloor \frac{n}{2} \rfloor} C(n, 2i) x^{n-2i} y^{2i}$. This could have been computed directly by recognizing $C^{\perp}$ as the code consisting of all vectors of even weights.

Section 11.10

(1) The following are eliminated: $6, 14, 21, 30$.

B.12   Problems of Chap. 12

Section 12.1

(1) YCPLTMMGDYLEOFFNIALCRUSHLETCDY.

(3) $\phi^{-1}(x) = 9x + 23$.

(5) $\phi(26)26 = 338$.

(7) Applying the rules of determinants we have

$$
\det(MM^{-1}) = \det(M)\det(M^{-1}) = \det(M)\det(M)^{-1} = 1.
$$

Then, since this is done modulo 26, the determinant of M must be relatively prime to 26, since it is a unit modulo 26.

Section 12.2

(1) 546999052092292800000.
(3) 32755471104142671992819712000000.

Section 12.3

(1) (a) $x \equiv 60 \pmod{77}$ (b) $x \equiv 211 \pmod{221}$ (c) $x \equiv 411 \pmod{418}$, (d) $x \equiv 1211 \pmod{1560}$.

(3) $L_2(1) = 0, L_2(2) = 1, L_2(3) = 4, L_2(4) = 2, L_2(5) = 9, L_2(6) = 5, L_2(7) = 11, L_2(8) = 3, L_2(9) = 8, L_2(10) = 10, L_2(11) = 7, L_2(12) = 6.$

(5) $\phi(2) = 1, \phi(3) = 2, \phi(4) = 2, \phi(5) = 4, \phi(6) = 2, \phi(7) = 6, \phi(8) = 4, \phi(9) = 6, \phi(10) = 4, \phi(11) = 10, \phi(12) = 4, \phi(13) = 12, \phi(14) = 6, \phi(15) = 8, \phi(16) = 8, \phi(17) = 16, \phi(18) = 6, \phi(19) = 18, \phi(20) = 8, \phi(21) = 12, \phi(22) = 10, \phi(23) = 22, \phi(24) = 8, \phi(25) = 20.$

(7) $n = 187, M = 23^7 \pmod{187} = 133, d = 23, 133^{23} \pmod{187} = 23.$

Section 12.4

(1) Applying the rules of determinants we have

$$\det(MM^{-1}) = \det(M)\det(M^{-1})\det(M)\det(M)^{-1} = 1.$$

Therefore, $\det(M) = 1.$

(3) 1.

(5)
$$\begin{pmatrix} 0 & 0 & 0 & 0 & 1 & 0 & 0 \\ 0 & 0 & 1 & 0 & 0 & 0 & 0 \\ 0 & 0 & 0 & 0 & 0 & 1 & 0 \\ 0 & 1 & 0 & 0 & 0 & 0 & 0 \\ 0 & 0 & 0 & 0 & 0 & 0 & 1 \\ 1 & 0 & 0 & 0 & 0 & 0 & 0 \\ 0 & 0 & 0 & 1 & 0 & 0 & 0 \end{pmatrix}.$$

B.13   Problems of Chap. 13

Section 13.1

(1) Let the first two moves be arbitrary, then show that no matter where the first player puts his second move, the second player can counter the move by making it impossible for the first player to have a winning strategy.

(3) Show that given the number of plays that the second player has, there is no way to have a point on each of the lines in these planes.

Section 13.2

(1) For the plane of order 2, the first move can be made arbitrarily. Then given any move by the second player, the next move by the first player gives two points on some line and so it is a win for the first player. Therefore, there is a winning strategy for the first player. For the plane of order 3, after the first three moves by the first player, there are three lines with two marked points between these points and the second player has only been able to block two. Therefore, the first player has a winning strategy. For the plane of order 4, there is a drawing strategy.

(3) For the 1-net, the second player can simply place points on the same line as the first player giving a draw. For the other nets, the second move for the first player can give a win.

Glossary

Affine plane An affine plane is a set of points $\mathcal{A}$ and a set of lines $\mathcal{M}$ and an incidence relation $\mathcal{J} \subseteq \mathcal{A} \times \mathcal{M}$ such that through any two points there is a unique line incident with them. More precisely, given any two points $p, q \in \mathcal{A}$ there exists a unique line $\ell \in \mathcal{M}$ with $(p, \ell) \in \mathcal{J}$ and $(q, \ell) \in \mathcal{I}$. If p is a point not incident with ℓ then there exists a unique line through p parallel to ℓ. More precisely, if $p \in \mathcal{A}$, $\ell \in \mathcal{M}$ with $(p, \ell) \notin \mathcal{J}$ then there exists a unique line $m \in \mathcal{M}$ with $(p, m) \in \mathcal{J}$ and ℓ parallel to m. There exists at least 3 non-collinear points. It is a $2\text{-}(n^2, n, 1)$ design.

Arc A k-arc is a set of k points in a plane such that no three are collinear.

Association scheme Let X be a finite set, with $|X| = v$. Let R_i be a subset of $X \times X$, for all $i \in I = \{i \mid i \in \mathbb{Z}, 0 \le i \le d\}$ with $d > 0$. We define $\mathfrak{R} = \{R_i\}_{i \in I}$. We say that $(X, \mathfrak{R})$ is a d-class association scheme if the following properties are satisfied:

1. The relation $R_0 = \{(x, x) : x \in X\}$ is the identity relation.
2. For every $x, y \in X$, $(x, y) \in R_i$ for exactly one i.
3. For every $i \in I$, there exists $i' \in I$ such that $R_i^T = R_{i'}$, that is we have $R_i^t = \{(x, y) \mid (y, x) \in R_i\}$.
4. If $(x, y) \in R_k$, the number of $z \in X$ such that $(x, z) \in R_i$ and $(z, y) \in R_j$ is a constant p_{ij}^k.

Automorphism An automorphism of an incidence structure $D = (\mathcal{P}, \mathcal{B}, \mathcal{I})$ is a bijection $\psi : D \to D$, where ψ is a bijection from the set of points to the set of points, a bijection from the set of blocks to the set of blocks and

$$(p, B) \in \mathcal{I} \iff (\psi(p), \psi(B)) \in \mathcal{I}.$$

Baer subplane A Baer subplane is a subplane of order m of a projective plane of order m^2.

Binomial coefficient The binomial coefficient is defined as $C(n, k) = \frac{n!}{k!(n-k)!}$.

© The Editor(s) (if applicable) and The Author(s), under exclusive license
to Springer Nature Switzerland AG 2020
S. T. Dougherty, *Combinatorics and Finite Geometry*,
Springer Undergraduate Mathematics Series

Bipartite graph Let $G = (V, E)$ be a graph where A and B are subsets of the set of vertices V, with $A \cup B = V$ and $A \cap B = \emptyset$. The sets A and B are said to be a partition of V. The graph G is said to be bipartite graph if $v \in A$ implies $w \in B$, whenever $\{v, w\} \in E$.

Biplane A biplane is a set of points $\mathcal{P}$ and a set of lines $\mathcal{L}$ and an incidence relation $\mathcal{I} \subseteq \mathcal{P} \times \mathcal{L}$ such that through any two points there are two lines incident with them and any two lines intersect in two points.

Blocking set A blocking set $\mathcal{B}$ in a projective plane is a subset of the points of the plane such that every line of Π contains at least one point in $\mathcal{B}$ and one point not in $\mathcal{B}$.

Catalan number Catalan numbers satisfy the recursion

$$C_{n+1} = \sum_{k=0}^{n} C_k C_{n-k}, C_0 = 1.$$

Their closed form is $C_n = \frac{1}{n+1} \frac{(2n)!}{n!n!}$.

Chromatic number The chromatic number of a graph is the least number of colors needed to color the vertices such that no two vertices connected by an edge have the same color.

Chromatic polynomial The chromatic polynomial of a graph is the number of possible ways to color a graph using n colors or less.

Code A code over an alphabet A of length n is a subset of A^n.

Coloring of a graph A coloring of a graph G is a map from V to a set C such that if $(v, w) \in E$ then $f(v) \neq f(w)$.

Complete graph The complete graph on n vertices, K_n, is the graph for which any two distinct vertices are connected by an edge.

Connected graph A connected graph is a graph in which for any two vertices v and w there is a tour from v to w. Otherwise we say that the graph is disconnected.

Cubic number Cubic numbers satisfy the recursion $a_{n+1} = a_n + 3n^2 + 3n + 1$. Their closed form is $a_n = n^3$.

Cycle graph A cycle graph on n vertices, C_n, is the connected graph on n vertices consisting of a single Euler cycle with no repeated vertices.

Design A t-(v, k, λ) design is $D = (\mathcal{P}, \mathcal{B}, \mathcal{I})$ where $|\mathcal{P}| = v$, every block $B \in \mathcal{B}$ is incident with exactly k points, and every t distinct points are together incident with exactly λ blocks.

Degree of a vertex The degree of a vertex in a graph is the number of edges incident with that vertex.

Difference Set A difference set in a group $(G, +)$ is a subset D of G such that each non-identity element g of G can be written in exactly λ different ways of the form $x - y$ with $x, y \in D$.

Directed graph A directed graph $G = (V, E)$ is a set of vertices V and edges E, where each edge is an ordered pair of the form (v_1, v_2) where $v_i \in V$.

Discrete graph The discrete graph on n vertices, D_n, is the graph consisting of n vertices and no edges.

Euler tour A tour in a graph $G = (V, E)$ is an ordered k-tuple $(e_1, e_2, \ldots, e_k)$, $e_i = \{v_i, w_i\} \in E$, where $w_i = v_{i+1}$ and $e_i \neq e_j$ if $i \neq j$. An Euler tour is a tour such that each edge of the graph appears in the tour. An Euler cycle is an Euler tour where the starting vertex of the tour is also the ending vertex, i.e. $v_1 = w_k$.

Fibonacci sequence The sequence is defined as $\mathcal{F}_0 = 0, \mathcal{F}_1 = 1, \mathcal{F}_n = \mathcal{F}_{n-1} + \mathcal{F}_{n-2}$.

Field A finite field $(F, +, *)$, is a set F with two operations $+$ and $*$ such that $(F, +)$ is a commutative group; $(F - \{0\}, *)$ is a commutative group; and for all $a, b, c \in F$, we have $a * (b + c) = a * b + a * c$ and $(b + c) * a = b * a + c * a$.

Generalized Hadamard matrix A generalized Hadamard (GH) matrix $H(q, \lambda) = (h_{ij})$ of order $n = q\lambda$ over $\mathbb{F}_q$ is a $q\lambda \times q\lambda$ matrix with entries from $\mathbb{F}_q$ such that for every $i, j, 1 \leq i < j \leq q\lambda$, each of the multisets $\{h_{is} - h_{js} : 1 \leq s \leq q\lambda\}$ contains every element of $\mathcal{F}_q$ exactly λ times.

Graph A graph $G = (V, E)$ is a set of vertices V and edges E where each edge is of the form $\{v_1, v_2\}$ where $v_i \in V$.

Group A group $(G, *)$ is a set G with an operation $*$ such that if $a, b \in G$ then $a * b \in G$; for all $a, b, c \in G$, $(a * b) * c = a * (b * c)$; there exists an $e \in G$ with $e * a = a * e = a$ for all $a \in G$; and for all $a \in G$ there exists $b \in G$ with $a * b = e$.

Hadamard matrix A Hadamard matrix is a square n by n matrix H with $H_{ij} \in \{1, -1\}$ such that $HH^t = nI_n$, where I_n is the n by n identity matrix.

Hadamard Design A Hadamard design is a $3\text{-}(4n, 2n, n - 1)$ design.

Hamiltonian cycle A Hamiltonian cycle hits every vertex exactly once except for the final vertex which was also the beginning vertex.

Hamiltonian tour A Hamiltonian tour is a tour that hits every vertex exactly once.

Hyperoval A hyperoval is an arc of maximal order in a plane of even order, that is, a set of $n + 2$ points such that no 3 are collinear.

Hyperplane A hyperplane is an $n - 1$ dimensional subspace of an n dimensional space.

k-partite graph Let $A_1, A_2, \ldots, A_k$ be a partition of the set of vertices V. The graph G is said to be a k-partite graph if $v \in A_i$ implies $w \notin A_i$, whenever $\{v, w\} \in E$. That is, if two vertices are connected then they must be in different sets of the partition.

Latin hypercube A Latin k-hypercube $L_{j_1, \ldots, j_k}$ of order n is a k dimensional array of size n, whose elements come from the set $\{0, 1, 2, \ldots, n - 1\}$. If $j_i = j_i'$ for all $i \neq \alpha$ and $j_\alpha \neq j_{\alpha'}$ then $L_{j_1, j_2, \ldots, j_k} \neq L_{j_1', j_2', \ldots, j_k'}$. That is, in any direction each "row" has each element from the set exactly once.

Latin square A Latin square L is an n by n array such that each element of an alphabet A with n elements, appears exactly once in every row and every column.

Latin subsquare Let L be a Latin square of order n, indexed by the elements of $\mathbb{Z}_n$. If there are subsets $R, C \subset \mathbb{Z}_n$ with $|R| = |C| = m$, where $L_{ij}, i \in R, j \in C$, is a Latin square of order m, this is said to be a subsquare of order m.

Lattice A partially ordered set is a lattice if and only if for every two elements a and b in the partially ordered set the least upper bound $l.u.b.(a, b)$ and the greatest lower bound $g.l.b.(a, b)$ always exist.

Linear graph The linear graph on n vertices, L_n, is the connected graph on n vertices consisting of a single Euler tour with no repeated vertices.

MDS (Maximum Distance Separable) code An $[n, k, d]$ code is called MDS if $d = n - k + 1$.

Main class If L is a Latin square, then any Latin square that can be obtained by any combination of permuting the rows, permuting the columns, permuting the symbols, taking the row adjugate, taking the column adjugate, and taking the transpose is said to be in the main class of L.

Magic square A magic square of order n is an n by n matrix where each element of the set $\{0, 1, 2, \ldots, n^2 - 1\}$ appears once and the sum of each row and column is the same.

Multinomial Coefficient The multinomial coefficient is defined as:

$$\binom{n}{k_1, k_2, \ldots, k_t} = \frac{n!}{k_1! k_2! k_3! \cdots k_t!}.$$

Multigraph A multilgraph is a structure $G = (V, E)$ where V is a set of vertices, E is a set of edges, and each edge is represented by $(\{a, b\}, i)$ where a and b are vertices connected by the edge and i is used to distinguish multiple edges. We say that two vertices are adjacent if they are connected by an edge.

Mutually Orthogonal Latin Squares (MOLS) If $L = (\ell_{ij})$ is a Latin square of order n over an alphabet A and $M = (m_{ij})$ is a Latin square of order n over an alphabet B then L and M are said to be orthogonal if every pair of $A \times B$ occurs exactly once in $\{(\ell_{ij}, m_{ij})\}$. If $L_1, L_2, \ldots, L_k$ are Latin squares of order n, then they are a set of k Mutually Orthogonal Latin Squares if L_i and L_j are orthogonal for all $i, j, i \neq j$.

Near field A nearifield $(Q, +, *)$, is a set Q, with two operations $+$ and $*$, such that the following hold:

1. (additive closure) If $a, b \in Q$ then $a + b \in Q$.
2. (additive associativity) For all $a, b, c \in Q$ we have $(a + b) + c = a + (b + c)$.
3. (additive identity) There exists an element $0 \in Q$ such that $0 + x = x + 0 = x$ for all $x \in Q$.
4. (additive inverses) For all $a \in Q$ there exists $b \in Q$ with $a + b = b + a = 0$.
5. (additive commutativity) For all $a, b \in Q$ we have $a + b = b + a$.
6. (multiplicative closure) If $a, b \in Q$ then $a * b \in Q$.
7. (multiplicative associativity) For all $a, b, c \in F$ we have $(a * b) * c = a * (b * c)$.
8. (multiplicative identity) There exists an element $1 \in Q$ such that $1 * x = x * 1 = x$ for all $x \in Q$.
9. (multiplicative inverse) For all $a \in Q - \{0\}$, there exists $b \in Q$ with $a * b = 1$.

10. (unique solution) For all $a, b, c \in Q$, $a \neq b$ there is exactly one solution to $xa - xb = c$.

11. (right distributive) For all $a, b, c \in Q$ we have $(b + c) * a = b * a + c * a$.

Nets A k-net of order n is an incidence structure consisting of n^2 points and nk lines satisfying the following four axioms: every line has n points; parallelism is an equivalence relation on lines, where two lines are said to be parallel if they are disjoint or identical; there are k parallel classes each consisting of n lines; any two non-parallel lines meet exactly once. It is a 1-(n^2, n, k) design.

Oval An oval is an arc of maximal size in a plane of odd order, that is $n + 1$ points in a plane of odd order such that no 3 points are collinear.

Parallel lines Two lines are parallel if they lie on a plane and are either equal or disjoint.

Parity check matrix A matrix H is said to be a parity check matrix for the code C, if $C = \{\mathbf{v} \mid H\mathbf{v}^\mathsf{T} = \mathbf{0}\}$.

Partial Latin square A partial Latin square of order n is an n by n array, where cells may be empty or from a symbol set A of size n, such that each symbol occurs at most once in any row or column.

Perfect code A perfect code is a code with equality in the sphere packing bound, namely if C is a code of length n over an alphabet of size p with minimum distance $2t + 1$ then $|C|(\sum_{s=0}^{t} C(n, s)(p - 1)^s) = p^n$.

Partially ordered set A partially ordered set is a set A together with a relation R on A such that R is reflexive, antisymmetric, and transitive.

Permutation A permutation of set A is a bijection from A to A.

Permutations The number of permutations of k objects chosen from n objects is defined as $P(n, k) = \frac{n!}{(n-k)!}$.

Planar graph A graph is said to be a planar graph if it can be drawn in $\mathbb{R}^2$ such that no two edges (when viewed as curves in $\mathbb{R}^2$) intersect at a point that is not a vertex.

Primitive root Let p be a prime. An element $a \pmod{p}$ is a primitive root $\pmod{p}$ if $a^{p-1} \equiv 1 \pmod{p}$ but $a^i \not\equiv 1 \pmod{p}$ for $0 < i < p - 1$.

Probability Given a sample space S and an event $E \subset S$, the probability of event E is $P(E) = \frac{|E|}{|S|}$.

Projective plane A projective plane is a set of points $\mathcal{P}$ and a set of lines $\mathcal{L}$ and an incidence relation $\mathcal{I} \subseteq \mathcal{P} \times \mathcal{L}$ such that through any two points there is a unique line incident with them. More precisely, given any two points $p, q \in \mathcal{P}$ there exists a unique line $\ell \in \mathcal{L}$ with $(p, \ell) \in \mathcal{I}$ and $(q, \ell) \in \mathcal{I}$. Any two lines meet in a unique point. More precisely, given any two lines $\ell, m \in \mathcal{L}$ there exists a unique point $p \in \mathcal{P}$ with $(p, \ell) \in \mathcal{I}$ and $(p, m) \in \mathcal{I}$. There exists at least four points no three of which are collinear. It is a 2-$(n^2 + n + 1, n + 1, 1)$ design.

Quasigroup A quasigroup $(Q, *)$ is a set G with an operation $*$ such that if $a, b \in G$ then $a * b \in G$; for all $a, b \in G$ there exists a unique x with $a * x = b$; and for all $a, b \in G$ there exists a unique y with $y * a = b$.

Regular graph A graph is said to be regular if the degree of every vertex is the same.

Residual design A residual design is $D' = (\mathcal{P}', \mathcal{B}', \mathcal{I}')$ where $D = (\mathcal{P}, \mathcal{B}, \mathcal{I})$ is a design and for a block v in D we define $\mathcal{P}' = \mathcal{P} - \{p| (p, v) \in \mathcal{I}\}$, $\mathcal{B}' = \mathcal{B} - \{v\}$ and a point is incident with a block in D' if it was incident in D.

Self-dual code A code $C \subseteq \mathbb{F}_q^n$ is said to be self-dual if $C = C^{\perp}$.

Self-orthogonal code A code $C \subseteq \mathbb{F}_q^n$ is said to be self-orthogonal if $C \subseteq C^{\perp}$.

Singer cycle A singer cycle is a map $\psi : \Pi \to \Pi$ such that ψ maps points to points, lines to lines and preserves incidence, that is an automorphism, such that for any two points p and p' in Π there exists an i such that $\psi^i(p) = p'$ and ψ^{n^2+n+1} is the identity.

Skew lines Two lines that do not lie on a plane but do not intersect are called skew lines.

Sphere Packing Bound A code C of length n over an alphabet of size p with minimum weight $2t + 1$ satisfies $|C|(\sum_{s=0}^{t} C(n, s)(p - 1)^s) \leq p^n$.

Steiner Triple System A Steiner Triple System is a 2-$(v, 3, 1)$ design.

Square numbers Square numbers satisfy the recursion $a_{n+1} = a_n + 2n + 1$. Their closed form is $a_n = n^2$.

Stirling numbers of the first kind Stirling numbers of the first kind satisfy the recursion $\begin{bmatrix} n \\ k \end{bmatrix} = (n - 1) \begin{bmatrix} n - 1 \\ k \end{bmatrix} + \begin{bmatrix} n - 1 \\ k - 1 \end{bmatrix}$, $\begin{bmatrix} n \\ 1 \end{bmatrix} = (n - 1)!$, $\begin{bmatrix} n \\ n \end{bmatrix} = 1$.

Stirling numbers of the second kind Stirling numbers of the second kind satisfy the recursion $\begin{Bmatrix} n \\ k \end{Bmatrix} = k \begin{Bmatrix} n - 1 \\ k \end{Bmatrix} + \begin{Bmatrix} n - 1 \\ k - 1 \end{Bmatrix}$, $\begin{Bmatrix} n \\ 1 \end{Bmatrix} = \begin{Bmatrix} n \\ n \end{Bmatrix} = 1$.

Symmetric design A symmetric design is a 2-(v, k, λ) design where the number of points equals the number of blocks and the axioms are symmetric for points and blocks.

Syndrome Let C be a code in $\mathbb{F}_q^n$ with parity check matrix H. Then the syndrome of a vector $\mathbf{v} \in \mathbb{F}_q^n$ is $S(\mathbf{v}) = H\mathbf{v}^T$.

Transversal A transversal of a Latin square L is a set of n coordinates in the square such that each symbol appears once in the set and the set intersects each row and column exactly once.

Transversal design A transversal design D of order n, block size k, is a set of points V and blocks B such that V has kn points; V is partitioned into k classes, each of size n; and every pair of points is either contained in the same set of the partition or in a block. It is a 1-(kn, k, n) design.

Triangular numbers Triangular numbers satisfy the recursion $a_{n+1} = a_n + n + 1$. Their closed form is $a_n = \frac{n(n+1)}{2}$.

Unit Let n be a positive integer. An element $a \in \mathbb{Z}_n$ is a unit if there exists an element $b \in \mathbb{Z}_n$ such that $ab \equiv 1 \pmod{n}$.

Wheel graph The wheel graph, W_{n+1} is formed from the cycle graph C_n by adding an additional vertex that is connected to ever other vertex.

Zero divisor Let n be a positive integer. A non-zero element $c \in \mathbb{Z}_n$ is a zero divisor if there exists a $d \neq 0$ with $cd \equiv 0 \pmod{n}$.

References

1. Appel, K., Haken, W.: Every planar map is four colorable. Part I. Discharging. Ill. J. Math. **21**, 429–490 (1977)
2. Appel, K., Haken, W.: Every planar map is four colorable. Contemp. Math. **98**, (1989)
3. Appel, K., Haken, W., Koch, J.: Every planar map is four colorable. Part II. Reducibility. Ill. J. Math. **21**, 491–567 (1977)
4. Assmus Jr., E.F., Key, J.D.: Designs and Their Codes. Cambridge University Press, Cambridge (1992)
5. Assmus Jr., E.F., Mattson Jr., H.F.: New 5-designs. J. Comb. Theory **6**, 122–151 (1969)
6. Bammel, S.E., Rothstein, J.: The number of 9×9 Latin squares. Discret. Math. **11**, 93–95 (1975)
7. Bannai, E., Ito, T.: Algebraic Combinatorics I: Association Schemes. The Benjamin/Cummings Publishing Co., Inc, Menlo Park (1984)
8. Berlekamp, E.R., Conway, J.H., Guy, R.K.: Winning Ways for Your Mathematical Plays. Academic, New York (1982)
9. Boole, G.: An Investigation of the Laws of Thought. Prometheus Books, New York (1854)
10. Bose, R.C., Shimamoto, T.: Classification and analysis of partially balanced incomplete block designs with two associate classes. J. Am. Stat. Assoc. **47**, 151–184 (1952)
11. Bose, R.C., Shrikhande, S.S.: On the construction of sets of mutually orthogonal Latin squares and the falsity of a conjecture of Euler. Trans. Am. Math. Soc. **95**, 191–209 (1960)
12. Bose, R.C., Shrikhande, S.S., Parker, E.T.: Further results on the construction of mutually orthogonal Latin squares and the falsity of Euler's conjecture. Can. J. Math. **12**, 189–203 (1960)
13. Bourbaki, N.: Corps communatifs, 12th edn. Hermann, Paris (1967)
14. Brown, E.: The fabulous $(11, 5, 2)$ biplane. Math. Mag. **77**, 2 (2004)
15. Brualdi, R.A., Ryser, H.J.: Combinatorial Matrix Theory. Encyclopedia of Mathematics and Its Applications, vol. 39. Cambridge University Press, Cambridge (1991)
16. Bruck, R.H.: Finite nets I. Numerical invariants. Can. J. Math. **3**, 94–107 (1951)
17. Bruck, R.H., Ryser, H.J.: The nonexistence of certain finite projective planes. Can. J. Math. **1**, 88–93 (1949)
18. Bruen, A.: Baer subplanes and blocking sets. Bull. Am. Math. Soc. **76**, 342–344 (1970)
19. Carroll, M.T., Dougherty, S.T.: Tic-Tac-Toe on a finite plane. Math. Mag. **260–274**, (2004)
20. Carroll, M.T., Dougherty, S.T.: Fun and Games with Squares and Planes. MAA Press, Resource for Teaching Discrete Mathematics (2009)

21. Carroll, M.T., Dougherty, S.T.: Tic-Tac-Toe on Affine Planes. The Mathematics of Various Entertaining Subjects. Princeton University Press, Princeton (2016)
22. Catalan, E.: Note sur une equation aux différences finies. Journal de mathématiques pures et appliquées **3**, 508–516 (1838)
23. Cayley, A.: On the colourings of maps. Proc. R. Geogr. Soc. **1**, 259–261 (1879)
24. Colbourn, C.J., Dinitz, J.H. (eds.): CRC Handbook of Combinatorial Designs. CRC Press, Boca Raton (1996)
25. Denes, J., Keedwell, A.D.: Latin Squares and Applications. Academic, New York (1974)
26. Diffie, W., Hellman, M.E.: New directions in cryptography. IEEE Trans. Inf. Theory IT **22**(6), 644–654 (1976)
27. Dirac, G.A.: Some theorems on abstract graphs. Proc. Lond. Math. Soc. 3rd Ser. **2**, 69–81 (1952)
28. Dougherty, S.T.: A coding-theoretic solution to the 36 officer problem. Des. Codes Cryptogr. **4**(1), 123–128 (1994)
29. Dougherty, S.T.: Algebraic Coding Theory over Finite Commutative Rings. SpringerBriefs in Mathematics. Springer, Cham (2017)
30. Dougherty, S.T., Vasquez, J.: Amidakuji and games. J. Recreat. Math. **37**(1), 46–56 (2008)
31. ElGamal, T.: A public-key cryptosystem and a signature scheme based on discrete logarithms. IEEE Trans. Inf. Theory **31**(4), 469–472 (1985)
32. Erdös, P., Selfridge, J.L.: On a combinatorial game. J. Comb. Theory **14**, 298–301 (1973)
33. Euler, L.: From the problem of the seven bridges of Königsberg, Commentarii. St. Petersbg. Acad. **8** (1736)
34. Euler, L.: Observationes Analyticae, Novi Commentarii Academiae Scientiarum Imperialis Petropolitanae **11**, 53–69 (1765). Reprinted in Leonardi Euleri Opera Omina Ser. I **15**, 50–69 (1923), Teubner, Berlin-Leipzig
35. Euler, L.: Recherches sur une nouvelle espece des quarres magiques. Leonardi Euleri Opera Omina Ser. I **7**, 291–392 (1923), Tuebner, Berlin-Leipzig
36. Fisher, R.A.: The Design of Experiments. Macmillan, London (1935)
37. Fisher, R.A., Yates, F.: The 6×6 Latin squares. Proc. Camb. Philos. Soc. **30**, 492–507 (1934)
38. Fraleigh, J.B.: A First Course in Abstract Algebra, 5th edn. Addison Wesley, Boston (1994)
39. Gauss, C.F.: Disquisitiones Arithmeticae (1801)
40. Golay, M.J.E.: Notes on digital coding. Proc. IEEE **37**, 657
41. Guthrie, F.: Tinting maps. The Athenæum **1389**, 726 (1854)
42. Hadamard, J.: Résolution d'une question relative aux déterminants. Bull. Sci. Math. **2**, 240–246 (1893)
43. Hall, M.: Projective planes. Trans. Am. Math. Soc. **54**, 229–277 (1943)
44. Hall, M.: Combinatorial Theory. Ginn-Blaisdell, Waltham (1967)
45. Hamilton, W.R.: Memorandum respecting a new system of roots of unity. Philos. Mag. **12**, 446 (1856)
46. Hamilton, W.R.: Account of the icosian calculus. Proc. R. Ir. Acad. **6**, 415–416 (1858)
47. Hamming, R.W.: Error detecting and error correcting codes. Bell Syst. Tech. J. **29**, 147–160 (1950)
48. Hill, R.: A First Course in Coding Theory. Oxford University Press, Oxford (1990)
49. Huffman, W.C., Pless, V.S.: Fundamentals of Error-Correcting Codes. Cambridge University Press, Cambridge (2003)
50. Hulpke, A., Kaski, P., Östergard, P.: The number of Latin squares of order 11. Math. Comput. **80**(274), 1197–1219 (2011)
51. Huygens, C.: De Ratiociniis in Ludo Aleae (The value of all chances in games of fortune) (1657)
52. Järnefelt, G., Kustaanheimo, P.: An Observation on Finite Geometries. Den 11te Skandinaviske Matematikerkongress, Trondheim, 166–182 (1949)
53. Jungnickel, D.: Latin Squares, Their Geometries and Their Groups. A Survey. IMA Volumes in Mathematics and Its Applications, vol. 21. Springer, Berlin

54. Keedwell, A.D., Dénes, J.: Latin Squares and Their Applications, 2nd edn. Elsevier/North-Holland, Amsterdam (2015)
55. Kharaghani, H., Tayfeh-Rezaie, B.: A Hadamard matrix of order 428. J. Comb. Des. **13**(6), 435–440 (2005)
56. Kirkman, T.P.: On a problem in combinatorics. Camb. Dublin Math. J. **2**, 191–204 (1847)
57. Knuth, D.E.: Axioms and Hulls. Lecture Notes in Computer Science, vol. 606. Springer, Berlin (1992)
58. Lam, C.W.H., Thiel, L., Swiercz, S.: The non-existence of finite projective planes of order 10. Can. J. Math. **41**, 1117–1123 (1989)
59. Laywine, C.F., Mullen, G.L.: Discrete Mathematics Using Latin Squares. Wiley-Interscience Series in Discrete Mathematics and Optimization. Wiley, New York (1998)
60. Lucas, É.: Récréations mathématiques, four volumes. Gauthier-Villars, Paris (1882–1894). Reprinted by Blanchard, Paris (1959)
61. MacWilliams, F.J.: Combinatorial problems of elementary abelian groups, Ph.D. Dissertation, Harvard University (1962)
62. MacWilliams, F.J., Sloane, N.J.A.: The Theory of Error-Correcting Codes. North-Holland, Amsterdam (1977)
63. McEliece, R.J.: A public-key cryptosystem based on algebraic coding theory. DSN Prog. Rep. **44**, 114–116 (1978)
64. McKay, B., Rogoyski, E.: Latin squares of order 10. Electron. J. Comb. **2**, Note 3 (1995)
65. McMahon, L., Gordon, G., Gordon, H., Gordon, R.: The Joy of SET. The Many Mathematical Dimensions of a Seemingly Simple Card Game. Princeton University Press, Princeton; National Museum of Mathematics, New York (2017)
66. Mullen, G.L., Purdy, D.: Some data concerning the number of Latin rectangles. J. Comb. Math. Comb. Comput. **13**, 161–165 (1993)
67. Pickert, G.: Projective Ebenen. Springer, Berlin (1955)
68. Póyla, G.: On picture writing. Am. Math. Mon. **63**, 689–697 (1956)
69. Qvist, B.: Some remarks concerning curves of the second degree in a finite plane. Ann. Acad. Sci. Fennicae (A,I) **134** (1952), 27 pp
70. Rapoport, A.: Two-Person Game Theory. University of Michigan Press, Ann Arbor (1966)
71. Ray-Chaudri, D.K., Wilson, R.M.: Solution of Kirkman's schoolgirl problem. In: Combinatorics (Proceedings of the Symposium in Pure Mathematics, vol. XIX, University of California, Los Angeles, California, 1968), pp. 187–203. American Mathematical Society, Providence (1971)
72. Reidemeister, K.: Topologische Fragen der Differentialgeometrie. Gewebe und Gruppen, Math. Z. **29**, 427–435 (1928)
73. Reiss, M.: Uber eine Steinersche Combinatoirische Aufgabe Welche in 45sten Bande Dieses Journals, Siete 181, Gestellt Worden Ist, 0J. reine angew. Math. **56**, 36–344 (1859)
74. Richardson, J.T.E.: Who introduced Western mathematicians to Latin squares? Br. J. Hist. Math. **34**(2), 95–103 (2019)
75. Rivest, R.L., Shamir, A., Adleman, L.: A method for obtaining digital signatures and public-key cryptosystems. Commun. ACM **21**(2), 120–126 (1978)
76. Rosenhouse, J.: The Monty Hall Problem. The Remarkable Story of Math's Most Contentious Brainteaser. Oxford University Press, Oxford (2009)
77. Segre, B.: Ovals in a finite projective plane. Can. J. Math. **7**, 414–416 (1955)
78. Shannon, C.E.: A mathematical theory of communication. Bell Syst. Tech. J. **27**, 379–423 and 623–656 (1948)
79. Shor, P.W.: Polynomial-time algorithms for prime factorization and discrete logarithms on a quantum computer. SIAM J. Comput. **26**(5), 1484–1509 (1997)
80. Singer, J.: A theorem in finite projective geometry and applications to number theory. Trans. Am. Math. Soc. **43**, 377–385 (1938)
81. Stanley, R.: Catalan Numbers. Cambridge University Press, Cambridge (2015)
82. Steiner, J.: Combinatorische Aufgabe. J. reine angew. Math. **45**, 181–182 (1853)
83. Stevenson, F.: Projective Planes. Freeman, San Francisco (1972)

84. Stinson, D.R.: A short proof of the non-existence of a pair of orthogonal Latin squares of order six. J. Comb. Theory **A36**, 373–376 (1984)
85. Stinson, D.R.: Cryptography. CRC Press Series on Discrete Mathematics and Its Applications, Theory and Practice (1995)
86. Stinson, D.R.: Combinatorial Designs. Constructions and Analysis. Springer, New York (2004)
87. Stirling, J.: Methodus Differentialis. London (1730)
88. Straffin, P.D.: Game Theory and Strategy. The Mathematical Association of America, New York (1993)
89. Sylvester, J.J.: Thoughts on inverse orthogonal matrices, simultaneous sign-succession, and tessellated pavements in two or more colours, with applications to Newton's rule, ornamental tile-work, and the theory of numbers. Lond. Edinb. Dublin Philos. Mag. J. Sci. **34**, 461–475 (1867)
90. Tarry, G.: Le probleme des 36 officers. Compte Rendu Ass. Franc. Pour l'avacement des sciences **2**, 170–203 (1901)
91. The Thirteen Books of Euclid's Elements, translated by Thomas L. Heath. Cambridge University Press, Cambridge
92. Trappe, W.: Introduction to Cryptography with Coding Theory, 2nd edn. Pearson Prentice Hall, Upper Saddle River (2006)
93. Weibel, C.: Survey of non-Desarguesian planes. Not. AMS **54**(10), 1294–1303

Index

© The Editor(s) (if applicable) and The Author(s), under exclusive license
to Springer Nature Switzerland AG 2020
S. T. Dougherty, *Combinatorics and Finite Geometry*,
Springer Undergraduate Mathematics Series

Printed in the United States
By Bookmasters